LAND USE CONTROL
Geography, Law, and Public Policy

Rutherford H. Platt
University of Massachusetts at Amherst

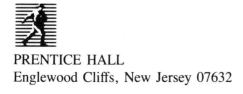
PRENTICE HALL
Englewood Cliffs, New Jersey 07632

Library of Congress Cataloging-in-Publication Data

Platt, Rutherford H.
  Land use control : geography, law, and public policy / by
Rutherford H. Platt.
    p. cm.
  Includes bibliographical references and index.
  ISBN 0-13-523457-3
  1. Land use—Law and legislation—United States. 2. Land use—
United States. 3. Physical geography—United States. I. Title.
KF5698.P59   1991
346.7304'5—dc20
[347.30645]   90-49075
                                                              CIP

Editorial/production supervision: Cyndy Lyle Rymer
Cover design: Joe Di Domenico
Prepress buyer: Kelly Behr
Manufacturing buyer: Susan Brunke

© 1991 by Prentice-Hall, Inc.
A Division of Simon & Schuster
Englewood Cliffs, New Jersey 07632

All rights reserved. No part of this book may be
reproduced, in any form or by any means,
without permission in writing from the publisher.

Printed in the United States of America

10 9 8 7 6 5 4 3 2 1

TEXT CREDITS

Page 96, F. L. Olmsted, Jr. and T. Kimball, *Forty Years of Landscape Architecture Central Park*, 1928, reprinted by permission of The MIT Press; pages 136–37, J. Bollens, *Special District Governments in the United States*, copyright © 1957 Regents of the University of California, © renewed 1985 John C. Bollens; page 162, R. C. Wood, *1400 Governments*, 1961, reprinted by permission of the Regional Plan Association, Inc.; page 226, R. M. Smith, "From Subdivision Improvement Requirements to Community Benefit Assessments and Linkage Payments, *Law and Contemporary Problems*, 50 (1): 5–30, copyright © 1987 Duke University School of Law; page 313, Table 12.3 adapted from R. G. Healy and J. A. Zimm, "Environment and Development Conflicts in Coastal Zone Management," reprinted by permission of *Journal of the American Planning Association*, vol. 51 (3), summer 1985; page 319, Table 12.4 R. H. Platt, "Floods and Man: A Geographer's Agenda," in R. W. Kates and I. Burton (eds.), *Geography, Resources and Environment*, vol. II, reprinted by permission of the University of Chicago Press; page 324, Table 12.5 R. J. Burby and S. P. French, "Action Instruments Used by Communities to Manage Flood Hazard Areas" in *Water Policy & Management*, no. 5, reprinted by permission of Westview Press; page 372, Table 13.1 "Status of State Coastal Management Programs" reprinted by permission of the Coastal States Organization.

ISBN 0-13-523457-3

Prentice-Hall International (UK) Limited, *London*
Prentice-Hall of Australia Pty. Limited, *Sydney*
Prentice-Hall Canada Inc., *Toronto*
Prentice-Hall Hispanoamericana, S.A., *Mexico*
Prentice-Hall of India Private Limited, *New Delhi*
Prentice-Hall Japan, Inc., *Tokyo*
Simon & Schuster Asia Pte. Ltd., *Singapore*
Editora Prentice-Hall do Brasil, Ltda., *Rio de Janeiro*

This book is dedicated with admiration and affection to two staunch friends of this beleaguered planet:

Gilbert F. White
and
Anne U. White (1919–1989)

CONTENTS

Preface   vi

1/Overview of Land Resources and Uses in the United States   1

    Introduction   *1*
    Cropland   *4*
    Forest Land   *7*
    Grasslands   *8*
    Recreation Land   *8*
    Wetlands   *9*
    Urban Land   *9*
    Conclusion   *15*

2/The Interaction of Law and Geography   17

    The Geographical Landscape   *18*
    The Legal Landscape   *22*
    The Legal Process   *25*
    Law as an Agent of Land-Use Morphology   *28*
    A General Model of Land-Law Interaction   *33*

3/English Roots of Modern Land Use Controls   39

    The Feudal Commons   *39*
    Municipal Corporations   *45*
    The Common Law   *51*
    Building Laws: The Act for Rebuilding London   *53*
    Private Land-Use Restrictions   *58*
    Improvement Commissions   *60*
    Conclusion   *62*

4/Urban Reforms of the Nineteenth Century   65

    Introduction   *65*
    Urban Growth During the Nineteenth Century   *66*

Regulation: Public Sanitary Reform   *68*
Redevelopment: A Century of Municipal Improvements   *76*
Relocation: The Ideal Communities Movement   *99*
Conclusion   *110*

5/Private Ownership of Land   113

Introduction   *113*
What Is Real Property?   *113*
Green Sticks: Rights of Real Property Ownership   *115*
Red Sticks: Duties of Real Property Ownership   *117*
Types of Owners and Legal Interests   *119*
Acquisition and Disposition   *124*
Land-Use Conversion   *125*
Legal Boundary Descriptions   *126*
Conclusion   *133*

6/American Municipal Geography and Law   135

Overview   *135*
Origins and Diffusion   *137*
Municipal Jurisdiction and Powers   *153*

7/Euclidean Zoning: Origins and Practice   165

Introduction   *165*
Precursors to Zoning   *166*
The Nuts and Bolts of Euclidean Zoning   *177*

8/Policy Issues of Zoning   195

Introduction   *195*
The Taking Issue   *195*
Reasonableness   *199*
The Role of Planning   *201*
Metropolitan Needs and Exclusionary Zoning   *204*
Aesthetics and Historic Preservation   *212*

9/Beyond Zoning: Innovations in Local Land Use Control   219

Introduction   *219*
Subdivision Regulations and Exactions   *221*
Growth Management   *227*
Transfer of Development Rights   *234*

Floodplain and Wetland Regulation *237*
Public Acquisition *246*

10/Management of the Federal Lands   256

Introduction *256*
The Public Domain *257*
Conflict and Management in the Twentieth Century *263*
Conclusion *270*

11/Federal Land Use Programs Before 1970   275

Introduction *275*
National Planning in the New Deal *275*
Post-War Federal Programs *278*
Rumblings of Dissent *279*
Outdoor Recreation *282*
Urban Housing *285*
Metropolitan and Regional Planning *288*
River Basin Commissions *293*
Conclusion *294*

12/The Federal Environmental Decade of the 1970s   299

Introduction *299*
The National Environmental Policy Act *302*
Coastal Zone Management *307*
The National Flood Insurance Program *314*
Federal Wetlands Programs *327*
Agricultural Land *334*
Solid and Hazardous Waste Management *343*
Surface Mine Reclamation *354*

13/State Land Use Programs   362

State Parks *362*
"The Quiet Revolution" *365*
Special Area Programs: The Pinelands Case *366*
Special Issue State Programs *369*
Conclusion *373*

Conclusion   375

Index and List of Cases   386

PREFACE

REFLECTIONS ON A CROSS-COUNTRY FLIGHT

To geographers and their fellow travelers, there are few greater treats than to fly a considerable distance over land on a clear day with a view unobstructed by the airplane wing. In June, 1987, I made such a flight nonstop from San Francisco to Boston. Between the sourdough vendors and live lobster purveyors of those two airports stretches about 2,700 miles of air distance. Along this trajectory, the route traverses a succession of geographic regions marked by vivid contrasts in both physical and human characteristics. Even the casual observer can scarcely fail to notice and perhaps to wonder about the diversity of the perceived landscape—its physical landforms, land cover, and patterns of rural and urban land use. If the movie is really boring, the window-gazer may attempt to annotate the passing scene by assigning causative factors and implications—some definite, others hypothetical—to what is seen or imagined in the landscape below. This is thinking geographically.

    The aircraft ascends over the crowded East Bay cities of Oakland and Berkeley, riddled by seismic faults (where the Nimitz Freeway collapsed in the October 17, 1989 earthquake). The flight continues over the geometric patterns of irrigated fields of the Central Valley (handsomely subsidized by the federal taxpayer), over the snowy peaks and steep declivities of the Sierra Nevada (where John Muir battled Gifford Pinchot over damming the Hetch Hetchy Valley to provide water supply for San Francisco), to the empty, somewhat radioactive wastes of the arid Great Basin. Cities and towns briefly reappear in the Mormon settlements east of Great Salt Lake (which in 1987 was pumped into nearby salt flats to reduce record flood levels). The Wasatch Range, crisscrossed by ski slopes and clearcutting, gives way to the upper Colorado River Plateau, another sparsely inhabited region of high desert, sagebrush, and spectacular landforms.

    Across the snowcapped Rockies lies the Front Range Urban Corridor, a chain of cities extending from Greeley to Pueblo, anchored by smog-bound Denver. For the next hour, one flies over the vast checkerboard of the High Plains dominated by green or brown circles of center-pivot irrigated fields within the square quarter-sections laid out under the Federal Land Survey a century ago. (On subsequent flights, it was observed that the green circles are increasingly turning brown as irrigation is suspended due to high costs of pumping from the declining Ogallala Aquifer and perhaps also the effect of federal land retirement payments.)

One crosses the Missouri River in the vicinity of the fabled "100th Meridian" (which roughly corresponds to the 20-inch average annual rainfall contour) where irrigation yields to rain-dependent agriculture. Towns reappear as beads on a string along mainline railroad lines and old section-line highways. Increasingly larger "central places" occasionally appear, culminating with the economic and governmental colossus of the Midwest: Chicago. The alternation of town and farm across the nation's heartland is a totally human-dominated landscape. Few natural or unused areas of land are observed until the Appalachian Upland is reached in Pennsylvania and New York State.

Flight attendants collect headsets as the aircraft descends in humid summer twilight over the northeastern urban corridor for which geographer Jean Gottmann coined the name "Megalopolis." This "stupendous monument erected by titanic efforts" (in Gottmann's words) is curiously dominated by forests, within which several of the nation's largest cities are situated. The failing daylight permits a glimpse of Quabbin Reservoir, Boston's primary water supply, surrounded by an "accidental wilderness" (as described by the Massachusetts Audubon Society) resulting from the forced abandonment of farms and villages when the state purchased the watershed in the 1930s. The plane swings over "the Nation's Technology Highway" (Route 128) where the arms race fostered the illusion of prosperity during the 1980s. It banks over the built-up coastal barriers south of Boston where hundreds of homes were damaged in the Northeaster of February 3–4, 1978, and lands at Logan International Airport which, like our departure point at San Francisco, is constructed on a filled wetland.

The cross section of the North American landscape just described is viewed by thousands of travelers on any clear day, albeit without the running commentary. To many observers, the scene below is a pleasant but seemingly random series of abstract images, like the geometric patterns produced by a kaleidoscope. But just as those may be explained through an understanding of optics, so too the patterns of human land use are by no means random. To one with geographic training or interests, the variation in the landscape offers not only aesthetic but intellectual stimulation. The geographer seeks to discern order, process, and coherency in the seemingly haphazard sequence of images.

The perennial question of geography is: *Why* is this place the way it is, and like or unlike other places? This leads to additional questions, for example: *What* benefits or costs arise from specific practices or ways of using land, air, and water and to whom do they accrue? and *How* can we better manage the use of land and other resources to promote the public welfare, however defined, and reduce social costs? These questions ultimately frame the central question of our time: *How may global resources be managed to sustain a world population that has more than doubled since 1950?*

Global resource crisis is at hand according to the Worldwatch Institute, World Resources Institute, Conservation Foundation, Smithsonian Institution, National Academy of Sciences, and a host of international organizations. Global resource problems include widespread deforestation, atmospheric warming and ozone depletion, loss of biodiversity, land degradation, food shortages, energy shocks, accumulating wastes, and surface and groundwater pollution.

Geographers of course do not claim any special monopoly on wisdom, nor do they offer ready solutions to these travails. But they do offer the perspective of the "Why" question. They seek lessons from the experience of the past and present, which may profitably be applied to the exigencies of the future. If we can better understand how we got to where we are in our inhabitation of the earth, or portions of it that we label regions or nations, we may have some valuable insight into how to deal with the challenges ahead.

Unlike more narrowly focused disciplines, geographers view the earth and its regions holistically and seek explanations (or solutions) through synthesis of diverse phenomena and the formulation of theories regarding the interaction of these variables. From time to time, certain classes of spatially distributed phenomena have been vested with greater importance than others as explanations of human settlement patterns and uses of resources. In the 1920s, the theory of environmental determinism sought to explain human actions in terms of the influence of climate and physical characteristics of regions. In the 1960s, central place theory and gravity models attributed human spatial behavior to economic forces operating through the private land market.

This book takes a different tack. While not discounting the role of physical, economic, and other spatial variables, the primary focus of this book is the role of law as a major factor in the way humans use their resources and design their patterns of settlement. In the phrase of James E. Vance, Jr., law is viewed as a "morphogenetic agent" in the shaping of the human environment.

Admittedly, law is not entirely an independent variable: Laws are products of institutions reacting to their perceptions of exogenous circumstances such as the threat of fire, flood, drought, famine, pollution, or crime. Yet the rules for human activities established by law differ according to the rule maker's perception of external circumstances, and often comprise an imperfect, partial, or even counterproductive response to the actual problem. Furthermore, laws established to address one problem may compound others. And laws have a habit of remaining in effect long after changes in circumstances have rendered them moot or even pernicious.

This book, then, is offered as an exploration of the influence of law over human use of land from the perspective of a geographer. The specific rules, doctrines, and practices discussed are drawn from the American context, including its common law roots in England. But recognition of law as a factor for better or worse in the use of land is believed to be of general applicability.

OVERVIEW OF LAND RESOURCES AND USES IN THE UNITED STATES

INTRODUCTION

A quarter-century ago, Marion Clawson (1963, p. 1) noted that the average American, viewed nationally, then enjoyed the products and services of 12.5 acres of land. In 1920, this figure stood at 20 acres per capita; it would further decline, in Clawson's estimate, to 7.5 acres in the year 2000. Clawson's prediction is on track. In 1984, U.S. land area per capita was 9.3 acres and shrinking. It nevertheless remained 14 times higher than that of the world as a whole and was at least three times the total for most other industrialized nations (Conservation Foundation, 1984, p. 144).

Acreage per capita, however, is not a very meaningful figure. In the first place, it masks regional variation and is a poor measure of social well-being. At a state level, the citizens of Connecticut have one acre per capita (dividing its land area by its 1984 population). New Jerseyites have only two-thirds of an acre each; while the 686,000 residents of North Dakota "claim" 64 acres apiece. But does this mean that the people of North Dakota are better off than those of Connecticut? Clearly not in economic terms: Connecticut ranks second in income per capita while North Dakota ranks 33rd. In terms of quality of life, that is a matter of personal judgment.

Second, most of the nation's wealth of land resources is distant from the everyday habitat of most of us. Three-quarters of the U.S. population (172 million) live in the nation's 337 metropolitan statistical areas (MSAs), which comprise about 16 percent of the nation's land area. Land area per capita within MSAs, viewed in the aggregate, amounts to 2.1 acres per capita. But, except for those who live on oversized lots in affluent suburbs or on the many farms that remain within MSAs, this figure clearly is meaningless. Most metropolitan area residents live, work, and seek nearby recreation under conditions of extreme and worsening crowding. So while there is no absolute shortage of land in the United States, most people have little direct contact with the vast expanses of rural and wild land located far from their metropolitan homes. They do of course benefit indirectly from such lands through the availability of food, fiber, forest products, water, energy, and, for the fortunate, occasional vacation trips.

A third limitation of the acres-per-capita measure of land wealth is the diversity of physical capabilities and use categories into which land resources may be classified. Overall totals of land area reveal little about the sufficiency of land for particular purposes such as production of food and fiber, forest products, watershed, and built-up purposes. Table 1–1 summarizes broad land-use purposes in 1959 and 1982 and Figure 1–1 provides a longer term perspective. Even these raw data reveal little about the sufficiency of land for the purposes listed. Assessments must be tempered by the potential for interchange among various categories, that is, the degree of reversibility of land-use changes. Also, the growing importance of the global economy, the rise of the U.S. trade deficit, the effect of currency exchange rates, and the flow of commodities and people across national borders vastly complicate the task of appraising the adequacy of U.S. land resources.

Finally, land differs as to ownership status. About one-third of the U.S. land

Table 1–1 Utilization of land in the United States, 1959 and 1982 (in millions of acres).

	1959	1982	1982 % of U.S.
Cropland	392	404	17.8
Pasture/range	699	662	29.1
Forest[a]	728	655	28.8
Wetlands[b]	N.A.	99*	
Urban/built-up	24	47**	2.0
Transportation	25	27	1.1
Recreation	44	116	5.1
Wildlife	17	95	4.1
Military	26	24	1.0
Farmsteads	10	8	0.3
Other[c]	306	234	10.3
Total	2,271	2,271	100.0

SOURCES: *U.S. Statistical Abstract: 1986.* Tables 336, 339.
[a]Excludes forested land in parks
[b]Overlaps other categories; not included in total
[c]Includes marshes, open swamps, bare rock, tundra and desert
*National Wetlands Inventory
**1982 National Resources Inventory

Introduction

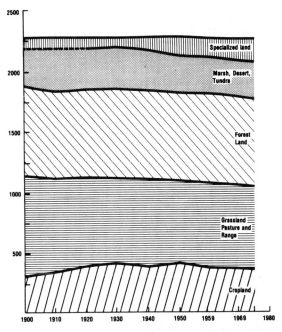

Note: *Forest Land* excludes reserved forest land in parks, wildlife refuges, and other special-use areas. *Specialized Land* is urban and built-up areas (including cities and towns, rural highway and road rights-of-way, railroads, airports, and public institutions in rural areas) and nonurban special-use areas (including Federal and State parks and other rural parks, recreational areas, Federal and State wildlife refuges, national defense sites, flood-control areas, Federal industrial areas, farmsteads, and farm roads).

Source: U.S. Department of Agriculture, Economics, Statistics, and Cooperatives Service, *Major Uses Of Land in the United States Summary for 1974*, draft, June, 1978, Table 3, p. 3a; U.S. Department of Argiculture, Economic Research Service, *Major Uses of Land and Water in the United States, Summary for 1959*, Agricultural Economic Report No. 13, pp. 10, 11; U.S. Department of Agriculture, Economics, Statistics, and Cooperatives Service, unpublished data.

Figure 1.1 Major uses of land in the United States: 1970–74.
Source: U.S. Council on Environmental Quality, 1979.

area is owned by the federal government and is thus removed from the operation of the private land market, although federal lands do accommodate a variety of private activities (Figure 1–2). (Federal lands are the subject of Chapter 10.)

With these shortcomings in mind, the rest of this chapter briefly reviews the status of land usage and related issues according to the major categories listed in Table 1–1. This discussion also serves as a synopsis of several topics treated in more detail in later chapters.

About three-quarters of the total U.S. land area is devoted to three primary categories of rural land usage: (1) cropland; (2) grazing land (pasture and range); and (3) forest land. Each of these represents, by definition, a productive and economically beneficial class of land resource, although the more remote forests and more arid grazing lands may be seldom utilized.

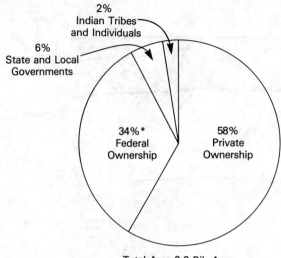

Total Area 2.3 Bil. Acres
*94 percent is in the eleven western-most states and Alaska. About 50 percent is in Alaska.

Figure 1.2 Land ownership in the 50 states, 1969.
Source: U.S. Department of Agriculture, 1974.

CROPLAND

Cropland is the most sensitive and valuable of the nation's rural land resources. The protection of the cropland base, especially those portions of it deemed "prime land," has been the subject of lively discussion and debate over the past two decades, both among scholars and in the public media (Platt, 1985). Despite widely expressed anxiety about the "loss" of cropland, Table 1–1 indicates that the total of cropland harvested in 1982 was actually larger by 12 million acres than in 1959. Furthermore, the average productivity of U.S. agriculture has risen by 46 percent in a quarter-century from a farm output index of 76 in 1960 to 111 in 1984 (U.S. Bureau of the Census, 1985, Table 1165). This increase, which has been achieved through the massive use of fertilizers, pesticides, irrigation water, and genetic research (the Green Revolution), has expanded U.S. agricultural production by the equivalent of an additional 186 million acres of cropland. Indeed, the rising productivity of American agriculture has had the ironic result of achieving such an abundance of farm commodities that low prices have prevailed over the past decade and farmers have widely suffered economic distress.

Despite the apparent sufficiency—even glut—of cropland, the need for careful stewardship of this resource remains a public policy concern. In the first place, the apparent increase in harvested acreage between 1959 and 1984 is not a totally

Cropland 5

reliable statistic (as is the case with most land-use data at a national scale). Even in 1982, cropland acreage totals were not derived from satellite imagery or other standardized remote sensing methodology. Instead, such data were obtained from the estimates of individual agricultural extension agents in most of the nation's 3,041 counties and equivalent units. There is thus a considerable margin of error due to variability in the accuracy of individual cropland estimates. Also, the inventories for 1959 and 1982 were not strictly comparable due to changes in definitions, sampling techniques, and data management (microcomputers were unknown in 1959).

A second important qualification is that totals of harvested cropland give no indication of the average quality of land under cultivation. As documented by the U.S. Soil Conservation Service (SCS) (1977), much land is converted from one

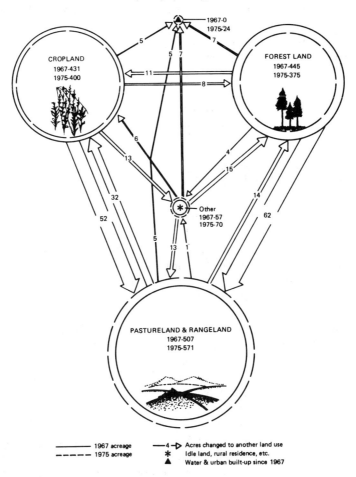

Figure 1.3 Land-use conversion between 1967 and 1975.
Source: U.S. Soil Conservation Service, 1977.

rural category to another, as well as into urban and built-up status over time. The 1982 National Resources Inventory indicated that about 900,000 acres per year are converted to urban and built-up purposes, essentially an irreversible process (U.S. Council on Environmental Quality, 1984, p. 283). Since harvested cropland has remained relatively constant for three decades, it is apparent that replacement land for cultivation has been drawn from other rural categories, for example, forest, range, pasture, and wetlands. (The U.S.D.A. estimated in 1982 that some 127 million acres of potential cropland could be found in these other categories of land use.) But the replacement land is not necessarily equivalent in quality to the cropland converted to nonagricultural purposes. Heavier application of irrigation water, chemical fertilizers and pesticides, and labor is required to render such marginal lands productive in cultivation. Furthermore, the continued drainage of wetlands for agriculture in the south central states, Florida and California—once consistent with national policy—now is viewed as threatening to the ecological values and functions of wetlands in their natural state.

There is growing evidence that increased use of fertilizer has reached a point of diminishing returns. According to Crosson and Brubaker (1982, p. 68):

> Per-acre applications of fertilizer grew much more slowly after 1972 as farmers moved toward more land-using technologies after that date. . . . The slower increase in per-acre applications of fertilizer after 1972 was consistent with the post-1973 slowdown in yield growth. Correspondingly, our belief that the trend of yields to 2010 will more nearly resemble the post- than the pre-1972 experience is based in good part on our belief that per-acre applications of fertilizer will grow more slowly.

These writers estimate that an actual increase of 64 million acres of cropland will be required to meet agricultural demand by 2010.

Similarly, the use of irrigation to reclaim arid lands for cultivation may also have peaked. Between 1959 and 1982, irrigated land in 17 arid western states plus Louisiana increased by 35 percent from 31 million to 42 million acres. In the latter year, the national total of irrigated land was 49.4 million acres, which represented 15 percent of total cropland. Continued expansion of the area under irrigation is doubtful as new ground and/or surface water supplies are increasingly difficult and expensive to develop. Environmental objections and economic costs have deterred the further expansion of irrigation facilities in recent years. Furthermore, the mining of ground water in the Ogallala Aquifer of the High Plains and the salinization of ground water supplies in the Colorado Basin threaten to reduce the acreage presently receiving irrigation water.

Soil erosion is another source of concern about the future adequacy of the nation's cropland base. The U.S.D.A. estimates that 44 percent of all cropland nationally is eroding at rates exceeding the normal rate of replacement through natural processes (which differ from one location to another). Especially high levels of soil erosion, ranging from 5 to 14 tons of soil per acre per year, were identified in most of the north central region—the Corn Belt—as well as in dissected uplands of the Southeast, Texas, and the Southwest (U.S. Council on Environmental Quality, 1984, Figure 5–10). The precise effects of soil erosion are difficult to quantify but represent a long-term loss of natural soil productive capacity.

The available supply of cropland is more than adequate for the present domestic and export needs of the nation. However, the expansion of outputs since 1960 through technological inputs and the use of marginal land for cultivation may be nearing the point of diminishing returns. Environmental objections to the use of certain chemicals, shortage of water for irrigation, and declining soil quality due to erosion all suggest the need for wise management of land naturally suited to cultivation of crops. (See further discussion in Chapter 12.)

FOREST LAND

Land covered by forest in the United States totalled 655 million acres in 1982, a decline of 72 million acres or 10 percent from 1959 (Table 1–1). Some of this decline represents clearance for cropland usage as expected from the foregoing discussion. Some was converted to national park or wildlife refuge status, which excluded it from definition as "forest land." Still other forest land was converted to water bodies, highways, and urban development. The exact extent of these conversions is unknown.

About three-fourths of forested land is classified by the U.S. Forest Service as "commercial timberland," which is defined as being capable of yielding at least 20 cubic feet per acre per year of commercial wood products and is not closed to cutting by governmental prohibition (e.g., wilderness status). Contrary to popular impression, 76 percent of New England is forested and all but five percent of this is designated as commercial timberland, although much of it is unmanaged. The preponderance of noncommercial forest lands is in the West, especially Alaska.

During the nineteenth century, most of the major forest lands accessible to loggers were cleared without consideration of adverse effects such as soil erosion, forest fires, and loss of regeneration capability. Out of concern for maintaining adequate forests for the nation's future needs, the National Forest System was initiated in 1891 at the urging of Gifford Pinchot, who was appointed by President Theodore Roosevelt as the first Director of the U.S. Forest Service in 1905. The national forests now comprise a total land area of 230 million acres, which has barely changed in the past 40 years. (About 190 million acres are actually forested.) National forests are managed to ensure a sustained yield of forest products while serving other public needs such as water supply, natural habitat, recreation, and mining.

Despite the decline of 10 percent since 1959, there appears to be slight cause for concern about the adequacy of forest land, particularly in light of the increasing reliance on imports. There is, however, some imbalance between hardwood stocks in the Northeast, which are increasing rapidly in relation to harvesting levels, and softwoods of the South and West, for which growth is about equivalent to cutting.

Also, there are serious regional problems with insect pests such as the gypsy moth, which has been identified in 30 states, and the spruce budworm, which has caused much damage to the pulp forests of Maine. In Oregon and Northern California, cutting of old growth timber is opposed by wildlife activists who seek to protect natural habitat for the spotted owl.

GRASSLANDS

Grasslands suitable for grazing have remained relatively unchanged over the past three decades and in 1982 amounted to about 662 million acres, the largest single land-use category. This figure, however, includes two very different subclasses: (1) cropland used for pasture (65 million acres), and (2) rangeland (597 million acres). While the former comprises only 10 percent of total grazing lands, it is generally located in the more humid regions of the country and yields a very large share of total forage production (U.S. Department of Agriculture, 1974, p. 8). Rangeland is substantially located in the Mountain and High Plains states and receives rainfall of only 10–20 inches annually. Such lands produce very little forage and must be grazed on a very land-extensive basis.

Ownership of western range lands is split between the federal government, states, and private owners. Federal grasslands total about 160 million acres, of which 130 million acres are administered by the Bureau of Land Management (BLM) and the remainder by the U.S. Forest Service. Grazing in the West has always involved joint usage of both private and public lands by private ranchers. Before 1934, private use of federal range was generally unregulated, illegal, and the source of disputes among competing interests, for example, cattle versus sheep and "sodbusting" versus grazing. The Taylor Grazing Act of 1934 authorized the establishment of federal grazing districts under the general land office and the issuance of permits for private grazing rights on such lands, bringing a semblance of order to the prior chaos. Since 1950, both the Bureau of Land Management and the Forest Service have sought to reduce grazing pressure on federal rangelands to protect their productive capacity over the long term (Clawson, 1983, p. 67).

RECREATION LAND

Recreation land appeared to increase by 163 percent from 44 million acres in 1959 to 116 million acres in 1982. This increase is misleading however—on December 1, 1978, about 40 million acres of federal land in Alaska were added to the National Park System, thereby doubling that system to a total of 80 million acres. This represented merely an administrative transfer of the lands concerned from one federal agency to another without any actual increase in accessibility or recreational value thereof. Excluding the Alaska lands, the recreational total amounts to 76 million in 1982, of which about 40 million were in the National Park System, 10 million in state parks, and the balance in county, regional, and local parks and other facilities.

On the other hand, these figures understate recreation resources on federal lands. Virtually all lands administered by the Bureau of Land Management and the U.S. Forest Service are managed according to the principle of multiple use, which includes recreation. (See Chapter 10.) Similarly, water resource projects of the Army Corps of Engineers, the Tennessee Valley Authority and the Bureau of Reclamation all involve provision for recreation. Much of the recreational benefit provided by these facilities is in the form of water-based activities, including boating, swimming, and fishing, for which land-acreage data are not an adequate measure.

Finally, there is a large but unquantifiable amount of private land devoted to recreation. This would include intensive-use facilities such as golf courses, tennis clubs, and private campgrounds, as well as more extensive facilities such as private nature sanctuaries and membership camping and hiking parks. In terms of land-use data, much of the foregoing would be classified as forest, urban, or "other."

A final qualification is that raw acreage data does not adequately measure the potential value of a recreational site. Location, site design, amenities, and natural characteristics are usually more important than mere size in determining the functional utility of a recreational facility.

WETLANDS

Wetlands comprise an important subset of the total land and water resources of the United States. Wetlands are generally characterized by: (1) the presence of water at or close to the surface; (2) predominance of saturated hydric soils; and (3) prevalence of vegetation adapted to wet conditions (Mitsch and Gosselink, 1986, pp. 15–16). Depending on their physical nature, size, and location, wetlands perform various natural functions such as providing natural habitat, flood storage, concentration of nutrients, absorption of pollutants, buffering of coastlines from storm waves, recharge of ground water aquifers, and scenic amenity.

The National Wildlife Inventory conducted by the U.S. Fish and Wildlife Service estimated total wetlands in the early 1980s to comprise about 99 million acres nationally (Frayer et al., 1983). A study by the U. S. Department of Agriculture (1981) estimated nonfederal wetlands to amount to 70.5 million acres, with an apparent loss of 14 percent between 1956 and 1980. Annual loss of wetlands nationally due to dredging, filling, drainage, and conversion to agricultural or urban purposes is roughly estimated to be 300,000 acres (U.S. Senate, 1985, p. 1). (See further discussion in Chapters 9 and 12.)

URBAN LAND

Estimation of urban land usage nationally is difficult. Rural land cover tends to occur in sizable, relatively homogeneous spatial units that are easy to identify. The urban landscape by contrast is a finely variegated mosaic of buildings, paved areas, parks, vacant land, private yards, and even residual agriculture. How much of this crazy-quilt pattern of land use is "urban" and where the boundary between urban and nonurban should be drawn are matters of definition and subjective judgment of the estimator.

Another problem is the scarcity of national-level data on urban land usage. In the case of agriculture and forestry, large federal agencies are charged with administering programs that require accurate and frequently revised data on the nature and extent of those land uses. By contrast, the federal government, particularly since the 1970s, is little involved with urban land planning, which it leaves to state, regional, and municipal authorities. Accordingly, the federal government does not assemble or publish detailed statistics on urban land as it does for rural natural resource regions. Even the U.S. Department of Housing and Urban Development, which administers urban housing assistance programs, devotes little attention to the growth of urban areas (notwithstanding the second part of its title).

By any estimate, urban and built-up areas comprise a very small proportion of the nation's total land resources. The 1982 National Resources Inventory (NRI) conducted by the U.S. Soil Conservation Service (undated) estimated urban and built-up areas to total 47 million acres, or about 2 percent of the total U.S. land area (Table 1–1). This figure in fact comprised a reduction from 69 million acres of urban land estimated in the 1977 NRI, also conducted by the SCS. This reduction of course does not mean that urban areas diminished between 1977 and 1982. Instead, the later estimate was based on "better maps, more time for data collection, many more sample points, and better quality control" (U.S. Council on Environmental Quality, 1984, p. 283).

Far more important than the absolute amount of land devoted to urban usage at a given time is the rate of growth of urban areas over time. Clearly, the NRI data do not support such intertemporal comparison. Two intertemporal sets of data relating to urban and urbanizing land, are provided by the federal Bureau of the Census for Metropolitan Statistical Areas (MSAs) and Urbanized Areas (UAs). These units, for which many kinds of statistics are published after each decennial census, serve different purposes.

Metropolitan growth involves the spatial expansion of older urbanized cores into surrounding rural areas. Over time, such expanding urban areas tend to coalesce with their neighbors to form even larger metropolitan areas encompassing multiple central cities together with their suburbs and nearby urban fringe areas. The political geography of such urbanizing regions is thus very complex, involving (1) central cities, (2) incorporated suburbs, and (3) unincorporated areas under county jurisdiction, which are now, or likely soon will, incur development pressure. To document regional trends in population, housing, land use, and other variables, without having to add up all the separate jurisdictions, the Bureau of the Census since 1950 has designated metropolitan regions for which it publishes statistics as a whole. (See Figure 1–4.)

MSAs consist of clusters of one or more counties (or cities and towns in New England) that contain or are economically related to one or more core cities of at least 50,000 people each. The number of metropolitan units designated by the

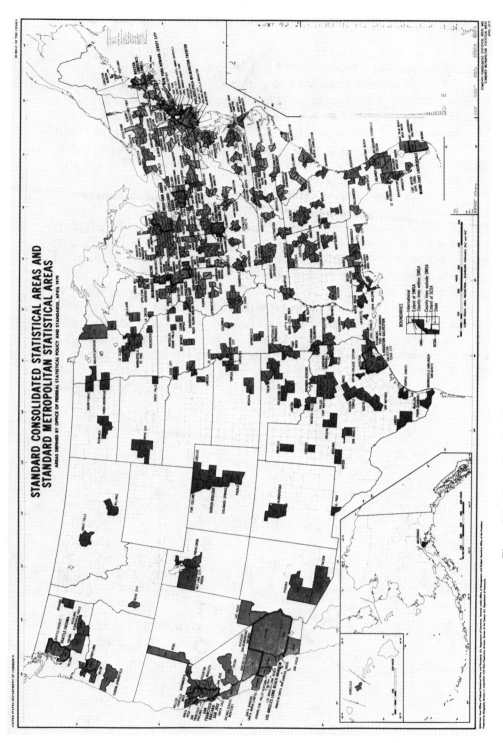

Figure 1.4 Metropolitan Statistical Areas, 1979. *Source*: U.S. Bureau of the Census.

Bureau of the Census has grown from 169 in 1950 to 337 in 1985. Total metropolitan population has increased from 84 million (55 percent of U.S. population) to 179 million (76 percent) in the latter year. (MSAs are often enlarged by adding counties or New England towns to reflect ongoing growth—so care must be taken in comparing population over time to hold area constant.) Land area within these units has risen from about 207,000 square miles in 1950 (6 percent of the United States) to 571,000 square miles (15 percent). Figure 1–5 displays the relative distribution and numbers of people living in central cities, metropolitan suburbs, and nonmetropolitan areas in 1960 and 1980.

The latter figures however vastly overstate the quantity of land actually devoted to urban purposes. MSAs, being defined according to county jurisdictions, contain extensive rural areas. The U.S. Department of Agriculture (1987, pp. 3–12) estimates that metropolitan areas contain 20 percent of the nation's cropland and 20 percent of the land with high and medium potential for conversion to cropland (currently in forest, pasture, or other rural use).

Urbanized Areas (UAs), unlike MSAs, are not based strictly on county or municipal units, but instead are drawn to conform approximately to the actual urban or built-up area within and surrounding a core city. According to the U.S. Bureau of the Census (1984, p. 5): "A UA comprises an incorporated place and adjacent densely settled surrounding area that altogether have a minimum population of 50,000." Political units are divided where necessary to distinguish urban and nonurban land. Inevitably designated boundaries err both as to over- and underinclusion

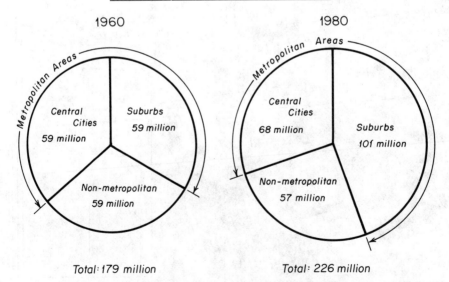

Figure 1.5 Distribution of U.S. population, 1960 and 1980.

of urban land. Nevertheless, the UA is the best national-level estimate of the extent and rate of increase of urbanized land in the United States.

As shown in Table 1–2, between 1920 and 1960, UAs exceeding 50,000 inhabitants doubled in number, grew from about 40 million to 100 million in population, and nearly quadrupled in land area (Pickard, 1967). This was accompanied by a decline in average density from 6,160 to 4,090 persons per square mile. By 1980, there were 366 UAs, which collectively contained 139 million inhabitants and covered 52,000 square miles. Average density by then had fallen to only 2,177 persons per square mile.

Thus between 1960 and 1980, 41 million persons were added to UAs of more than 50,000 inhabitants, while the total land area of such UAs expanded by 27,906 square miles. This amounts to an average density of 1,469 persons per square mile, or 2.2 persons per acre. Each new UA resident therefore accounted for 0.45 acre of additional urbanized land (including residential, nonresidential, and transportation uses).

The population of the United States is expected to stabilize (i.e., reach zero population growth) at about 289 million early in the next century (Brown, 1987). This would represent an increase in total population of 63 million above the 1980 level. Assuming conservatively that the proportion of this increase that will accrue to UAs is the same as in 1980 (61 percent of total population), this would represent an increase of 38.4 million UA residents. Applying the 1960–80 rate of land consumption per additional UA resident of 0.45 acres, this new population would require about 17 million additional acres of urbanized land or about 27,000 square miles. That would raise the total UA land area to about 79,000 square miles (about 50 million acres). This would represent an addition of new urban land between 1980 and "ZPG year" exceeding in size the total area of UAs in 1960 (24,111 square miles) and about equivalent to the increase in that area between 1960 and 1980 (27,906 square miles). In other words, the total land area of UAs exceeding 50,000 inhabitants in 1960 doubled over the following 20 years and will be tripled by the time the nation reaches ZPG. Barring economic collapse or other calamity, the United States appears to be well on its way to replicating for a second time the total

Table 1–2 Urbanized areas exceeding 50,000 in population: census data (1920–1980)

	No. of Areas	Population (millions)	Land Area (sq. mi.)	Average Density
1920	131	38.9	6,318	6,160
1940	190	58.1	10,452	5,560
1960	269	98.7	24,111	4,090
1980	366	139.1	52,017	2,675
(1980 central cities)		67.0	18,876	3,551
(" rest of UA)		72.1	33,141	2,177

SOURCES: Pickard, 1967. U.S. Bureau of the Census, 1984.

area of land urbanized by 1960. (Public response to urbanization during the 1960s and 1970s is considered in Chapters 9, 11, and 13.)

Even the achievement of national ZPG would not terminate the expansion of all UAs. Both intra- and interregional shifts of population, which were prominent in the 1970s, may be expected to continue. Thus, while net national population may level off at approximately 289 million, some UAs will continue to attract migrants from other regions. Urban land requirements in such growing UAs (e.g., Tampa, Phoenix, San Diego) would continue to rise, but land already urbanized in UAs losing population (e.g., New York, Cleveland, Detroit) would remain in urban condition, albeit underutilized. Thus, total urbanized land could continue to expand indefinitely.

The rate of such future expansion is difficult to predict. Earlier discussion simply extrapolated from rate of land consumption per additional UA resident in 1980 (0.45 acres). This rate might prove to be too high in light of the trend toward smaller household size (including single parent households) and toward multiple-family construction (rental and condominium) in place of single-family houses. A recurrence of the 1970s oil shocks coupled with the substantial completion of the federal Interstate Highway System also might suggest a lower rate of urban land expansion in the future.

Problems of undercounting and overcounting may be expected. Infilling of residual pockets of undeveloped land already included in UAs would not add to the statistical size of such units, even though appreciable effects from such conversions might be experienced at a local or metropolitan scale. On the other hand, population growth in smaller towns and cities, which presently do not qualify as UAs of more than 50,000 inhabitants, will promote many of them to that status. At that point, their entire land areas would be added to the national UA total. That would overstate the amount of newly urbanized land since long-standing urban land in newly recognized UAs would be included as well.

A number of issues for public policy arising from the urbanization of land may be identified in general terms at this point, with more detailed consideration of some of them deferred until later chapters.

- **Inefficient use of land**
  waste of cropland
  loss or pollution of wetlands
  overextension of public services
  visual blight
- **Energy waste**
  lengthy journeys to work
  traffic congestion
  decline of public transportation
  heating and air conditioning of small structures
- **Natural hazards**
  urban flooding

seismic risk
soil and slope instability
coastal storm hazards
- **Solid Wastes**
rising volume of wastes
shortage of landfill capacity
siting of new landfills and incinerators
- **Public recreation and open space**
spatial imbalance of supply and demand
multiple functions and constituencies
deterioration of older facilities
- **Affordable housing**
exclusionary zoning
inadequate public financing
conversion of rental units to condominiums
deterioration of older housing

CONCLUSION

This chapter has summarized the status and usage of land resources in the United States as of the mid-1980s. A clear dichotomy exists between rural land uses on the one hand and urban and built-up uses on the other. Rural land, predominantly used for cropland, grazing, or forestry, is abundant in quantity to meet anticipated domestic needs. Adequacy in terms of international demand is difficult to assess in light of fluctuating currency exchange rates and global trade patterns. The overriding goal of public policy toward rural land should be to preserve the productive capacity or "sustainability" of such resources to meet future domestic and foreign demand. In particular, those lands deemed most or least suitable for specific uses should be identified, designated, and managed accordingly by public and private land managers. Reversible conversion of rural land from one use to another is a normal response to changing economic circumstance. Irreversible transformation of productive rural land, either to a degraded condition (e.g., due to soil erosion, salinization, or inundation) or to urban or built-up condition pose important public policy issues.

The spatial growth of urban land is the mirror image of the loss of rural land to development. But the implications of such growth are not limited to the loss of productive or potentially productive rural land. Urbanization involves a spectrum of public issues including environmental quality, adequacy of water supply, equity in housing and economic opportunity, energy consumption, traffic congestion, visual blight, natural hazards, and rising public costs per capita for providing utilities and services to a vastly expanded region of urban habitation.

REFERENCES

BROWN, L. et al. 1987. *State of the World-1987*. New York: Norton.

CLAWSON, M. 1963. *Land for Americans: Trends, Prospects, and Problems*. Chicago: Rand McNally.

———. 1983. *The Federal Lands Revisited*. Baltimore: Johns Hopkins Press.

CONSERVATION FOUNDATION. 1984. *State of the Environment: An Assessment at Mid-Decade*. Washington: Conservation Foundation.

CROSSON, P. R. AND S. BRUBAKER. 1982. *Resource and Environmental Effects of U.S. Agriculture*. Washington: Resources for the Future.

FRAYER et al. 1983. *Status and Trends of Wetlands and Deepwater Habitats in the Conterminous United States: 1950s to 1970s*. St. Petersburg, FL: U.S. Fish and Wildlife Service.

MITSCH, W. J. AND J. G. GOSSELINK. 1986. *Wetlands*. New York: Van Nostrand Reinhold.

PICKARD, J. P. 1967. *Dimensions of Metropolitanism*. Research Monograph 14. Washington: Urban Land Institute.

PLATT, R. H. 1985. "The Farmland Conversion Debate: NALS and Beyond." *Professional Geographer* 37(4): 433–442.

U.S. BUREAU OF THE CENSUS. 1984. *Population and Land Area of Urbanized Areas . . . 1980 and 1970*. Supp. Report. Washington: U.S. Government Printing Office.

———. 1985. *Statistical Abstract of the U.S.—1986*. Washington: U.S. Government Printing Office.

U.S. COUNCIL ON ENVIRONMENTAL QUALITY. 1979. *Environmental Quality—1979*. Washington: U.S. Government Printing Office.

———. 1984. *Environmental Quality—1984*. Washington: U.S. Government Printing Office.

U.S. DEPARTMENT OF AGRICULTURE. 1974. *Our Land and Water Resources*. Misc. Pub. no. 1290. Washington: U.S.D.A. Economic Research Service.

———. 1981. *America's Soil and Water Conditions and Trends*. Washington: U.S.D.A.

———. 1987. *The Second RCA Appraisal: Soil, Water, and Related Resources on Nonfederal Land in the U.S.* (Review Draft). Washington: U.S.D.A.

U.S. SENATE. 1985. *Oversight Hearings on Section 404 of the Clean Water Act*. Committee on Environment and Public Works, Subcommittee on Environmental Pollution (99th Congress, 1st Session). Washington: U.S. Government Printing Office.

U.S. SOIL CONSERVATION SERVICE. 1977. *Potential Cropland Study*. Stat. Bull. no. 578. Washington: U.S. Government Printing Office.

———. Undated. "Preliminary Data, 1982 National Resources Inventory: Executive Summary." (mimeo).

2

THE INTERACTION OF GEOGRAPHY AND LAW

The fields of geography and law to some people may seem to have as much in common as astronomy and neurology, or mathematics and poetry. But as applied to problems of managing land resources, the two fields are logical and necessary allies, with much to contribute to each other. Geography, if defined as the science of spatial organization of human activities, seeks to explain the functional interrelation of units of land area to each other and to larger systems of land use, for example, communities, regions, nations, and the planet.

The law of land use is concerned with the process by which land is allocated for various purposes through the recognition and exercise of private rights and public powers to influence the use of land. A rational society uses its laws to facilitate the efficient and productive use of land. According to the philosopher John Locke (1689/1952, p. 71):

> I cannot count upon the enjoyment of that which I regard as mine, except through the promise of the law which guarantees it to me. It is law alone which permits me to forget my natural weakness. It is only through protection of law that I am able to enclose a field, and to give myself up to its cultivation with the sure though distant hope of harvest.

Law thus confirms property rights in a capitalist society. If Locke had lived in the twentieth century rather than the seventeenth, he would surely have mentioned the additional role of law as a means to control harmful externalities among users of land. Both of these functions of law—facilitating productive use of land and abating externalities—are fundamentally influenced by the geographical context of the land in question.

This book examines the interactive nature of geography and law as reflected in the evolution of societal controls over land use. Law is viewed both as a product of societal perception of the physical environment and as an independent variable that itself shapes that environment in sometimes unexpected ways. The model that describes this interactive relationship is discussed later in this chapter. First, however, the opening section contrasts some fundamental differences between the two fields which have impeded recognition of their interdependence. After that, we consider the role of law in the shaping of land-use patterns or morphology and briefly summarize the nature and powers of land-use managers (which will be considered in more detail in later chapters).

THE GEOGRAPHICAL LANDSCAPE

The field of geography is inherently eclectic. Its two major branches, physical and human geography, are in turn subdivided into a number of subfields which respectively contribute to an understanding of the nature and use of land resources. These subfields include on the physical side, geomorphology, hydrology, biogeography, and climatology; and on the human side, urban, political, economic, cultural, and social geography, among other specialties. Some geographical topics bridge the physical-human dichotomy, as in studies of human response to natural hazards (James and Martin, 1981).

It is sometimes charged that geographers have more in common with their colleagues in cognate disciplines (e.g., geology, economics, political science, urban planning) than they do with one another. Geographers have sometimes wondered if they have a field of their own. In the role of devil's advocate, the noted geographer Richard Hartshorne (1939, p. 125) has written: "Defined not in terms of a particular set of facts, but in terms of causal relationships presumed to exist, [the field of geography] could have but a parasitic character."

Its interest in causal relationships (the perennial "Why"), however, as Hartshorne demonstrated, indeed defines geography as a separate discipline. Certain organizing themes, models, and concepts characterize the geographic perspective and method. For present purposes, the following concepts are pertinent:

1. spatial organization,
2. scale,
3. function, and
4. externalities.

Spatial Organization

A common denominator that unites geographers of all specialties is a fundamental concern with the organization or distribution of diverse phenomena in space (James and Martin, 1981, Ch. 15). Spatiality is to geography what spirituality is to the ministry and health is to the physician. The identification and explanation of areal differentiation of diverse phenomena, notably including land use, is a long-standing geographical pursuit (Morrill, 1970).

In short, geographers are concerned with the where and the why of spatial patterns on the earth's surface. To delineate the where, geographers utilize maps, photographs, graphs, and other means of representing spatial data. Cartography, the development and preparation of maps suitable for particular tasks, is itself a subfield of geography. To explain the why of perceived spatial patterns of phenomena requires the application of various sources of inference including statistical analysis, empirical field research, modeling, and scholarly intuition. Geographical analysis often involves the identification of cross-relationships between different systems of spatial data.

Spatial differentiation of land use may be interpreted in terms of the interaction of three overlapping categories or macrosystems of spatial data: (1) physical phenomena; (2) human socioeconomic activities (referred to by geographers as *cultural phenomena*); and (3) patterns of legal and political authority over land. (See Figure 2–5 and accompanying discussion later in this chapter.)

Under the first category, the physical landscape may be described and interpreted in terms of the spatial characteristics of bedrock and landforms, soils, hydrology, natural vegetation, wildlife, and climate. Each of these subclasses of physical data represents a major branch of natural science. The geographer draws on the findings of the appropriate field to the level of detail necessary to resolve the problem under consideration.

The geography of "cultural" human activity patterns includes systems of rural land use—agriculture, forestry, and mining—as well as urban settlements ranging from hamlets to metropolitan regions. Aggregate human activity patterns reflect systems of economic enterprise. It has been argued that human geography is basically economic geography (Hartshorne, 1939). The spatial analysis of economic systems involves the identification of "nodes" or points of activity and "linkages" or connections between nodes, for example: transportation corridors, pipelines, and communications networks. Such linkages serve as conduits between activity nodes for a variety of dynamic commodities, for example, people, materials, nutrients, energy, goods, information, and capital.

Spatial patterns of legal and political authority may be analogized to a series of jurisdictional templates overlying land, which represent respectively: (1) ownership (private or public); (2) minor civil divisions (municipalities, special districts, counties); (3) states; (4) the federal government. Units of authority at each level are bounded by precise, if irregular, territorial limits that define the geographic reach of their legal power over land. Moreover, units at different levels in the hierarchy influence the use of land within their jurisdictions in different ways according to their respective legal, political, and fiscal capabilities. The use of individual parcels of land thus reflects a complex interaction among the various levels of land managers who share jurisdiction over a given site. Broader patterns of land use result from the aggregation of use characteristics of individual parcels.

Legal and political boundaries of course are invisible to observation unless marked by a sign, fence, or other visual indicator. But the presence of institutional boundaries may often be inferred from observation of abrubt changes in land-use patterns, as from a high-income, low-density residential district to an adjoining area of congestion, commerce, clutter, and blight. The fortuitous location of legal and political boundaries often exerts substantial influence on the arrangement of human activities and perceived land-use patterns (Whittlesey, 1935; Cohen and Rosenthal, 1971; Clark, 1985). Of course, institutional patterns are not totally independent variables. Particularly in the case of suburban jurisdictions established in this century, political boundaries tend both to reflect and in turn to reinforce the spatial arrangement of housing markets, economic activity, and locational preferences of the wealthy.

Scale

Fundamental to geographic analysis of spatial organization is the concept of scale or hierarchy. Certain kinds of spatial variability may be classified according to position within a hierarchy of nested subsets of phenomena that comprise the overall spatial system under consideration. In the physical context, for instance, a riverine system may be divided into a hierarchy of main stem, major tributaries, subtributaries, and headwaters, each with its associated area of surface drainage or watershed. For purposes of land planning and water management, the position of a tract of land in relation to this hierarchy of drainage is crucial in relation to quality and quantity of surface flow and degree of flood risk.

Patterns of economic activity and human settlement are also hierarchical. Commercial centers may be classified in terms of size and complexity, for example: (1) crossroads gas pump and convenience store; (2) traditional village center; (3) conventional postwar shopping center; (4) medium-sized city center; (5) regional shopping center; (6) large city center; (7) primary metropolitan center (e.g., New York, Chicago, San Francisco).

The position of a particular settlement in this hierarchy is not accidental. Geographers have developed numerous theories and models to account for the spatial organization of urban systems. Central place theory, in particular, relates the size and spacing of commercial centers to the distance consumers are willing to travel to obtain certain goods and services. Thus, beer, bread, and milk are purchased from the nearest among a myriad of neighborhood or convenience outlets. New cars and banking, legal, and health services are obtained from larger ("higher-order") centers, while rare art objects, fur coats, large-scale financing, and open heart surgery are likely to be sought in a major metropolitan city. Commercial centers at each higher level thus provide a wider array of goods and services to consumers from a broader geographic "hinterland." Concomitantly, the size and diversity of a commercial center is limited by its proximity to competing centers of the same or higher order (Berry and Garrison, 1958; Berry, 1967; Palm, 1981, Ch. 8).

The geographic location of a tract of land in relation to the hierarchy of urban places, like its position in a drainage system, bears an important relationship to its suitability for particular uses. The history of American land settlement is replete with examples of speculative land ventures that failed to achieve the shining prosperity predicted by their promoters, due in part to locational disadvantages. For every Chicago or St. Louis, there have been many disappointed Michigan Citys (Indiana) or Cairos (Illinois). The equivalent in contemporary metropolitan America is the proliferation of regional shopping centers and office parks, regardless of the need or economic viability of such uses in a particular location. Furthermore, municipal zoning for such sought-after development typically precludes uses more suited to the location, such as lower-cost housing. As discussed in Chapter 8, the exercise of public land-use control powers incompatibly with geographic reality may be constitutionally unsound.

Function

A closely related notion to that of scale is the concept of function. The raison d'etre of a settlement, and its size and rank in the central place hierarchy, are directly related to the function(s) that it performs. An urban place without a function is a virtual nullity, regardless of what laws or promoters may say. Colonial legislatures of Virginia and Maryland attempted to encourage the growth of towns by laying out sites adjoining rivers and granting them special port privileges. The region, however, shipped its products directly from river landings at each plantation and needed no port towns. In the pithy words of Thomas Jefferson (1784/1944, p. 227), a geographer among his many talents:

> ... the laws have said there shall be towns; but nature has said there shall not, and they remain unworthy of enumeration.

Urban economic functions may take the form of primary (agriculture, forestry, mining); secondary (manufacturing); or tertiary (retail and service) activities, or a combination of them. Some urban places are characterized by a dominant function, such as ports, transportation hubs, governmental centers, academic towns, recreational resorts, and retirement communities. Urban functions change over time with technological innovation, patterns of economic investment (both intra- and international), demographic and lifestyle trends, shifts in political power, and changing popular perception. Thus urban places may lose or gain in their range of functions and correspondingly in their size and importance within the national system of settlements.

The notion of function may also be applied to individual parcels of land or real property. This concept relates each discrete unit of land to the larger physical and human spatial systems of which it is a component. In ecological terms, land in its natural state may facilitate recharge of ground water, store surface runoff in soil moisture or ponding, support vegetation, concentrate energy through photosynthesis, and provide habitat for wildlife. Land within the agricultural system functions as cropland, pasture, fallow, horticulture, or farmstead. Urban land functions as a site for residential, commercial, industrial, institutional or public activities.

The notion of function is related to, but not synonymous with, land use. Function refers to the relationship between a parcel of land and the wider physical and socioeconomic spatial systems to which it belongs. Even vacant land that is unused in a market sense may function in a physical and socioeconomic sense, and such functions may be either positive or negative witn respect to the surrounding area. Thus, a vacant urban lot may have no formal use but may function as visual amenity, perceptual buffer between urban activities, a play space for children, an informal parking area, a habitat for wildlife. Negatively, it may serve as a dumping site for trash, junk cars or hazardous wastes, or as a refuge for drug abusers or other illicit activities.

Externalities

The concept of function suggests that no parcel of land is an island unto itself. The use or condition of any unit of land generates external effects or "externalities" on surrounding areas. Such effects, which may be either positive or negative, result from the inconsistency of the various patterns of spatial organization of land—physical, socioeconomic, and institutional. While the authority of land managers is confined to the geographic space defined by their jurisdictional boundaries, the physical and socioeconomic consequences of their actions are distributed according to the spatial pattern of the variable considered. (See Figure 2–1.)

Physical externalities may take such forms as air and water pollution; flooding; impacts on fisheries, birds and other wildlife; depletion of water supplies; littering and dumping of wastes; noise; and visual blight. Socioeconomic externalities due to spatial shifts in economic activity result in loss of jobs, retail sales, and taxes and increased burdens on schools and other public services in adversely affected jurisdictions. Traffic congestion in neighboring jurisdictions is a frequent concomitant of a development decision.

While the nature, extent, and economic consequences of externalities vary according to the scale at which they arise, the fundamental problem is the same: How can favorable externalities be encouraged and adverse externalities suppressed or mitigated? The geographer, having thus framed the problem, refers its solution to the law.

THE LEGAL LANDSCAPE

The legal landscape is very different from that of the geographer. While the latter is a composite of several interacting types of spatial phenomena—physical, economic, social, and institutional—the law is primarily concerned with the last category, legal and political authority over land, and only secondarily with the others.

EXTERNALITIES

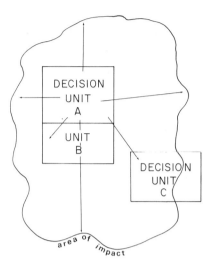

Figure 2.1 Diagram of spatial impacts of externalities on adjacent and nonadjacent land.

Where the geographical model of the land economy identifies systems of nodes, linkages, flows, hierarchies, and functions, the legal counterpart may best be described as a "battlefield" on which private property interests struggle against each other (the private law context) as well as against governmental constraints imposed on behalf of the "public interest" (the public law context). This landscape is crisscrossed with earthworks of entrenched legal interests and is littered with the shell craters of past legal salvos.

The adversarial and pragmatic perspective of law contrasts with the positivist orientation of geography. Land-use patterns to the lawyer are the collective outcome of a myriad individual cases, conflicts, appeals, administrative rulings, and political actions to which geographical notions of scale, function, and central place theory may seem irrelevant abstractions. The holistic, systematic perspective of geography yields to the particularistic, adjucative focus of the law. To the latter, substantive outcomes, particularly broad scale, long-term applications, are secondary to constitutional issues of fairness and reasonableness of the process by which conflicts are resolved. As will be discussed in Chapter 8, however, the constitutionality of public land use measures is strongly related to the reasonableness of their impact, which raises geographical questions.

A fundamental dichotomy in land use law lies between *ownership* (private or public) and *jurisdiction* (public). The former represents a set of rights and duties that inure to the party(ies) that hold a legally recognized interest in specific real property. The latter represents the power of a governmental body to regulate private activities including land usage within its political boundaries. Confusingly, both the public authority and the geographic territory within which it is exercised are each referred to as "jurisdiction."

Figure 2–2 represents the templates of ownership and public jurisdictions that overlie land in the United States. It is the interaction of the relevant units at each level of the hierarchy that determines the usage of any specific tract of land.

The recognition of competing interests is itself an important difference between geography and law. The former identifies spatially differentiated "clusters" of common interest defined by diverse criteria (for example, immediate neighbors, neighborhood, ethnic community, business district, watershed, metropolitan housing market, and so on). In contrast to this fluid and dynamic approach, the law rigidly confines its recognition of legitimate "parties in interest" in land use controversies to those deemed to possess legal "standing" which qualifies them to seek legal redress. Classes of parties with legal standing in land use disputes normally are limited to property owners directly affected by a decision, their immediate neighbors, the municipality, the state, and the federal government. Other geographically relevant constituencies such as the neighborhood, downstream watershed occupants, or the metropolitan region lack legal standing except to the extent that an organizational surrogate may successfully claim to represent affected interests (for example, a community, environmental, or civil rights organization).

The complex "nested hierarchies" of political geography are collapsed into a monolithic "public" for purposes of land use litigation that commonly takes the form of private property owner versus "the public." This blurs or erases geographical distinctions between diverse "publics" with very different interests—neighbors, community, region, state, nation, globe.

HIERARCHY OF LAND USE MANAGERS

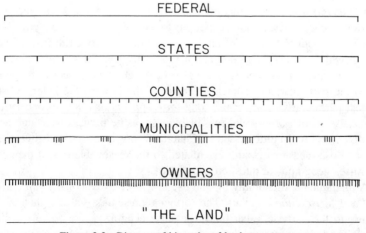

Figure 2.2 Diagram of hierarchy of land-use managers.

Actually, three classes of land use disputes may be identified: 1) owner *versus* owner; 2) owner *versus* public (just mentioned); and 3) public *versus* public. Disputes of the first type are usually settled by a court through application of common law doctrines such as nuisance and trespass derived from past judicial decisions in the United States and Great Britain (see Chapters 3 and 5). The second class of disputes involves *constitutional* issues governing the respective rights and powers of private and governmental interests (see Chapter 8). The third category may involve disputes between peer units of government on the same level (horizontal) or between units at different levels (vertical). Disputes between the federal government and one or more states present issues of *federalism,* while disputes between a state and a local government involve the extent of municipal autonomy under express or implicit grants of state authority or *home rule* (Clark, 1985). (See Chapter 6 for further discussion.) Federal–local disputes are also possible.

Land use disputes of all three classes listed above are prevalently concerned with harmful externalities inflicted by one land use manager on one or more other units (of the same or different level in the hierarchy). Like private owners, units of public authority seek to minimize internal benefits of land use within their own jurisdiction, to ignore adverse externalities inflicted on neighbors, and to resist such adverse impacts being inflicted on themselves. In this "beggar thy neighbor" game, there is no real "public interest," regardless of the constitutional rhetoric. There is only the interest of the "publics" represented by the parties to the litigation. In the absence of a representative of a genuine public interest, courts are often at a loss to determine which "sub-public" to protect, for example, private owner, municipality, or region. (See contrasting views of the Federal District Court and Court of Appeals in the *Petaluma* case in Chapter 9.)

With all the contrasts indicated above, can it be said that land use geography and law occupy any common ground? The answer is strongly affirmative. It was suggested above that the problem of externalities is an area of overlap between the geographical and legal frames of reference. Externalities, which represent friction among components in the geographer's macro view of the land economy, are in fact the central concern of the lawyer's micro view. If there were no externalities among land use management units, there would be no need for land use law. The central problem of land use law is that of externalities, and that problem is fundamentally geographical.

THE LEGAL PROCESS

Before proceeding further, it is important that the nonlegal reader understand the context in which the American law of land use is articulated. The starting point is the U.S. Constitution—the "supreme law of the land." The federal Constitution and its state counterparts establish fundamental principles governing the relationship between government (federal, state, municipal) on the one hand, and the private individual on the other. Of particular importance to the land-use context, the Fifth Amendment to the U.S. Constitution declares:

> No person shall be . . . deprived of life, liberty, or property without due process of law, nor shall private property be taken for public use without just compensation.

Responsibility for the establishment of laws concerning land use lies with the individual states, which in turn have delegated most of their authority to local municipalities. Pursuant to state enabling laws, local municipalities (and counties for many unincorporated areas) may adopt land-use zoning laws and other measures regulating private use of land. In recent years, many states have adopted additional laws establishing statewide regulations for particular types of sensitive or significant areas such as flood plains, wetlands, historic sites, or agricultural land (Kusler, 1985; DeGrove and Stroud, 1987). State and municipal laws may further delegate authority to administrative bodies to promulgate specific rules that elaborate on the broader provisions of the statute.

Statutes and administrative regulations concerning land use are subject to judicial review upon suit by an aggrieved party against the public entity whose action is in question. In such cases, the court is asked to resolve whether the measure is constitutional and fairly applied to the plaintiff. Many suits are decided on procedural grounds without reaching the merits of the case. Others are found to be "fairly debatable" on their merits and the public action is presumed to be valid. In a third set of decisions, courts are persuaded by the plaintiff that the public measure is "discriminatory, arbitrary, or capricious" and the measure is held invalid, at least as applied to the plaintiff. The latter two classes of decisions thus involve some degree of judicial scrutiny of the facts of the case, including its geographic context.

Only a very small proportion of land use disputes are the subject of a lawsuit and even fewer are appealed by the unsuccessful party in the trial court to a higher court, cases of major significance occasionally reach the U.S. Supreme Court which has issued only about 20 land-use opinions during the 1970s and 1980s. The state supreme courts collectively account for the preponderance of significant decisions in the field of land use.

Published decisions on land-use cases in the state and federal court systems comprise a rich archive of judicial perspectives on the relationship of law and geography. Judicial opinions result from the application of legal authorities (e.g., prior case law, statutes, principles of "black letter law," learned treatises, etc.) to the facts of the case. It is the role of the attorneys for each party to portray applicable legal authority and the facts of the case favorably to their respective positions. The court in turn forms its own opinion of the state of the law and the facts of the case and reaches a decision accordingly. The legal outcome therefore reflects in part the court's perception of the geographic context.

Judges, being human, do not necessarily view the circumstances of a land-use issue in the same way (Clark, 1985). Judicial disagreement may arise: (1) between individual judges on a multi-judge court (as expressed in dissenting opinions); (2) between a lower and higher court reviewing the same case; (3) between courts in different states or federal jurisdictions reviewing similar cases; and (4) between courts considering a similar issue at different points in time. The last category is particularly important in weighing the role of geographical perspective in the judicial process. Law is a flexible and dynamic institution. The adjudicative process permits reinterpretation of legal principles over time in response to actual or perceived changes in society and its needs. As stated by the U.S. Supreme Court in *Village of Euclid v. Ambler Realty Co* (272 U.S. 365, 1926):

> . . . while the meaning of the constitutional guarantees never varies, the scope of their application must expand or contract to meet the new and different conditions which are constantly coming within the field of their operation (272 U.S., at 386).

Leading decisions on similar land-use issues over time may thus reflect shifts of legal response to new and different conditions among which geographical circumstances loom large.

A NOTE ON LEGAL CITATIONS IN THIS BOOK

Many statutes and judicial opinions are cited in this book, particularly in Chapters 7, 8, 9, 11, and 12. For law review articles, legal briefs, and treatises, the standard forms of legal citations are set forth in *A Uniform System of Citation* ("Blue Book") available in any law library. In the interest of simplification, this book does not adhere strictly to those rules of style. Federal statutes are usually cited by their Public Law (P.L.) number, which refers to the text of a law as originally adopted. For example, the National Environmental Policy Act of 1969 (P.L. 91–190) was the 190th act to be passed by the 91st Congress. Public Laws are published in two forms: the *U.S. Code Congressional and Administrative News* (an annual series of very large volumes found in any law library) and *Statutes at Large* (less frequently used and sometimes not available).

The Public Law form of a statute of course does not reflect earlier statutes that it may have amended or subsequent laws that amend it. The current version of a federal law, reflecting all additions and deletions by various public laws passed at different times, is "codified" by subject matter in the *U.S. Code* (U.S.C.) and *U.S. Code Annotated* (U.S.C.A.). The former contains only the statutes themselves, while the latter provides a wealth of additional information on legislative history, changes in language, court decisions, and law review articles that discuss the statute, and so on.

Discussion in this book is primarily historical, focusing on the evolution of public response to perceived environmental needs. The use of Public Law citations is appropriate for this purpose, since we are interested in the language of a law as adopted at a particular moment in time. But for anyone who wishes to research the current status of a federal law, the U.S.C. or U.S.C.A. are indispensable. [State laws follow the same two-fold form of citation: (1) by order of adoption and (2) codified by subject matter].

Judicial opinions (court decisions) are published in series of "reporters" by West Publishing Co., which are available in law libraries. The standard form of citation which is used in this book is as follows:

Plaintiff v. Defendant, Vol. no.   Reporter   First Page no., (State, Year).

For example, *Just v. Marinette County,* 201 N.W.2d 761 (Wis., 1972). (N.B. "N.W.2d" refers to the Northwest Reporter, Second Series. Consult a law librarian for explanation of other reporter abbreviations.)

LAW AS AN AGENT OF LAND-USE MORPHOLOGY

The influence of law in the pattern or morphology of landscape was noted by the Harvard geographer Derwent W. Whittlesey (1935, p. 85):

> Political activities leave their impress upon the landscape, just as economic pursuits do. Many acts of government become apparent in the landscape only as phenomena of economic geography; others express themselves directly. Deep and widely ramified impress upon the landscape is stamped by the functioning of effective central authority.

The "impress" of law, however, differs from place to place, from one historical period and social order to another, and among different types of land use. As will be illustrated in the case of Boston, the role of law may be discerned in different ways among districts of the same city built at different periods of social and economic development. The urban geographer James E. Vance, Jr. (1977, p. 24) draws a rough dichotomy between "organic cities" which reflect individual rather than centralized control over their origin and growth, and "preconceived cities" which display clear evidence of an imposed land use plan (Figure 2–3). Most urban landscapes lies somewhere between these two extremes. Law as an instrument of both private rights and public authority is a ubiquitous agent in the evolution of urban form, and is often discernible in the rural landscape as well. Law, along with other factors such as topography, regional context, economic function, and culture, is an agent of land-use morphology.

To further illustrate this point, this section first considers the most conspicuous and familiar example of a preconceived or imposed land-use pattern: the grid street plan. It then takes a longitudinal look at the evolution of a particular city, Boston, in terms of the relative influence of law over time.

Law as an Agent of Land-use Morphology

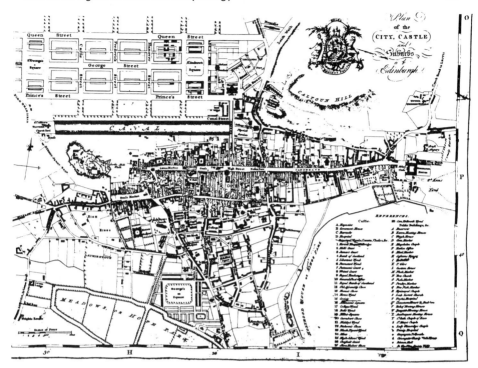

Figure 2.3 Edinburgh, Scotland: medieval Old Town and Craig's baroque New Town, 1766. Reproduced by permission of the Scottish Records Office.

Imposed Plans: The Ubiquitous Grid

The use of the grid as a basis for land allocation and street layout has been traced back to the third millenium B.C. in the Indus Valley by Stanislawski (1946), who cites such a plan, wherever found, as evidence of centralized control:

> This pattern is not conceivable except as an organic whole. If the planner thinks in terms of single buildings, separate functions, or casual growth, the grid will not come into being; for each structure considered separately the advantage lies with irregularity. History is replete with examples of the patternless, ill-formed town that has been the product of growth in response to the desires of individual builders.

Similarly, Castagnoli (1971, p. 124) stresses the influence of governing authority in establishing a preconceived pattern of land allocation and usage:

> The regularity or irregularity of town forms depends entirely on the presence or absence of spontaneity in their birth and growth. The irregular city is the result of development left entirely to individuals who actually live on the land. If a governing body divides the land and disposes of it before it is handed over to users, a uniformly patterned city will emerge.

The grid plan serves several administrative goals: equitability of land allocation, defensibility, convenience of survey, and ease of expansion into later settled areas. Its disadvantages include incompatibility with irregular terrain, problems of defensibility, excessive street length in relation to built-up land, and (to contemporary observers at least) monotonous regularity.

Grid street plans have been employed in diverse sociocultural contexts and periods of urban history. Beijing, Kyoto, Mexico City, Berlin, Buenos Aires, New York, and most North American cities are orthogonal, at least in part. The grid has been particularly suited to the layout of colonial outposts and new settlements, from the cardo and decumanus of Roman enclaves in conquered territories (Grimal, 1954/1983), to thirteenth-century bastide towns in France, Wales, and Ireland (Beresford, 1967), to Spanish and French settlements in the New World following Roman precedents (Stanislawski, 1947). The grid also appears where a newly planned sector has been added to an earlier, unplanned city, as in London's seventeenth- and eighteenth-century West End estate developments. James Craig's 1766 plan for Edinburgh's New Town stands in stark contrast with the organic clutter of the old medieval town (Figure 2–3). Pombal's reconstruction of central Lisbon following the 1755 earthquake similarly was rigidly orthogonal (Mullin and DeMello, unpublished ms.).

The grid served as a standard pattern for new settlements established by land proprietors in colonial America such as Penn's 1686 checkerboard plan for Philadelphia, Oglethorpe's 1733 plan for Savannah, and the 1820 Mount Auburn subdivision in Boston (Reps, 1969). The pervasive rectangularity of rural and urban land use in the United States west of the Appalachians results from one of the first acts of the new national government upon its assumption of sovereignty, the federal cadastral survey initiated in the Land Ordinance of 1785 (Clawson, 1983, p. 20).

The grid plan thus unmistakably reflects the influence of supervening authority—crown, military governor, proprietor, national or local government—upon the use of land by individuals. So also do other geometric urban forms that conspicuously reflect an imposed plan, as in the case of round cities (Johnston, 1983), baroque cities (Sitte, 1945), or composite cities such as Washington, D.C. (which combines a grid plan and a baroque radial plan).

But the urban-shaping role of legal constraints on land use is not necessarily as obvious or as areally extensive as in these cases. The influence of law on urban form may operate more subtly and prosaically, as for instance in the width of streets; the size, spacing, construction, and use of buildings; and in the balance of built and unbuilt space. How are physical differences among sections of the same city explained? To paraphrase landscape architect Grady Clay (1973), we must "read the city" in terms of legal and institutional influences as well as economic, cultural, and geomorphic factors.

Evolution Over Time: Boston

A short walk through downtown Boston traverses an archive of different stages of public involvement in the city-shaping process (Figure 2–4). Starting in the

Law as an Agent of Land-use Morphology

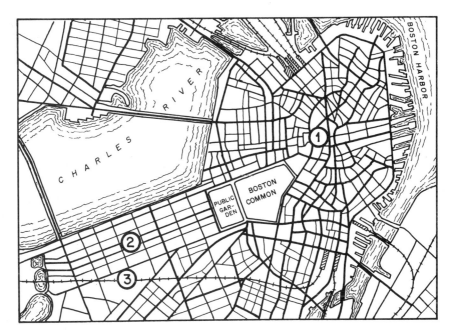

Figure 2.4 Three eras of legal geography in Boston, Massachusetts: 1) the North End; 2) Back Bay; 3) Prudential Center.

vicinity of Commercial Wharf, one wanders through the irregular, narrow street pattern of the North End and financial district dating from the mid-seventeenth century. These lead eventually to the city's open core, the Common and Public Garden, set aside from private development by public action respectively in the 1630s and the 1830s. Adjoining the Public Garden to the west are the rectilinear streets and bowfront brick rowhouses of Back Bay, Boston's mid-nineteenth-century expansion onto newly filled land bordering the Charles River. Crossing Boyleston Street from Back Bay, one enters the high-rise complex of multiple-use structures in Prudential Center and adjacent areas.

How does one account for the perceived differences in morphology between the North End, Back Bay, and Prudential Center? Contrasts between the three districts may be conventionally explained in terms of their functional relationships to the economic circumstances of Boston and the demand for land at the time of their development. Thus, the North End outwardly reflects its origins as the core of a crowded, mercantile settlement oriented chiefly to its wharves (McManis, 1975, pp. 81–82). Back Bay originated, and in part still serves, as an elite residential quarter reflecting Boston's nineteenth-century prominence as a center of education, finance, and culture (Whitehill, 1968). The Prudential Center involved the revitalization of a blighted site to enhance Boston's function as a corporation headquarters and convention site (Conzen and Lewis, 1976, p. 50). One may also interpret the morphology of the three districts in terms of changes in transportation technology, from foot and sailing vessel, to carriage and streetcar, to automobile and elevator. In construction technology, one moves from the era of clapboard and shingle; to brick and slate; to reinforced concrete, glass, and steel.

There is an additional dimension of contrast between the three districts, namely the extent and form of public intervention in the private building process. Boston's early growth was largely organic. Actions by the town selectmen to constrain individual freedom in building were limited to measures concerned with fire, as in specifying materials to be used in roofing and chimneys and in requiring the possession by householders of fire-fighting implements (Bridenbaugh, 1964, pp. 55–61). Otherwise, the town placed no restrictions on the layout of individual structures.

Back Bay, by contrast, was a totally preconceived expansion of Boston. New land created through filling of the malodorous fens vested legally in the Commonwealth of Massachusetts, which historically holds tidelands in trust for the public. In 1856, a multipartite indenture to govern the filling and development of the fens was executed between the Commonwealth, the City of Boston, and various private proprietors (Whitehill, 1968, p. 151). Pursuant to this agreement, the Commissioners on the Back Bay, a legal entity created by the legislature in 1852, exercised plenary control over the layout of streets and disposition of parcels. Purchasers of building lots were required to accept deed restrictions limiting the use, height, and external appearance of structures. Municipal land-use zoning would not appear in Boston until after World War I. Meanwhile, deed restrictions provided strict legal control to ensure harmonious development of Back Bay.

Prudential Center legally resembles the Back Bay project to the extent that public authorities facilitated the development of an underutilized site (in this case a railroad yard) and controlled the physical result through deed restrictions. But Prudential Center involved other public roles as well. The site (including the newly recognized property interest in "air rights") had to be acquired from the private owner using public eminent domain power. It was then reconveyed at a lower cost to the redevelopment corporation. The latter was required to provide an auditorium and convention hall, as well as public ways and parking spaces, as a condition to constructing private commercial space. A special state law deferred certain real estate taxes for up to 45 years as a subsidy to the development's future profitability (Haar, 1963, pp. 181–2). Prudential Center thus resulted from a complex interaction of public and private initiatives characteristic of much contemporary metropolitan development.

Legal influences on land use assume many forms and operate in subtle and sometimes contradictory ways. These interact with nonlegal constraints, such as site and situation, economic conditions, culture, and technology, in various combinations. The relative influence of legal and other constraints differs from one time period to another and from site to site. A further tour along the Boston waterfront, for instance, reveals a highly differentiated pattern of selective public intervention and private investment as one proceeds from the revitalized Faneuil Hall Market towards Charlestown or South Boston. Historic restoration tax credits, flexible zoning provisions, tidelands building lines, National Park Service programs, accel-

A General Model of Land-Law Interaction

erated depreciation, urban development action grants—these and many other legal measures interact with physical and economic factors to shape the visible morphology of the Boston waterfront.

A GENERAL MODEL OF LAND-LAW INTERACTION

The complexity of land-use management, in terms of the types of decision makers, the powers they exert, and the interactions between them, tends to impede perception of an overall process: The forest cannot be seen for the trees. Furthermore, the legal perspective on land use, as stated earlier, views the decision process as adversarial between private and public interests. In fact, both private and public managers may be viewed as "partners in equity" (to use a term suggested by resource economist Mason Gaffney) which jointly, although not necessarily amicably, determine for better or worse the way in which a society puts its land resources to use.

Figure 2–5 depicts a simple model in which all relevant decision makers,

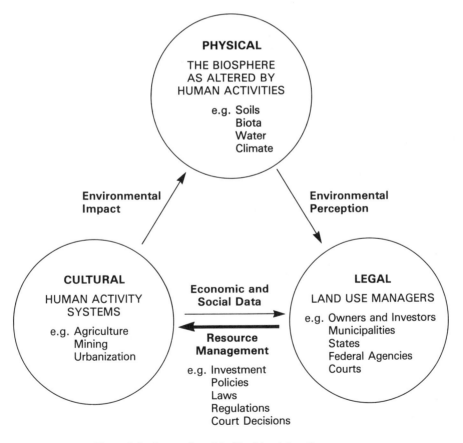

Figure 2.5 A general model of land-law interaction.

public and private, are represented holistically by the "legal" circle. The other two circles represent the "physical" resource and "cultural" or human activity systems. Each circle thus depicts a different set of spatially distributed phenomena or geographies—of the land itself, of human activities on the land, and of legal/political authority that directs those activities. Each circle may be thought of as a template overlying the land surface. For any given location on that surface, a specific set of physical, cultural, and legal characteristics apply. And these are interrelated with each other as suggested by the vectors in the model.

The most important vector for purposes of the present discussion is labeled *resource management,* a term that includes but is not limited to management of land use or "land-use control" (air and water management also fall within the broad parameters of the model). This vector represents the aggregate influence of public and private authority over land use as expressed through the various management instruments available to each type of decision maker. The nature and form of human activity systems such as urbanization are determined by the collective guidance of private and public decisions articulated via the "resource-management" vector.

Two inputs to the "legal" circle inform the content of the resource-management output. One of these is *environmental perception.* This term is used broadly here to refer to the flow of information or cues concerning the state of the physical resource as perceived by managers relevant to their respective areas of property rights or jurisdiction. Environmental cues include data concerning the sustainability of particular land uses, risks posed by natural or technological hazards, and, in general, the positive and negative implications of particular ways of using land.

Environmental perception applies both to cues originating within the land area of direct interest as well as to cues derived from activities external to such land area (externalities). By definition, harmful externalities tend to be ignored by the manager (usually the owner) that generates them, unless forced by the victims to abate them (e.g., through a nuisance action) or by higher public authority through administrative or judicial intervention. The resource-management vector represents the net effect of all these various signals issued by different managers having a legal interest in the use of land in question.

The other input vector represents socioeconomic feedback on the functioning of each activity system to responsible or affected decision-makers.

Environmental and socioeconomic cues are interpreted in terms of the goals or expectations of each manager. Different types and levels of managers may differ in their objectives. Furthermore, different public agencies of municipal, state, or federal governments may seek different objectives, for example, economic development versus environmental quality. Also, each manager differs with respect to receptivity to environmental cues in relation to the second set of inputs—economic and social data derived directly from the activity systems themselves. The relative weight of these two vectors may change over time. Until the 1960s, economic profitability prevailed over resource degradation in determining resource-management actions. With the rise of the environmental movement in the late 1960s, the "environmental-perception" vector gained greater weight.

Clearly the processing of environmental and socioeconomic inputs within the "legal" circle and the interaction of resultant management outputs or signals is very complex. The latter may include conflicting signals from different levels or agencies of government that modulate in different ways the management actions of individual users of land. Also, the use of land is modified in response to changes in official or social priorities over time. For example, the design of cities has historically been influenced by many objectives. From time to time and from place to place, these have included such goals as military defensibility, religious piety, royal or imperial vanity, colonial orderliness, public healthfulness, civic attractiveness, and functional efficiency.

The interactive loop of the model is completed by the vector labeled *environmental impact:* the modification of the physical resource by human activities. In the agricultural context, row cropping on hilly terrain hastens soil erosion and causes sedimentation and pollution of downstream water bodies; irrigation may lead to salinization of the soil mantle or, where based on ground water, to a lowering of aquifer levels. Such practices, if continued unchecked, may lead to a loss of productivity in the areas affected, and eventually to a destruction of a portion of the physical resource. If and when the counterproductive tendency of existing practices is environmentally perceived by a legal entity having jurisdiction over the area in question, new resource-management signals may be issued to modify the status quo. In the case of soil erosion, for instance, Congress has twice acted to modify the destructive practices of farm operators: first in the establishment of the Soil Conservation Service in 1935 and, second, in the creation of a "conservation reserve" in the Food Security Act of 1985. (See Chapter 12.)

Thus the model depicts a dynamic process whereby the way in which land is used in human activity systems may be modified by a new set of resource-management signals emitted by the "legal" circle in response to new awareness of the impacts of existing practices on the physical resource. This new awareness may result from a single dramatic catastrophe, which instantly leads to revision of the prevailing rules, as when the Fire of London of 1666 led to a dramatic reform of building practices in the Act for Rebuilding London of 1667 (as discussed in Chapter 3). Or it may result from a change in social values regarding an existing state of affairs which, previously acceptable or ignored, becomes intolerable and leads to legislative reform. Thus, the Sanitary Reform Movement of the 1840s and the Progressive Movement in the early twentieth century focused public attention on conditions of urban squalor and overcrowding and prompted the adoption of sanitary codes and zoning laws, respectively. (See Chapters 4 and 7.)

The past two decades have witnessed a proliferation of new environmental laws and programs in response to the perception of environmental deterioration inspired by the writings of Rachel Carson, Barry Commoner, Joseph Sax, William H. Whyte, Gilbert F. White, and many others during the 1960s. (See Chapters 11 and 12.) At this writing, the feedback loop of our model is laboring to formulate legal response to the growing recognition of new threats to the biosphere in the form of global warming, stratospheric ozone depletion, and deforestation. As in the past, the development of new institutions—additions to the "legal" circle—may be required to supply the needed resource-management signals to modify human actions.

A number of writers have commented on the relationship of law and society in terms consistent with the model formulated in the foregoing. For example, the political geographer Derwent Whittlesey (1935) wrote:

> Laws affect both the tempo and the direction of settlement in all new countries, although in the process, the law itself is much modified, or where it conflicts too stridently with its new-found environment, abrogated.

As illustration, Whittlesey cited the evolution of the appropriation doctrine of water rights in the arid western parts of the United States in place of the common law riparian doctrine which applies in the humid eastern states. He further noted the appearance of land redistribution laws in the wake of social revolution. Thus, alteration of objectives through political or social upheaval may precipitate drastic modification of the legal status quo.

Relocation of an organized social community to a new physical environment may prompt alteration of legal institutions inherited from the country of origin. Concerning the Puritan migration to New England in the 1630s, McManis (1975, p. 41) writes:

> . . . Old World ways-of-doing were almost immediately modified as the colonists adopted native practices or adjusted to local soils and climatic conditions . . . . The individuality of colonial development was encouraged by the selective transfer of English practices to the colonies and by the relative freedom . . . from restraints of English traditions such as feudal rights and land entailments, from English local government which was a combination of feudalism and later innovation, and from many contemporary actions of the central government that were reshaping the landscape of Britain.

A case in point was the initial establishment of the open field system of land tenure in certain towns such as Sudbury, Massachusetts by settlers from areas of England where that practice still prevailed. Under pressure from population growth and scarcity of arable land, however, such open field settlements soon evolved into a proprietary form of land tenure (Powell, 1963).

Gradual, as opposed to sudden, social change may also be expressed in, and in turn furthered by, modification of legal doctrines concerning land, as in the case of the abolition of primogeniture in England in 1926. The long persistence of this doctrine, according to Cohen and Rosenthal (1971, p. 9):

> . . . maintained intact the great estates that were the source of wealth and power, not just for the landed nobility but for their familial offshoots who were impelled, encouraged, and assisted to apply their energies to the Church, the military, or the colonial service. . . . Without primogeniture, more farmers would have remained on the land

but in less affluent conditions, and the early enclosure laws could not have been used to transform agriculture from row crops to livestock.

Physical catastrophe, whether or not accompanied by social upheaval, may also serve as a catalyst to legal innovation in managing the redevelopment process. The Act for Rebuilding London after the Great Fire of 1666 involved no change of government but a major change in official policy regarding urban development. The 1755 Lisbon earthquake precipitated both a social revolution and a novel approach to the management of the city's reconstruction (Mullin and DeMello, unpublished ms.).

On the other hand, Hohenberg and Lees (1985, p. 36) suggest that institutional inertia impaired the viability of the late medieval city:

> A key to the declines that dogged cities through the postplague years and periodically for centuries beyond that is their inability to change when the environment did. In modern terms, medieval cities were institutionally closed systems, poorly attuned to changes in the external environment.

Over time, specific legal institutions and measures have thus emerged in response to the prevailing coalescence of political, social, and economic objectives regarding land use. As societal conditions and expectations changed, however, the broad legal concepts did not necessarily vanish, although specific applications may have been superseded. Like the woodstove and the windmill, the legal approaches of earlier eras remain available for subsequent reapplication.

REFERENCES

Beresford, M. 1967. *New Towns of the Middle Ages*. New York: Praeger.

Berry, B. J. L. 1967. *Geography of Market Centers and Retail Distribution*. Englewood Cliffs, N.J.: Prentice-Hall.

Berry, B. J. L. and W. L. Garrison. 1958. "The Functional Bases of the Central Place Hierarchy." *Economic Geography* 34: 145–154.

Bridenbaugh, C. 1964. *Cities in the Wilderness: Urban Life in America 1625–1742*. New York: Capricorn Books.

Castagnoli, F. 1971. *Orthogonal Town Planning in Antiquity*. Cambridge: The M.I.T. Press.

Clark, G. L. 1985. *Judges and the Cities*. Chicago: University of Chicago Press.

Clawson, M. 1983. *The Federal Lands Revisited*. Baltimore: The Johns Hopkins Press for Resources for the Future.

Clay, G. 1973. *Close-Up: How to Read the American City*. Chicago: University of Chicago Press.

Cohen, S. B. and L. D. Rosenthal. 1971. "A Geographic Model for Political Systems Analysis." *Geographical Review* 61: 5–31.

Conzen, M. P. and G. Lewis. 1976. *Boston: A Geographical Portrait*. Cambridge: Ballinger.

DeGrove, J. M. and N. E. Stroud. 1987. "State Land Planning and Regulation: Innovative Roles in the 1980s and Beyond." *Land Use Law* (March): 3–8.
Grimal, P. 1954/1983. *Roman Cities*. (G. M. Woloch, trans. and ed.). Madison: University of Wisconsin Press.
Haar, C. M. 1963. "The Social Control of Urban Space," in *Cities and Space* (L. Wingo, Jr., ed.). Baltimore: Johns Hopkins Press.
Hartshorne, R. 1939. *The Nature of Geography*. Lancaster, PA: Association of American Geographers.
Hohenberg, R. M. and L. H. Lees. 1985. *The Making of Urban Europe: 1000–1950*. Cambridge: Harvard University Press.
James, P. E. and G. J. Martin. 1981. *All Possible Worlds: A History of Geographical Ideas* (2nd. ed.). New York: Wiley.
Jefferson, T. 1784/1944. "Notes on Virginia," in *The Life and Selected Writings of Thomas Jefferson* (A. Koch and W. Peden, eds.). New York: The Modern Library.
Johnston, N. J. 1983. *Cities in the Round*. Seattle: University of Washington Press.
Kusler, J. A. 1985. "Roles Along the Rivers: Regional Problems Meet National Policy." *Environment* 27(7): 18–20; 37–44.
Locke, J. 1689/1952. *Second Treatise of Government*. New York: The Liberal Arts Press.
Matthews, O. P. 1984. *Water Resources: Geography and Law*. Washington: Association of American Geographers.
McManis, D. R. 1975. *Colonial New England: A Historical Geography*. New York: Oxford University Press.
Morrill, R. L. 1970. *The Spatial Organization of Society*. Belmont, CA: Wadsworth.
Mullin, J. R. and J. DeMello. Forthcoming. "Pombal and the Reconstruction of Lisbon Following the Earthquake of 1755: A Study in Despotic Planning."
Palm, R. 1981. *The Geography of American Cities*. New York: Oxford University Press.
Powell, S. C. 1963. *Puritan Village: The Formation of a New England Town*. Middletown, CT: Wesleyan University Press.
Reps, J. 1969. *Cities of the American West: A History of Frontier Urban Planning*. Princeton: Princeton University Press.
Sitte, C. 1945. *City Planning According to Artistic Principles*. New York: Random House.
Stanislawski, D. 1946. "The Origin and Spread of the Grid-Pattern Town." *Geographical Review*. 36(1): 105–20.
———. 1947. "Early Spanish Town Planning in the New World." *Geographical Review* 37(1): 94–105.
Vance, J. E., Jr. 1977. *This Scene of Man*. New York: Harpers College Press.
Whitehill, W. M. 1968. *Boston: A Topographical History* (2nd ed.). Cambridge: Harvard University Press.
Whittlesey, D. W. 1935. "The Impress of Effective Central Authority upon the Landscape." *Annals of the Association of American Geographers* 25: 85–97.

3

ENGLISH ROOTS OF MODERN LAND USE CONTROLS

The history of social organization of land use in the United States has been profoundly influenced by English precedents and practice. In part this derives from the transplanting of the common law of England to the American colonies during the seventeenth and eighteenth centuries. Judicial precedent from the courts of the mother country were thus directly applicable in the colonies, and most American states today still adhere to the common law tradition in matters relating to land. (A few states such as Louisiana, Florida, Texas, New Mexico, and California retain elements of the civil law imported from France or Spain.) After American independence, England continued to influence the development of legal doctrines affecting land use in the United States, as with the Chadwick Report on Sanitary Reform in the 1840s, Ebenezer Howard's garden cities idea at the turn of the century, and the post-World War II New Towns movement.

This chapter reviews the evolution of English legal practice concerning land use from the feudal commons of the medieval period (ca. 900–1400 A.D.) through the appearance of the improvement commission, forerunner of the modern special district, in the eighteenth century. The following chapter traces the expansion of urban institutions and forms of public response to the explosive growth of cities in Europe and the United States during the nineteenth century.

THE FEUDAL COMMONS

Private ownership of land is characteristic of a mature capitalistic society. In England and most other capitalist nations, the division of land into separately owned units replaced an earlier system of social control over land which we refer to as the *feudal commons*. Common rights in land differ markedly from private rights. In a commons, no individual has exclusive and permanent control over any particular land. Instead, rights to use available land are shared or exercised in common on a group basis (Figure 3–1). The group, ideally, is small and clearly defined, such as the inhabitants of a feudal manor. But the group may also be large and diffuse, such as the population of the world, which shares the common oceans and atmosphere. Clearly, the size and specificity of the group sharing a common resource affects the feasibility of its imposing collective rules on its members to manage the resource for the sustained benefit of all.

The biologist Garrett Hardin (1968) has formulated the gloomy axiom of "the tragedy of the commons." According to Hardin's proposition, if shared resources are not regulated through group self-restraint, individual users will inevitably maximize their own gains to their eventual mutual detriment:

> Ruin is the destination toward which all men rush, each pursuing his own best interest in a society that believes in the freedom of the commons. Freedom in a commons brings ruin to all.

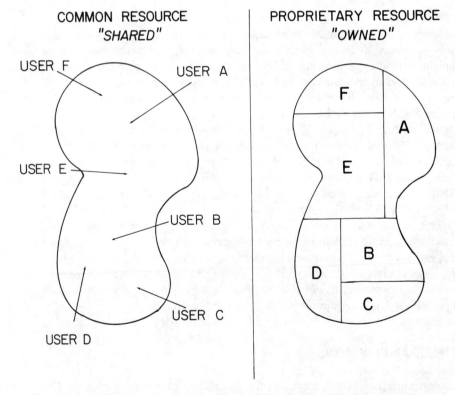

Figure 3.1 Diagram of common and proprietary resource management.

The Feudal Commons

Students of resource management since the publication of Hardin's article have, with good reason, identified the notion of the commons with the threat of environmental disaster, global warming, or nuclear war. But despite its contemporary implications, the commons proved to be a durable method of organizing land use under feudalism. In England, its origins predated the Norman Conquest; its period of dominance spanned the Middle Ages from approximately the twelfth to the fifteenth century. Thereafter it was eroded by the process of enclosure but persisted in some locations until well into the nineteenth and even the twentieth centuries. An institution for land management that lasts a thousand years deserves closer scrutiny.

The system of mutual interdependence known as the feudal commons arose, not from fiat or statute, but from practical necessity. The collapse of Roman urbanism, trade, and military protection after the sixth century A.D. caused a general reversion to a precarious agrarian existence. Mere survival against starvation, freezing weather, and hostile attack assumed paramount importance. In the words of Lewis Mumford (1961, p. 249):

> From the eighth century to the eleventh, the darkness thickened; and the early period of violence, paralysis, and terror worsened with the Saracen and the Viking invasions. Everyone sought security . . .

Security was achieved, to the extent possible, through the evolution of the feudal hierarchy of authority. Petty monarchs and local warlords assumed transient control over particular districts only to be overthrown from time to time by invading "barbarian hordes" or jealous neighbors. To sustain continuous preparations for war and to indulge in the pleasures of riotous living during peacetime, each local princeling exacted tribute from the inhabitants of his realm in the form of a portion of the product of the land, as well as monetary fines and manpower for military service. As feudalism developed in Saxon England, local lords or barons typically exercised authority within their fiefdoms by the grace of a higher overlord who assured the baron's status in exchange for rendition of a portion of his tribute. The overlord in turn had to secure his position by rendering tribute to the crown.

Thus feudalism involved the formation of a pyramid of alliances in which the price of security was the surplus product of the land. The entire system relied on the development of a socioeconomic institution that would ensure that land would be productively utilized to sustain both the local population and the superstructure of nobility. The institution that served this purpose was the feudal manor within which the commons prevailed. The manor and commons were well established in Saxon England at the time of the Conquest. William I consolidated his political control by replacing Saxon overlords with his own principal followers, thus installing himself at the apex of the feudal pyramid. Yet he did not tamper with the equilibrium of the manor and commons, which were essential to the viability of the entire system. Instead, he inventoried the assets and resources of all manorial units within his realm in the *Domesday Book* of 1086 (Trevelyan, 1953, pp. 171ff).

The manor was the fundamental economic, political, and social unit of English feudalism. It typically consisted of an extensive tract of land divided into (1) arable or cropland, (2) rough pasture or meadows, and (3) waste, including woods, ponds, wetlands, and uplands. The heart of the manor was a village settlement. This was no borough or town, but merely a cluster of dwellings huddled near the baron's hall, a parish church, and a water-powered mill. The manorial village had no legal or corporate status but served as the domicile and socioeconomic nexus of the baron, yeomen or free tenants, and villeins or unfree tenants. The baron's hall served as gathering place and occasionally as court for the village (Trevelyan, 1953, pp. 199ff.).

The feudal manor in its ideal form represented a balance between population and resources. Use of the land required the limitation of individual greed and desire for short-term gain in the interest of long-term productivity. This in turn required a state of legal equilibrium in which all parties, baron and peasantry alike, were bound by customary rules and constraints in the use of manorial resources. Without such constraints, the "tragedy of the commons" would soon have resulted.

The degree of control differed among the three classes of manorial lands—cropland, pasture, and "waste" (forest, wetlands, upland, etc.) (Figure 3-2). Apparently little regulation was needed regarding the use of waste. As long as population pressure was low, there was adequate fish, game, and firewood (although the killing of wildlife in protected royal forests was a capital offense). Pasture ("Green common") was normally subject to limitations on the number of livestock that each household, including the baron's, could graze.

Management of the cropland ("arable") was more complicated. A manor's cropland was usually divided into three large open fields. These three fields were

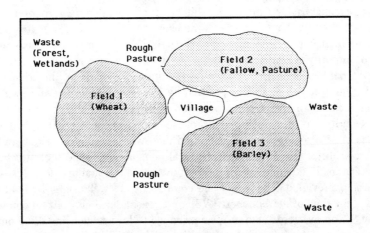

LAND ALLOCATION IN A FEUDAL MANOR

Figure 3.2 Diagram of land allocation in an English feudal manor.

rotated annually among wheat, oats, and fallow. This customary cycle allowed the soil to restore itself to ensure long-term productivity. Fertility was also maintained through the application of human and animal wastes. Each open field was internally divided into strips of approximately one acre each. Certain strips were reserved to support the baron's household and provide tribute to higher nobility. The remaining strips were allocated among the tenants as *commoners*. Each commoner household was assured the use of about 30 one-acre strips, from which it derived food and fiber for its own sustenance and for tribute to the baron. (Commoners also had to cultivate the baron's strips).

Allotments of strips within the open fields were interspersed side by side and end to end rather than being clustered in blocks under a single land holder. What this fragmentation lacked in productive efficiency, it gained in equity. According to Holdsworth (1927, p. 39):

> It is probable that the strips were scattered in this way in order to give each owner a little bit of the good land, a little bit of the indifferent, and a little bit of the bad. To allot to each owner a continuous area, compensating by extent of area for deficiency in quality, was beyond the powers of a primitive community.

Although internally apportioned, the open field was a species of commons. Individual strips were not fenced and were separated only by low ridges or turf balks which served as both boundaries and foot paths. During the fallow cycle, animals could be pastured without the need to be tethered within particular strips. Undoubtedly conflicts arose regarding trespass, vandalism, or encroachment on each other's strips. These were settled in a peculiarly feudal institution: the manorial court. Herein was an important antecedent to the future courts of equity in England and its colonies (Trevelyan, 1953, pp. 173–6).

Land management in feudal England was thus dominated by the commons. Much of the manorial land was literally shared in common, and even allocated cropland in open fields was subject to a high degree of collective mutual involvement. The use of one's own strips was dependent on the compatible use of surrounding land. No one, not even the baron, was free to break out of this system and introduce new crops or fence in his holdings. In the words of Maitland (1898, p. 32):

> Even in the boroughs the common bell called the commons to the town from the common streets and the green commons to the common hall and in common hall assembled a common seal to release of their common land, for which a fine is paid into their common chest. All is common; nothing is public.

The last sentence of this quotation is vital to the understanding of the development of land-use control. Under feudalism, there was no "public" regulation or management of land. The manorial system was symbolically subject to the power of lord, overlord, and crown. But, in fact, none of these could tinker with the manorial system of open fields and commons without toppling the entire delicately balanced structure. This system was perhaps the best example in all history of a land management system that was self-perpetuating. To quote Maitland once again:

> ... We underrate the automatism of ancient agriculture. .... So far as the arable land is concerned, the common-field husbandry, when once it has been started, requires little regulation. .... [By 1803 in Cambridge], for some centuries the common-field husbandry had needed no regulation; it had been maintaining itself (p. 25).

Why was the commons so durable as a land-management institution? In earlier times, prior to the seventeenth century, there was no feasible alternative to the commons as a means of organizing land use to supply a reliable, if not necessarily optimal, supply of food. According to Gonner (1966, p. 4):

> Common rights . . . were a necessary element in agricultural systems, they were involved in the ownership and cultivation of the land, and they were largely the source of the profits obtained from the land, the means of rendering its cultivation effective.

But stability does not necessarily imply efficiency. Beginning with the depopulation of the rural manors due to the Black Death in the mid-fourteenth century, changing demographic and technological conditions placed increasing pressure on the ancient ways of using land. Trade with the continent and the rise of the wool staple, for instance, generated attempts by landholding barons to fence off (enclose) common lands for sheep raising. Since common rights were protected by common law, they could be abridged only by statute. Beginning as early as 1235, Parliament adopted a long series of special acts authorizing specified tracts of land to be enclosed for animal husbandry or larger-scale agriculture to the exclusion of commoners who were forced to choose between working as hired laborers or seeking employment elsewhere (Gonner, 1966, pp. 43ff.). The commoners resisted this erosion of their livelihood and security as best they could, sometimes resorting to open violence, but with little success. The enclosure movement represented a gradual but distinct social and legal revolution in which common rights in land were slowly extinguished and replaced with the modern system of private rights in real property. This process was consistent with the general model of land-law interaction proposed in the preceding chapter.

The seventeenth-century settlement of New England coincided with the last stage of the open field or commons system of land tenure in certain parts of England. Settlers originating in various districts of England initially established the system of land tenure that prevailed in their place of origin, either open field or private proprietorship. Thus, Sudbury, Massachusetts followed the open field practice of Sudbury, England for its first few decades (Powell, 1963). Each family was granted a small house lot and garden for its personal use, together with rights to plant crops and graze livestock on the town's open or common fields. Eventually in Sudbury and elsewhere, the open field system, a legacy of the vanishing feudal manor, yielded to private proprietorship of enclosed farm units, as was the trend in England.

In "proprietary towns," individual holdings were initially allocated by the town council among the founding families. Each household was given a house lot in the village and certain parcels of nearby arable land; all initially shared in common the resources of forest, marsh, and water bodies within the larger town limits. In many cases, the town council reserved certain arable land and house lots for future disposition. The pattern of land tenure varied considerably from one town to another. Milford, Connecticut allocated several dispersed parcels of irregular size, shape, and quality to each founding family. Connecticut River towns such as Wethersfield and Springfield allocated long, narrow strips extending perpendicular to the river. The quantity of land alloted to each family was not necessarily equal—differences in position and wealth were recognized (and the more prominent citizens presumably served on the town council) (Meinig, 1986, p. 104). But there appears to have been an effort, as in the feudal manor of England, to allocate to each household a cross section of the available resources, including arable bottomland, upland pasture, and steep, rocky land for wood and game.

A type of commons persists to the present time in England and the northeastern United States in the form of patches of open space in the centers of old towns. But could a citizen of Boston today cut firewood or graze a cow on Boston Common? Clearly the legal status, purposes, and rules of usage of these open space legacies have changed. The village common is no longer common property of the inhabitants, but instead is owned and managed by the local municipal government.

The modern world, however, is replete with examples of common resources: the oceans, major rivers and lakes, the atmosphere, outer space. In a different sense, there are many urban commons: for example, streets, parks, subways. As society attempts to manage the use of these contemporary common resources, the feudal experience provides useful inspiration. During its period of ascendency, the feudal commons was a successful land-management institution that was self-perpetuating and, as long as population was limited, in harmony with the demands upon it, both human and ecological.

MUNICIPAL CORPORATIONS

The feudal commons was ill-suited to the governance of urban communities. As Maitland so vigorously stated, the commons involved no concept of "public." All transactions were based on custom and personal status, not laws adopted by a few on behalf of the many. The revival of towns in England and the continent starting in the eleventh century called for the development of a new legal institution more adapted to closely built, nonagricultural settlements: the municipal corporation.

Conditions for Urban Revival

For centuries after the Fall of the Roman Empire in the sixth century A.D., feudalism blanketed England and Europe like a miasma, smothering commercial and artistic exuberance and confining most of the population to a short, ignorant pastoral existence. Only the Church through its farflung network based in Rome was to preserve literacy and promote artistic impulse as reflected later in the great cathedrals erected between 1100 and 1400. A further legacy of the Church was the advent of universities in the Middle Ages such as Oxford, Cambridge, Paris, and Heidelburg, whose roots lay in the classical traditions of learning, which survived feudalism within the walls of monasteries and abbeys throughout Europe.

Towns would not long remain moribund, inhabited by monks, cats, and Roman ghosts. By the eleventh century, hints of a coming urban revival could be detected. According to Pirenne (1952), the prerequisite to this process was the revival of trade between regions, giving rise to the need for markets. This in turn would lead to the regrowth of a merchant class that would inhabit towns and give them political as well as functional importance. Broadly speaking, the urban revival was characterized by: (1) an increase in urban populations largely due to migration from rural areas; (2) reappearance of a middle class engaged in manufacturing and commerce; (3) physical growth of towns as reflected in new building, both within and outside the old fortified areas; (4) the emergence of the municipality as a new legal institution independent of the bonds of feudalism; and (5) the onset of urban problems such as water supply, disease, and fire.

The market function of towns involved both a physical space and a legal climate within which it could flourish. The physical marketplace was a central open space at the heart of the old walled city. Often the marketplace was adjoined by the cathedral, town hall, guild hall, or other civic buildings. The marketplace was multifunctional: besides its commercial role, it provided open space for ecclesiastic and civic ceremonies, social interaction, and games (Mumford, 1961). (Today, many European marketplaces retain these functions, plus outdoor cafes, political demonstrations, street life, and parking.)

For the central marketplace to function, it had to be accessible. Streets leading from the city gates to the marketplace had to be wide enough for people, animals, and carts to squeeze past each other. Given the scarcity of buildable land within the walls, streets and the marketplace itself were subject to chronic pressure of encroachment by adjoining property owners. This pressure was opposed, not by building laws, which were rarely effective where they existed at all, but literally by the throng of humanity and traffic. According to Saalman (1968, p. 30): "Streets will be as narrow as they can be while allowing for transit of goods and persons." (Figures 3–3 and 3–4).

In many cases, the demand for market space generated by growing trade simply outstripped the capacity of available land within the city walls. As the threat of hostile attack declined, development of new markets and accompanying houses and workshops appeared *outside* the gates of many new cities. These areas, known

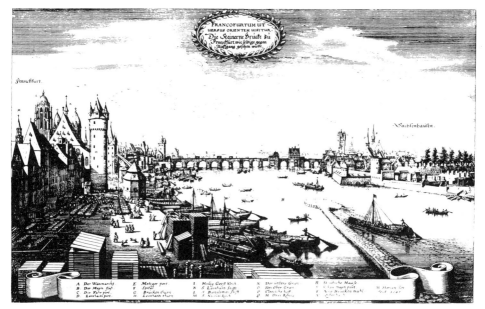

Figure 3.3 Riverfront view of Frankfurt Am Main, 1646. Source: Saalman, 1968.

Figure 3.4 Plan of Frankfurt Am Main, 1646, a classic late medieval walled city. Source: Saalman, 1968.

as *faubourgs,* were the original suburbs. Pirenne stresses that their commercial functions were not limited to periodic markets or fairs but assumed the continuous nature of modern commercial districts. It is likely that commerce outside the walls was promoted in some cases by a desire to escape the restrictions imposed upon trade within the walls.

Municipal Charters

The medieval city was as much a legal as a geographic phenomenon. Seldom of any great size in area or population, medieval cities nevertheless achieved a high degree of self-governance as virtual city-states. Legally independent of the onerous structure of feudalism, "the symbol of the city in the Middle Ages was eventually found in the sworn community which legally assumed the form of a corporation." (Weber, 1958, p. 105).

The origins of this "corporate revolution" are obscure. In England after the Conquest, certain older towns obtained charters or grants of privileges from the crown. Charters were either purchased or awarded as a token of royal favor. Some towns claimed the benefit of charters on the ground that they had been exercising certain powers of self-government since time out of mind and therefore such powers could not be abrogated.

The effect of a charter was to release the town and its inhabitants from traditional feudal obligations to render tribute in money, goods, or military service. Municipal courts replaced the whim of crown or overlord in petty judicial matters. Persons attending the weekly market would be excused from paying a market toll to the lord, which was regarded by merchants as a hindrance to trade. In place of these feudal obligations, the town was authorized to appoint its own sheriff, raise revenue from any available sources, and render annual tribute in the form of monetary payment to the crown. According to Lunt (1956, p. 178):

> Charters appeared at the close of the Eleventh Century and for the next two centuries they increased both in number and in the extent of the privileges granted. The acquisition of charters finally made the majority of towns communities with extensive rights of self-government and helped to make the townsmen a distinct element in the political, social, and economic life of the nation.

Broadly speaking, the privileges bestowed by charter included:

1. the right to hold a market;
2. the right to adopt municipal ordinances;
3. the right to establish a municipal court and for citizens to be tried before it;
4. the rights to organize a guild merchant;
5. freedom from certain taxes and obligations payable to the Sheriff on behalf of the crown or local lord;

6. the right to elect municipal officials; and
7. the right to coin money and to regulate weights and measures.

Citizens of towns (*burghers*) enjoyed not only commercial freedom but personal freedom as well. Even villeins or serfs who escaped from their manors and resided in towns for a year and a day were legally released from their feudal bonds and gained the status of freemen:

> The status . . . [of the individual under city law] was one of freedom. It is a necessary and universal attribute of the middle class. Each city established a "franchise" in this respect. Every vestige of rural serfdom disappeared within its walls. Whatever might be the differences and even the contrasts which wealth set up between men, all were equal as far as civil status was concerned. "The air of the city makes free," says the German proverb . . . (Pirenne, 1952, p. 193).

However, the medieval countryside and town were not alienated. Towns depended on their rural hinterland for the necessities of life as well as products to be traded in their markets. Amicable relations were often preserved with the local nobility and the church as well. In general, this period impresses the modern mind with its high degree of pragmatism and mutual interdependence among manor, aristocracy, church, and town.

Guilds

Guilds were organizations of merchants or craftsmen, which wielded great influence within the medieval town and its economy. The guilds' economic and political power arose from grants of monopoly status conferred on them by the crown. Thus, the wool traders guild could establish the place and hours of operation, standards of quality, weights and measurement, and terms of credit for all wool trading in the town. Nonmembers of the guild were either prohibited from wool trading in the town or were required to pay exorbitant fees to the guild. In addition, they could sell only to guild retailers: no nonmember middlemen were allowed. (Stenton, 1962, p. 178).

Guild members were the wealthiest and most influential elements of the medieval city's population. London in the early seventeenth century had over 50 craft guilds ranging from apothecaries to woodmongers, each with its own hall or meeting place. The leading members of the more important guilds were *ipso facto* leading citizens. The guilds provided a social and cultural dimension; their halls were the scene of banquets, plays, and ceremonies. They also contributed to the physical development of the community. Street maintenance, construction or replacement of bridges, additions to hospitals, repair of fortifications, and most permanent of all, the building of cathedrals—all were among the public-spirited works of guilds (Pirenne, 1952, p. 186). The phrase public-spirited is deliberately chosen; for the first time since the fall of Rome, there was emerging a sense of "public."

Ultimately, the influence of the guilds was reflected more enduringly in legal institutions than in bricks or mortar. Over time, medieval cities under their direction assumed a new legal status as "bodies politic" or "municipal corporations." By 1400, English municipal corporations had acquired the following characteristics: (1) perpetual existence; (2) power to own land and buildings; (3) power to sue and be sued; (4) power to adopt ordinances or bylaws; and (5) the right to formalize documents with a corporate seal (Holdsworth, 1927). These legal characteristics of the municipal corporation have remained fairly constant from the Middle Ages to the present time. (Chapter 6 further discusses municipal corporations in the American context.)

Municipal corporations exercise jurisdiction over a specific territory—the walled city in the medieval case. This concept of *jurisdiction* differed from *ownership:*

> A municipal corporation owns a few, but only a few, of the houses in the town. Over the whole town it exercises a certain governmental power. We have here two different ideas; they can be sharply contrasted. . . . There [was] a sort of lordship over the whole town, and of a few houses there will be landlordship (Maitland, 1898, p. 30).

The question logically arises as to what use municipal corporations made of their "lordship" or jurisdiction over land within the city walls. How in particular did they utilize their power to adopt ordinances and to what extent did they attempt to improve the quality of the deteriorating urban environment?

English municipal ordinances of the Middle Ages may be roughly divided into two classes. First were those concerned with public morals, health, and safety in the urban environment. London, for instance, had ordinances dealing with the removal of dung from stables, the lighting of streets, and "sweating houses, whereunto any lewd women resort" (Hearsey, 1965, p. 11). These ordinances were at best unevenly enforced and, at worst, totally ignored.

The second class of regulations dealt with offenses against trade and commerce, such as theft, overcharging, and sale of inferior goods. Penalties for these offenses, which struck directly at the economic welfare of the guilds, were swift and often terrible. One of the lighter punishments, as described by Pendrill (1937, p. 22), involved a baker caught selling bread of substandard weight who was " . . . strapped to a sort of low cart harnessed to a horse and dragged through the streets, accompanied by the City Minstrels playing on tabors and pipes, and finally brought back and released at his own door." Breaches of the "market peace," for example, theft or disorderly conduct, were subject to far more brutal punishments. According to Pirenne (1952, p. 200):

> This city peace was a law of exception, more severe, more harsh, than that of the country districts. It was prodigal of corporal punishments: hanging, decapitation,

castration, amputation of limbs. It applied in all its rigor the *lex talionis:* an eye for an eye, a tooth for a tooth. Its evident purpose was to repress derelictions, through terror.

Regulation of the urban environment was clearly of lower priority than deterring crimes against property. Petty theft of commercial goods in a public thoroughfare was a breach of the market peace and punished severely. But the permanent encroachment of private buildings into or above the same public way was likely to be ignored (Saalman, 1968, pp. 30–31).

The existence of corporate jurisdiction over the medieval city, therefore, did not necessarily mean that such power was used effectively to regulate the placement, height, construction, or use of private buildings. The resulting cityscape was characterized by narrow, twisting streets, overhanging upper stories, and prevalent use of wood as a construction material. Just as the casual disposal of human and animal waste contributed to periodic epidemics, the unregulated crowding of buildings posed a constant and growing danger of citywide fire. As in modern times, reform and progress were the result, not of enlightened foresight, but of bitter hindsight.

THE COMMON LAW

The thirteenth century marked the dawn of the modern era of land law in England. The new era was signified by the gradual replacement of feudal tenure with property ownership or "proprietorship," a process that extended over the next six centuries. This has been characterized by legal scholars as a transition from "status" to "estate" (Dukeminier and Krier, 1981, p. 362). As described earlier, feudal land was not "owned" in the modern sense, but instead was "held" by one party in subservience or fealty to another. The rights of the holder were essentially limited to mere right of usage in exchange for rendition of tribute to the immediately superior noble. A right of usage did not involve the right to sell, give, or devise the land to one's heirs. Nor, under the management constraints of the commons, could land be converted to nontraditional uses or simply removed from production.

The earliest evidence of transition from feudal tenure to freehold ownership of land occurred in the reviving cities and towns rather than the countryside. According to Pirenne (1952, pp. 194–5), the breakdown of feudal control over land was a concomitant of the comparable growth of personal freedom within city walls:

> With freedom of person there went on equal footing, in the city, the freedom of the land. In fact, in a merchant community, land could not remain idle and be kept out of commerce by unyielding and diverse laws that prevented its free conveyance and restrained it from serving as a means of credit and acquiring capital value. This was the more inevitable in that land, within the city, changed its nature—it became ground for building. It was rapidly covered with houses, crowded one against the other, and increased in value in proportion as they multiplied. Thus it automatically came about

that the owner of a house acquired in course of time the ownership, or at least the possession of the soil upon which it was built. . . . Cityhold thus became freehold.

Of course, even freehold was not totally free of obligation to higher authority. The medieval burgher owed tribute in the form of property taxes to the municipal authorities who collectively paid tribute when necessary to the crown. This was equivalent to the modern practice of property taxation.

The equivalent transformation of rural land from feudal tenure to proprietorship was facilitated by the Statute Quia Emptores in 1290. This statute prohibited the practice of "subinfeudation" which was causing proliferation of intermediate lordships and devaluation of tribute received by the crown. The statute however permitted the substitution of one holder for another, subject to the same obligations to the overlord as the earlier holder. This in effect legitimized conveyance of land from one party to another on a monetary basis, which is the essence of property ownership (Dukeminier and Krier, 1981, p. 358). Over time, this resulted in a withering away of intermediate lordships with the result that virtually all land came to be held by individual proprietors directly "of" the crown. That is the state of land tenure in Great Britain and its remaining colonies today.

By the eighteenth century, property ownership in both England and the American colonies was solidly established as a sacred institution with which even the crown was constrained from interfering. The concept that "every man's home is his castle" was most forcefully stated by William Blackstone in his commentaries on the common law in 1768:

> There is nothing which so generally strikes the imagination, and engages the affections of mankind as the right of property; or that sole and despotic dominion which one man claims and exercises over the external things of the world, in exclusion of the rights of any other individual in the Universe (Blackstone 1768/1863, p. 1).

Although tinged with hyperbole, Blackstone's view of private property stands the feudal tenure system on its head and exalts the proprietor over the crown.

But as Locke noted in the quote in Chapter 2, private property is valueless unless the owner is secure in reaping the "harvest" or other benefits of ownership. Blackstone's proposition declares that the owner is protected from the crown, but what about one's neighbors? The answer lay in the development of the common law doctrines of "trespass" and "nuisance" under which an aggrieved property owner could seek the protection of the courts from offensive conduct by other parties.

The common law consists of the accumulated wisdom of courts faced with specific disputes in which they invoke principles of law, drawn from earlier decisions and other sources, as a basis for decision. The common law evolves on a case-by-case basis and requires the aggrieved party (plaintiff) to bring an action against the alleged offender (defendant) to court at the former's expense. This reliance on past decisions in similar fact situations is referred to as *precedent* and has been the basis for the development of the common law in England and its present and former dependencies.

The common law has never been an efficient means of regulating urban land use. It was supplemented (but not replaced) in the nineteenth and twentieth centuries by public land-use regulations such as zoning. But before the advent of such regulations, adjudication of disputes between private parties under the common law provided a crude means for redressing flagrant abuses against property ownership in the preindustrial urban community.

The most fundamental protection afforded by the common law was against *trespassers*. Without security against physical invasion of the premises, property ownership meant nothing. The essence of trespass was a physical invasion or encroachment by an individual who had no right to enter the property in question. Over time, courts expanded the doctrine of trespass to hold parties liable who allowed livestock or even water to enter on the property of another with permission.

Beginning in the thirteenth century, trespass was broadly construed as a breach of the criminal law, punishable by fine or imprisonment. No proof of any actual damage was required since invasion of the plaintiff's rights was regarded as wrongful in itself. Furthermore, liability for trespass was imposed regardless of negligence: The defendant was "strictly liable" if proven to have encroached on the plaintiff's property (Prosser, 1971, pp. 357–64). The common law doctrine of trespass since medieval times has always reflected the principle that one's home and land are sacred.

The doctrine of *nuisance* afforded additional protection to property owners under the common law. Unlike trespass, nuisance did not require any physical invasion of the premises. Rather, it addressed the externalities of actions taken elsewhere, typically on adjoining land, which injured the beneficial enjoyment of the plaintiff's property. Typical forms of nuisance include blocking off a neighbor's light and air, causing bad odors and air pollution, loud continuing noises, and other externalities that impair the quiet enjoyment of nearby property. The medieval doctrine of nuisance—still cited today—was *sic utere tuo ut alienum non laedas:* use your property so as not to harm that of others.

This "golden rule" of nuisance has always been easier to express than to apply. In deciding nuisance cases, courts have traditionally attempted to "balance the equities," that is, weigh the social benefits, if any, of the conduct complained of against the degree and type of harm suffered by the plaintiff. Courts favor the plaintiff where the harm clearly outweighs the value of the activity. But, with the coming of industrialization, plaintiffs often lost cases in which the defendant was an industrial polluter which happened to supply a useful product and employ many people. Courts gradually became more creative in fashioning orders (injunctions) that limited the harmful effects of a particular enterprise without terminating it entirely. (See further discussion in Chapter 5.)

BUILDING LAWS: THE ACT FOR REBUILDING LONDON

While medieval municipal corporations had attempted to impose rules for building practices in urban areas on a piecemeal basis, most were doomed to futility. Part of the problem lay in the "tragedy of the commons:" Each property owner viewed the streets and marketplaces as common property to be encroached on for private gain to the maximum extent possible. Without a clear-cut purpose for such restrictions, backed up by a determined enforcement effort, the early attempts to regulate the urban environment were ineffectual (Saalman, 1968). A comprehensive building law did not arrive in London until the late seventeenth century, and its advent required the bitter experience of a citywide disaster: the Great Fire of 1666.

London Before the Fire

Between 1400 and 1666, London's population grew from 50,000 to about 400,000 inhabitants—an eight-fold increase. This drastic demographic expansion reflected both an influx of rural laborers displaced by enclosure of manors and immigration of persons fleeing persecution on the Continent. It was accompanied by physical expansion of London's housing stock both within and beyond the old Roman city walls. Queen Elizabeth I in 1580 attempted to halt the peripheral sprawl in a famous decree (anticipating the Greenbelt legislation of the twentieth century) which ordered all persons to:

> . . . desist and forbear from any new buildings of any house or tenement within three miles from any of the gates of the said city of London. . . . (quoted in Rasmussen, 1934/1967, p. 68).

This decree was a total failure as indicated by the continued growth of London outside its walls and gates. By 1666, the walled City of London was described as comprising only one-third of the total urbanized area of London: "The great urban spread had begun, and already a number of the better-off preferred to live outside the City where their work or business was" (Hearsey, 1965, p. 2). What is now the West End of London was then suburban in character, with large homes interspersed with residual common fields and districts of small cottages inhabited by migrants from the countryside and abroad (Rasmussen, 1934/1967, Ch. 4).

The walled City (now the financial district known simply as the City) remained in 1666 solidly medieval in character. Pre-fire London was a labyrinth of narrow, twisting streets with pervasive overhanging upper stories. Wood was the usual construction material. Since 1189 exterior walls were required to be of brick or stone, but "the precaution was very partially observed" (Bell, 1920/1971, p. 11). The City was connected to the surrounding countryside by gates on the landward side and by the famous London Bridge across the Thames. The walled City contained 109 parish churches and some 50 guild or livery halls. The ancient gothic pile of St. Paul's Cathedral, the largest in Europe (its tower reached 449 feet and its nave measured 585 feet in length), loomed above the smoky, crowded city (Hearsey, 1965, p. 60).

The Fire

The Great Fire of September 2–7, 1666, was perhaps the first major catastrophe of modern times to be fully described by competent eyewitnesses. Samuel Pepys' *Diary* relates:

> . . . Jane comes in and tells me that she hears that above 300 houses have been burned down tonight by the fire we saw [the night before] and that it is now burning down all Fish-street, by London Bridge. So I made myself ready presently, and walked to the Tower, and there got up upon one of the high places, . . . and there I did see the houses at that end of the bridge all on fire, and an infinite great fire on this and the other side. . . So down with my heart full of trouble to the Lieutenant of the Tower, who tells me that it begun this morning in the King's baker's house in Pudding-lane, and that it hath burned down St. Magnes Church [the first of 80 churches to be burned] and most part of Fish-street already. So I down to the water-side and there got a boat, and through bridge, and there saw a lamentable fire. . . . Everybody endeavoring to remove their goods, and flinging into the river, or bringing them into lighters that lay off; poor people staying in their houses as long as till the very flames touched them, and then running into boats, or clambering from one pair of stairs by the water-side to another . . . and the wind mighty high, and driving it into the City; and everything after so long a drought proving combustible, even the very stones of the churches . . . (Pepys, 1666/1898, pp. 392–3).

In the absence of an efficient water distribution system, little could be done to fight such a conflagration. The standard technique of medieval firefighting was to "pull down houses," thus creating a firebreak to stop the progress of the flames. The Lord Mayor of London vacillated in using this technique out of concern for potential liability to the house owners. An old law required that whoever pulled down a house in London had to bear the cost of replacing it (Hearsey, 1965, p. 125). In the absence of any effective effort to limit the spread of the fire, it burned unchecked for three days and consumed most of London within the walls and a considerable area outside. Within this area, 13,200 houses were destroyed in some 400 streets and alleys. Over 100,000 were homeless and left camping miserably in fields outside the city.

In terms of loss of life, the Great Fire of 1666 was vastly eclipsed by the outbreak of the plague in London the previous year. While 56,558 persons were reported to have died in the plague, only four deaths were attributed to the fire out of a population of 400,000. It is perhaps an ironic quirk of human nature that physical destruction such as occurred in the fire is given a higher place in history than human loss of life as occurred in the plague. In any event, the fire indeed eliminated medieval London, though fortunately not its population.

The First Modern Building Law

The fire epitomized Hardin's axiom: "Freedom in a commons brings ruin to all." Although London was scarcely a feudal commons, neglect of the urban environment under four centuries of municipal self-government had yielded disaster. Its public spaces were virtually an unregulated commons, with private structures clogging the narrow, twisting streets and blocking access to the Thames River. In the absence of effective regulation of building size, location, and construction materials, the fire was inevitable. Without access to water, it could not be halted.

The point was not missed by certain leading minds of the time. While the ruins were still smoking, plans, or at least concepts, for the rebuilding of London were being prepared by Sir Christopher Wren, the city's leading architect, and several others (Bell, 1920/1971, Ch. 13). Wren's plan proposed to transform the city into a monumental, baroque-style imperial capital, much as Haussman would later restructure Paris in the mid-nineteenth century. The prefire street alignments and property lines were to be abolished where necessary and replaced by an orderly, geometric network of major streets and open plazas with lesser streets leading into them. Churches, company halls, and other important buildings would be situated on the new squares or along the connecting arteries. The Thames embankment would be cleared and reserved for major buildings with open space between and in front of them. Dwellings would be confined to the lesser streets.

Such a heroic proposal for restructuring the city on baroque lines was incompatible with the political and economic temper of the times. Charles II, the recently restored Stuart, lacked the despotic powers that brought the downfall of his father Charles I at the hands of Oliver Cromwell. Furthermore, the government could not afford to pay property owners whose private lots would have been taken to implement the plan. The plan would take too long to implement; the City's commercial interests were anxious not to lose time in reestablishing their business activities. Finally, the plan was too grandiose for the English taste (Hearsey, 1965, p. 179).

A week after the fire had subsided, on September 13, 1666, Charles II addressed by proclamation the need for restraint and foresight in the rebuilding process, pending a full investigation of the causes of the disaster. Rasmussen (1934/1967, p. 117) refers to this declaration as "astonishingly modern . . . a true piece of town-planning, a programme for the development of the town." In its preamble, the proclamation combined seventeenth-century moralism and twentieth-century civic boosterism:

> And since it hath pleased God to lay this heavy Judgement upon Us . . . as evidence of His displeasure for Our sins We do comfort Our Self with some hope, that he will . . . give Us life, not only to see the foundations laid, but the buildings finished of a much more beautiful City than is at this time consumed. . .

The proclamation went on to address five practical aspects of the rebuilding process: (1) stone or brick was to be used for exterior facades in place of wood; (2) the width of streets was to be established in relation to their importance; (3) a broad quay or open area would be maintained along the Thames for access to water for firefighting; (4) public nuisance activities such as breweries or tanneries should be removed from central London to more suitable locations; and (5) reasonable com-

Building Laws: The Act for Rebuilding London

pensation should be determined and paid to property owners whose right to rebuild was curtailed by public restrictions (Rasmussen, 1934/1967, pp. 116–117).

He then appointed a special investigative commission (a forerunner of the modern blue ribbon committee) to draw up recommendations for future building practices in London. Christopher Wren served on the Commission and was to be a principal figure in the reconstruction of the city. Another interim measure of "the greatest aid in the rebuilding of London" was an act of Parliament that established a Court of Fire Judges to resolve disputes between landlords and tenants and to apportion rebuilding cost equitably between them (Bell, 1920/1971, pp. 242–3).

Based on the proclamation and the commission's report, the Act for Rebuilding London was adopted February 8, 1667, five months after the fire. What the law lacked in immediacy, it made up in detail. The Act for Rebuilding London has been described as London's first "complete code of building regulations" (Bell, 1920/1971, p. 251).

The Act was long and detailed. According to Summerson (1962, p. 53):

> The Act covered important aspects of the rebuilding program: first, the rearrangement of some of the worst features of the old plan, with its apparently wayward meanderings, jutting corners, and frequent bottle-necks; second, the partial standardization of the new buildings, particularly with a view to fire resistance; and third, the raising of money for the public . . . buildings by a tax on coal.

By far the most conspicuous and lasting of the Act's provisions dealt with the height and construction of dwellings to replace those burned in the fire. These consisted of four classes of structures, ranging from two-story, modest dwellings to houses of "the greatest bigness." The class of home that could be built on a site depended on the location and importance of the street or square on which it faced. The use of stone or brick for exterior walls was required. Thickness of walls, ceiling heights, and other architectural details were also specified. No overhang above the public way would be allowed.

The building provisions of the Act marked a significant departure from medieval style and construction practices, but those had long been abandoned in new construction. According to Summerson (1962, p. 54), the Act " . . . crystallized the best practice of the time. . . . " It was, in effect, a building code for the redevelopment of the burned area and a guide to new construction in surrounding areas. The Act was farsighted in its administrative provisions for permits and fines, a precedent for modern building codes. And its regulations regarding the banishment of smoky or noxious activities to specified locations anticipated modern zoning laws.

The Act was not uniformly effective, for indeed there was little experience or administrative structure to enforce its requirements in every case. Furthermore, it did not purport to change building patterns or land usage so as to eliminate overcrowded alleys and courts behind other buildings. Working-class London sprang back to life possibly safer from fire but otherwise densely crowded and deprived of light and air. Such conditions would become increasingly intolerable during the eighteenth and early nineteenth centuries. Meanwhile, the landed aristocracy were engaged in subdividing their properties north and west of the city walls into fashionable townhouses and mansions facing on verdant squares.

While the exact influence of the Act is difficult to discern from what might have occurred in its absence, it clearly marked a threshold between the medieval and the modern eras of urban land-use control. The lethal combination of overcrowding, use of flammable building materials, and abysmal sanitary conditions produced the twin perils of the medieval city: disease and fire, which struck London in tandem in 1665–1666. The fire swept away, not only the overhanging wooden houses and the rats they contained, but also the attitude of medieval neglect toward the urban environment. The royal moratorium on rebuilding and appointment of an expert investigative commission marked a vast change in governmental process from Elizabeth's useless edict of 1580. The Act of Parliament that resulted from the commission's findings represented the beginning of modern urban planning, although two centuries would elapse before public building codes were widely adopted in England and the United States. The Act for Rebuilding London, based on expert advice and perception of the causes of the disaster, epitomized the operation of the general model proposed in Chapter 2.

PRIVATE LAND-USE RESTRICTIONS

At the same time that the old walled City of London was being rebuilt under the 1667 Act, farms and estates beyond the walls were undergoing development for the first time. The building boom north and west of the city proper (including what is now the fashionable West End district) was not subject to the Act, but displayed nevertheless a remarkable uniformity in land-use pattern and architectural style. This uniformity was achieved, not through governmental intervention, but through private agreements between owners and occupants of the land in question, as expressed in deed and lease restrictions.

The architectural prototype for the seventeenth-century London square was the Place Royale in Paris, commissioned by Henri IV in 1615 (now the Place des Vosges). The Place Royale consisted of a small park surrounded by townhouses of equal height and similar facades, fronted by a street-level arcade. This design was soon introduced to London by Inigo Jones in 1630 in his design for Covent Garden. Although the "least English" of London's squares (Zucker, 1959, p. 200), Covent Garden in plan served as a paradigm for two centuries of residential baroque development in London.

The building of "Georgian London" (Summerson, 1962) in the seventeenth and eighteenth centuries responded to private demand for upper-class housing and residential neighborhoods. Several factors underlay this demand, including: (1) the increased overcrowding of London with its demonstrated hazards of plague and fire; (2) the accumulation of wealth due to the economic success of the City; and (3) the availability of buildable land in large estates at the outskirts of the City (Morris, 1979, Ch. 8). Some of these estates were held by nobility under feudal grants dating back to the Conquest; others were more recently acquired through the dissolution of monastaries and redistribution of church land by Henry VIII. In either case, the aristocratic landlords were eager for monetary gain.

A quirk of English legal history was to determine the scale and quality of the development of the London West End (Vance, 1977, p. 234). Land tenure was customarily in "fee tail" rather than "fee simple" which meant that land was always inherited and could not be sold on the open market. This accounts for the customary English practice of leasing land for a long term rather than selling it outright. This in turn permitted the careful planning and uniform execution of the neoclassic squares and residential terraces that characterize upper-class London to the present day.

The homogeneity of exterior appearance that characterized these developments was achieved through private restrictions or "covenants." A *covenant* is a promise made by the purchaser or lessee of real estate that the use of the premises will conform to conditions specified by the seller or lessor. Thus landlords, with the aid of architects such as Inigo Jones, John Woods, Thomas Cubitt, and John Nash, could absolutely control the use of their property by the lessee. This included the exact locations of streets, squares, and building lot lines as well as the size, appearance, and sometimes the interior layout of buildings.

The enforceability of covenants over time was seldom an issue where the original lessee or purchaser remained in possession. The landlord always had the legal right to enforce covenants against the party that had accepted them. However, many leases ran for periods of 99 years or longer and transfers from one party to another were common. This practice raised the question of whether the covenants would still be enforceable against subsequent parties in possession who had not specifically accepted the restrictions. The 1848 decision in *Tulk v. Moxhay* (41 Eng. Rep 1143) concerning London's Leicester Square established the rule that covenants may "run with the land" and bind the use of land indefinitely.

The neoclassical residential districts regulated through private covenants were the antithesis of medieval unplanned growth. In Vance's terms, they epitomized the "preconceived" as opposed to the "organic" urban pattern. They exuded wealth, conservatism, power, and control in contrast to the irregular, heterogeneous human scale of earlier districts. James Craig's 1766 plan for the New Town in Edinburgh stands in startling contrast to the medieval Old Town (See Figure 2–2). Although little of medieval vintage remains today in the Old Town, it retains a mysterious, romantic, and vaguely sinful atmosphere. Across a narrow linear park (formerly a fetid creek), the New Town stands rectilinear, symmetrical, and respectable, graced by its Adam doorways and leafy squares.

In the United States, the English-style residential square inspired several counterparts, notably Louisburg Square in Boston's Beacon Hill, laid out by private developers in 1826, and New York City's Gramercy Park, established by Samuel Ruggles in 1832.

Private covenants today are commonly used for effectuating and maintaining the provisions of a subdivision plan over time. No longer confined to aristocratic developments, they are employed in all sorts of residential and commercial subdivisions, including condominium developments. Deeds containing covenants and other restrictions are filed in public registries of deeds and are thus available to anyone purchasing property to which they apply. This ensures that the restrictions may be enforced by the original developer, his or her successor, or by other lot buyers in the same subdivision. (See Chapter 9.) Private land-use restrictions may exceed zoning and other public controls in their specificity and breadth of coverage. In the United States, a buyer and seller may agree to virtually any conditions imposed on the use of land, except for those such as racial occupancy, that are unconstitutional.

IMPROVEMENT COMMISSIONS

Private deed restrictions served the needs of the wealthy to implant their concepts of style on their elite districts. But working-class districts, which began to grow rapidly in the eighteenth century, were a different story. London grew from about 400,000 at the time of the fire to 864,000 in 1801 (Ashworth, 1954, p. 9). Other British and Scottish industrial cities, including Manchester, Birmingham, Liverpool, and Glasgow, grew at comparable rates. Most of this population increase, which anticipated even more massive growth in the following century, consisted of migration from the countryside and from Ireland and the Continent to the "satanic mills" of the industrial cities. This caused hideous overcrowding of existing dwellings and a proliferation of cheap, shoddy, unplanned tenement districts within walking distance to the mills.

To make matters worse, the old municipal corporations that nominally governed each city had ossified by the eighteenth century. Municipal offices were allocated according to status and privilege, not experience or interest in reform or public service. The corporations were unresponsive to what Ashworth (1954, p. 50) has termed "the increasingly lethal nature of the swelling towns." The result was a state of anarchy in the town building process throughout the eighteenth and well into the nineteenth century.

Not only the municipal corporations but also the courts were largely ineffectual in confronting the new circumstances. In earlier and simpler times, the common law doctrine of nuisance had afforded crude relief to property owners and the public against the more obnoxious abuses of public health, safety, and the common welfare. For instance, a builder could be prevented from blocking the "ancient lights" and ventilation from reaching pre-existing structure or from threatening an adjoining structure with collapse. Such actions had earlier provided limited, if cumbersome, protection against unreasonable or dangerous building practices.

But abatement of nuisance through court intervention was limited to case-by-case treatment upon petition of the wronged party. Normally, the harm complained about had to be in existence; it was rare that courts would enjoin a prospective nuisance. And the party seeking a remedy had to own the property, not be a mere tenant. Also legal action was expensive then as it is today. In theory nuisance actions remained available, but in practice, the courts in the eighteenth and nineteenth centuries exercised little restraint over the tenement-building process since the victims seldom were able to complain.

Sydney and Beatrice Webb (1963) in their classic history of English local government provide vivid images of the residential environment of early industrial England, for example:

> To begin with the houses—springing up on all sides with mushroom-like rapidity—there were absolutely no building regulations. Each man put up his house where and as he chose, without regard for building-line, width of street or access of light and air . . . 'Streets of projecting houses nearly meeting at the top rooms with small windows never meant to open; and dirt in all its glory, excluded every possible access for fresh air.' . . . The narrow ways left to foot and wheeled traffic were unpaved, uneven, and full of holes in which the water and garbage accumulated. Down the middle of the street ran a series of dirty puddles, which in time of rain became a stream of decomposing filth (p. 236).

This was mild in comparison with the reports of Edwin Chadwick and other public health reformers of the mid-nineteenth century. But the urban environment, apart from the elite districts, was fast disintegrating during the eighteenth century.

English society in this period, however, was dominated by conservatism, distracted by foreign wars, and generally disinterested in the fast-growing squalor within its cities. Creative and large-scale measures to address the crisis—for example, public sanitation and building laws, the public parks movement, urban redevelopment, and model planned towns—would not appear until the second half of the nineteenth century. Meanwhile, a simpler but practical stopgap was devised in response to the crisis: the improvement commission.

Improvement commissions were statutory bodies established by Parliament to perform certain functions in specified localities. They were the forerunners of contemporary special districts and authorities in England and the United States. Like their modern counterparts, improvement commissions could overlie general-purpose units of government and indeed could serve more than one municipality. The first of these new institutions was the Commissioners of Scotland Yard (1662), which encompassed London and Westminster. This was not a detective agency but rather was empowered to "make new sewers, enlarge old ones and to remove nuisances . . . to appoint public rakers or scavengers, who were to make daily rounds with 'carts, dungpots, or other fitting carriages' . . . and to remove encroachments upon public ways and to license hackney coaches" (Webb and Webb, 1963. p. 240).

Improvement commissions proliferated throughout England during the eighteenth century, numbering some 300 by the early 1800s. These bodies overlapped geographically and functionally with the old municipal corporations (which were substantially reformed by Parliament in 1835). During this period they assumed many of the functions of modern local governments, for example: "paving, cleansing, lighting, watching, and regulating" (Webb and Webb, 1963, p. 242). In addition, they engaged in activities of a more regional nature, for example: building bridges and canals, drainage improvement, enclosure of commons, erection of markets and slaughterhouses, water supply, and highway construction.

But like the municipal corporations and courts, the improvement commissions were hampered by institutional constraints in their efforts to stem the deterioration of English cities. They were strictly limited to specified functions and geographic areas as established by Parliament. They had no general jurisdiction or "home rule" authority to address a wider spectrum of urban needs without Parliamentary approval. Thus, a commission responsible for paving and lighting in a particular district had no authority to deal with drainage or water supply, no matter how obvious the need. Furthermore, they lacked any authority to plan or regulate new building; they were largely limited to dealing with harmful conditions after the fact, if at all,—not beforehand. By the early nineteenth century, the improvement commissions:

> . . . too often concentrated their attention solely on the middle-class districts of their towns, leaving the greater number of streets inhabited by the poorer classes wholly without essential services. However valiantly the improvement commissioners might struggle to cope with the flood-tide of urbanization—and few of them struggled very valiantly—they were fighting losing battles. [They were] constitutionally, financially, administratively, technically, and ideologically ill-equipped to cope with the frightening immensity of the task . . . (Flinn, 1965, p. 17).

Improvement commissions were thus a temporary expedient, a palliative to remedy certain kinds of urban ills on a piecemeal basis until more sweeping public approaches were devised. They totally failed to restrain the continued proliferation of slums and the lethality of the working-class residential environment. But the idea of the improvement commission, like the municipal corporation, was to become a permanent addition to the institutional fabric of urban and metropolitan government in Great Britain, the United States, and elsewhere. (See further discussion in Chapter 6.)

CONCLUSION

The creativity of the English people and their legal system thus yielded a series of institutional innovations over several centuries to meet perceived needs to better organize and control the use of land. The source of each device differs considerably. Some of the institutions discussed in this chapter, such as the feudal commons and the municipal corporation, arose spontaneously from the "invisible hand" of social necessity. The doctrines of trespass and nuisance and later the recognition of private covenants as restrictions that run with the land were products of the English judicial system. The Act for Rebuilding London and the improvement commissions were created by acts of Parliament. The latter two institutions—the courts and the legislature—have continued to play dominant roles in the fashioning of additional public responses to urban and environmental deterioration in the nineteenth and twentieth centuries. With the stage now set, we turn to the dramatic transformation of industrial cities and the forms of institutional response that ensued between 1800 and 1900.

REFERENCES

ASHWORTH, W. 1954. *The Genesis of Modern British Town Planning.* London: Routledge and Kegan Paul.
BELL, W. G. 1920/1971. *The Great Fire of London in 1666.* Westport, CT: Greenwood Press.
BLACKSTONE, W. 1768/1863. *Commentaries on the Laws of England* (Second Book). Philadelphia: J. B. Lippincott and Co.
DUKEMINIER, J. AND J. KRIER. 1981. *Property.* Boston: Little, Brown.
FLINN, M. W. (ed.). 1965. *Edwin Chadwick's Report on the Sanitary Condition of the Labouring Population of Great Britain.* Edinburgh: Edinburgh University Press.
GONNER, E. C. K. 1966. *Common Land and Inclosure* (2nd ed.). London: Frank Cass.
HARDIN, G. 1968. "The Tragedy of the Commons." *Science* 162: 1243–1248.
HEARSEY, J. E. 1965. *London and the Great Fire.* London: J. Murray.
HOLDSWORTH, W. S. 1927. *An Historical Introduction to the Land Law.* Oxford: Clarendon Press.
LUNT, W. E. 1956. *History of England.* New York: Harper and Row.
MAITLAND, F. W. 1898. *Township and Borough.* Cambridge: Cambridge University Press.
MEINIG, D. W. 1986. *Atlantic America: 1492–1800.* New Haven: Yale University Press.
MORRIS, E. E. J. 1979. *History of Urban Form: Before the Industrial Revolution* (2nd ed.). New York: Wiley.
MUMFORD, L. 1961. *The City in History.* New York: Harcourt, Brace and World.
PENDRILL, C. 1937. *Old Parish Life in London.* London: Oxford University Press.
PIRENNE, H. 1952. *Medieval Cities: Their Origins and the Revival of Trade.* Princeton: Princeton University Press.
PEPYS, S. 1666/1898. *Diary.* (H. B. Wheatley, ed.). London: George Bell and Sons.
POWELL, S. C. 1963. *Puritan Village: The Formation of a New England Town.* Middletown, CT: Wesleyan University Press.
PROSSER, W. L. 1971. *The Law of Torts* (4th ed.). St. Paul: West Publishing Co.
RASMUSSEN, S. E. 1934/1967. *London: The Unique City.* Cambridge: The M.I.T. Press.
SAALMAN, H. 1968. *Medieval Cities.* New York: George Braziller.
STENTON, D. M. 1962. *English Society in the Early Middle Ages.* Baltimore: Penguin Books.
SUMMERSON, J. 1962. *Georgian London.* Baltimore: Penguin Books.
TREVELYAN, G. M. 1953. *History of England,* Vol. 1. Garden City, NY: Doubleday Anchor.
VANCE, J. E., JR. 1977. *This Scene of Man.* New York: Harper's College Press.

WEBB, S. AND B. WEBB. 1963. *Statutory Authorities for Special Purposes.* Hamden, CT: Archon Books.
WEBER, M. 1958. *The City.* New York: The Free Press.
ZUCKER, P. 1959. *Town and Square. From the Agora to the Village Green.* New York: Columbia University Press.

4

URBAN REFORMS OF THE NINETEENTH CENTURY

INTRODUCTION

The modern industrial city came of age in Europe and North America during the nineteenth century. The population and geographic extent of many urban places mushroomed at unprecedented rates. With rapid expansion came a paroxysm of threats to life, health, and morality. In the early decades of the century, tenements proliferated, sanitation collapsed, crime and disease flourished, and life expectancy declined. Gradually, the horrors of uncontrolled urbanization were recognized, at first by a few perceptive individuals, and ultimately by a broader spectrum of society and its law-making bodies.

Beginning in the 1830s, three fundamental avenues of public and private response to urban deterioration evolved in England and America. Each involved legal and institutional innovation in addition to progress in engineering technology, urban design, and the social sciences. The first of these approaches was *regulation*. Beginning with the British Public Health Act of 1848, perception of squalor, overcrowding, and lack of basic sanitation yielded a series of public police power measures intended to gain some degree of control over the building of cities. These would lay a constitutional foundation for the proliferation of land-use and environmental regulations to appear in the twentieth century.

The second approach was *redevelopment*, the physical replacement of expansion of urban infrastructure in the form of wider streets, paving and lighting, water systems, mass transportation systems, and urban parks. This anticipated the urban renewal and revitalization efforts of the mid-twentieth century in Great Britain, the United States, and elsewhere.

Third, there were the advocates of *relocation* of people from existing overcrowded and unhealthy cities to newly built model towns in outlying locations. These new towns were to be carefully planned, both physically and socially, to uplift the spirit as well as to provide an honest living and healthful surroundings. The building of New Lanark, Pullman, and Letchworth Garden City involved new forms of finance and institutions of governance. The experimental communities of the nineteenth and early twentieth centuries inspired much larger-scale new town programs after World War II in Great Britain, France, Israel, the Soviet Union, India, China, Hong Kong, and (with less enthusiasm) the United States.

This chapter first summarizes the statistical growth of cities and then considers in turn each of these fundamental forms of response during the nineteenth century to the perils of laissez-faire urbanization.

URBAN GROWTH DURING THE NINETEENTH CENTURY

The advent and concentration of manufacturing activities in certain locations led to an astonishing increase in the level of urbanization in Europe and the United States during the nineteenth century. The growth of cities during that century was precisely documented in a classic study by Adna F. Weber (1899/1963), which an anonymous preface characterized as "the first really sound, comprehensive, and complete contribution to urban studies by an American."

Three aspects of urban growth during the nineteenth century that emerge from Weber's study are: (1) the rapidity of increase of urban population; (2) the emergence of large numbers of new urban places; and (3) the phenomenal expansion of very large cities such as London, New York, and Paris. These are interrelated facets of the prevailing movement of people to urban places from the countryside and from other countries.

In England and Wales, the aggregate population of places exceeding 20,000 grew from 1.5 million to 15.5 million between 1801 and 1891, a tenfold increase. This represented a rise in the urban portion of the total population of those regions from 16 percent to 53 percent (Table 4-1). Eighty percent of the total population growth of England and Wales between 1801 and 1891 was urban (Weber, 1899/1963, p. 43). Similarly, France experienced rapid growth of its urban population (living in places exceeding 10,000) from 2.6 million in 1801 to 9.9 million in 1891, with the urban share of national population rising from 9.5 percent to 25.9 percent. The United States, starting with a negligible urban population in 1800, had 8.2 million people living in places exceeding 8,000 by 1890, representing 29 percent of its population. (Urban population in the United States would exceed rural areas for the first time in 1920.)

The shift of energy source from running water to coal, and later to electricity, facilitated the geographic diffusion of manufacturing facilities during the nineteenth century. This was reflected in the rapid increase in the number of urban places. Cities exceeding 20,000 inhabitants, in England and Wales rose from 15 to 185

Table 4–1 Nineteenth century urban growth: England and Wales, France, and the United States

	England and Wales: Urban Places Exceeding 20,000 Inhabitants		
Year	No. of Places	Total Pop. (millions)	% of National Pop.
1801	15	1.5	16%
1851	63	6.2	35%
1891	185	15.5	53%
	France: Urban Places Exceeding 10,000 Inhabitants		
1801	90	2.6	9.5%
1851	165	5.1	14.4%
1891	232	9.9	25.9%
	United States: Urban Places Exceeding 8,000 Inhabitants		
1800	6	.13	3.3%
1850	85	2.9	12.5%
1890	448	18.2	29.0%

SOURCE: Adapted from Weber (1899/1963)

during the period 1801–1891. In France, places of more than 10,000 rose from 90 in 1801 to 232 in 1891. In the United States, places exceeding 8,000 increased from merely 6 in 1800 (Philadelphia, New York, Baltimore, Boston, Charleston, and Salem) to 448 in 1890 (Table 4-1).

But while urbanization was spreading to smaller towns and cities in outlying locations, the principal cities nevertheless attracted the major share of population growth, largely due to immigration from rural areas. London expanded by 625 percent to 5 million between 1801 and 1891. Its share of the population of England and Wales accordingly rose from 9.7 percent to 14.5 percent during that period. New York City grew tenfold from 1800 to 1850 and then tripled again in the next four decades, reflecting the arrival of large numbers of immigrants from the British Isles and Europe. Boston grew from a modest town of 24,900 in 1800 to a world-class city of 448,500 in 1890 (Weber, 1899/1963, p. 450). Paris quadrupled from 547,000 in 1801 to 2.4 million in 1891, rising from 2 percent to over 6 percent of national population. Berlin grew from 201,000 to 1.6 million during the period 1819–1890 (Table 4-2).

Second-tier cities also grew rapidly during the nineteenth century. In England and Wales, no cities except London had over 100,000 inhabitants in 1801. By 1851, there were eight cities of that size besides London with a total population of 1.6 million; by 1891, there were 23 such cities with a collective total of 5 million inhabitants, slightly more than London itself.

Table 4-2 Nineteenth century population growth of selected cities

	1800-1	1850-1	1890-1
London	860,000	1.7 mill.	5.0 mill.
Paris	547,000	1.0 mill.	2.4 mill.
Berlin	201,000	850,000	1.6 mill.
New York	62,500	660,000	2.7 mill.
Boston	25,000	137,000	448,500

SOURCE: Adapted from Weber (1899/1963).

In the United States, of 28 "great cities of 1890" containing 9.7 million people, only nine existed in 1800, with a total of 208,000 inhabitants (Weber, 1899/1963, Table II). (New York accounted for 2.7 million of 1890 total). Thus, a broader distribution of large-city growth is apparent in nineteenth-century America than in Great Britain, reflecting both the larger geographic size of the former and the westward migration of American settlement.

Weber's data suggest that Paris (with 2.5 million out of 4.5 million inhabitants of 12 "great cities" in 1891) dominated France as London did England. In the more decentralized German Empire, however, Berlin accounted for only 1.6 million out of 5.9 million inhabitants of 25 German "great cities" in 1890—comparable to the status of New York in the American system of cities.

REGULATION: PUBLIC SANITARY REFORM

Urban Squalor

The building of dwellings to accommodate the astronomic increase in urban populations in the industrializing nations during the nineteenth century lagged far behind demand. Overcrowding to inhuman levels was ensured by the prevailing building practices of the times. Unfettered by any public regulations, these practices were guided on the one hand by the need for proximity—housing had to be within walking distance to locations of employment—and, on the other hand, by the builder's greed. Thus, dwellings were minute in size and packed together with space left unbuilt only to the minimum extent necessary to provide physical access to each unit. A prevalent building pattern in English industrial cities during the first half of the nineteenth century was the "court system." Dwellings were constructed facing streets with a second row, back to back with the first row, which faced only onto an interior court or alley. Narrow tunnels connected these interior courts with the streets and outside world (Figure 4-1). In the absence of any means for removing sewage and refuse from the premises, the courts, alleys, and the streets served as waste receptacles. As exacerbated by the blocking out of sunlight and ventilation, the stench of accumulating bodily wastes and other refuse and the hazards that they posed to human health was unimaginable (Figure 4-2).

Regulation: Public Sanitary Reform

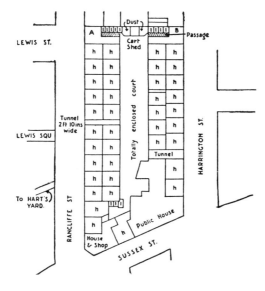

Figure 4.1 Diagram of a tenement courtyard in Nottingham, England, ca. 1840. *Source:* Benevolo, 1967. Reproduced by permission of Routledge & Kegan Paul Ltd.

Numerous eyewitness accounts of conditions in the mid-nineteenth century were provided by socialists and other reformers. Friedrich Engels described Manchester in 1845 as follows: (quoted in Benevolo, 1967, p. 23).

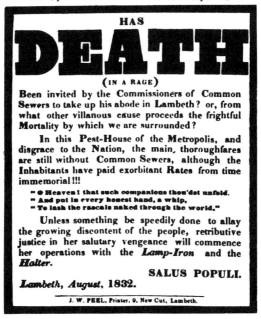

Figure 4.2 Broadside protesting inaction of improvement commissioners regarding sewerage, 1832. *Source:* Benevolo, 1967. Reproduced by permission of Routledge & Kegan Paul Ltd.

Here one is in an almost undisguised working-man's quarter, for even the shops and beer-houses hardly take the trouble to exhibit the trifling degree of cleanliness. But all this is nothing in comparison with the courts and lanes which lie behind, to which access can be gained only through covered passages, in which no two human beings can pass at the same time. Of the irregular cramming together of dwellings which defy all rational plan, of the tangle in which they are crowded literally one upon the other, it is impossible to convey an idea. And it is not the buildings surviving from the old times of Manchester which are to blame for this; the confusion has only recently reached its height when every scrap of space left by the old way of building has been filled up and patched over until not a foot of land is left to be further occupied. . . . he who turns [into the maze of passageways and courts] gets into a filth and disgusting grime the equal of which is not to be found . . . . In one of these courts there stands directly at the entrance a privy without a door, so dirty that inhabitants can pass into and out of that court only by passing through foul pools of urine and excrement. Below it on the river there are several tanneries which fill the whole neighborhood with the stench of animal putrefaction.

Not only were the dwelling units pitifully small to begin with, but they were hopelessly overcrowded. According to Rosen (1958, p. 206), Manchester in 1841 "had 1,500 cellars where three persons, 738 where four, and 281 where five slept in one bed." Nor were such conditions limited to Manchester: Liverpool, Bristol, Leeds, Glasgow, and London all contained sizable districts of similar nature. Liverpool in 1884 was reported to have certain districts with up to 1,210 persons per acre (Ashworth, 1954, p. 10). In the United States, high-density tenement districts flourished in ports of entry for European immigrants. New York's Lower East Side had a density of 272 persons per acre in 1860, which doubled in the next 30 years as further waves of Irish and Italian immigrants arrived (Weber, 1899/1963, p. 460).

Public Health Implications and Reforms

The pervasive overcrowding and lack of sanitation and basic hygiene, water supply, fresh air, and opportunity or provision for outdoor recreation, combined with long working hours in the factories and mills, inevitably magnified human misery and shortened life expectancy. Tuberculosis or "consumption" was the leading cause of death in urban England during the nineteenth century. TB was inevitably associated with undernourishment, poor ventilation, and general debilitation (Flinn, 1965, p. 11). TB, however, attracted little social consternation before the 1840s since it was viewed as an inevitable, albeit undesirable, aspect of the working-class existence. Also, statistics on TB were unreliable due to the difficulty of diagnosing the disease and the absence of a central governmental office for collecting statistics on morbidity and mortality. As long as people of influence were unthreatened by tuberculosis and its companion urban killer, typhus, nothing was done about it.

Cholera was another story. This Asian import struck London in 1831–1832, and reappeared in 1848–1849, 1854, and 1867. In terms of numbers of deaths and chronic level of threat, cholera was far less important than tuberculosis or typhus. But its effects were not confined to poor districts. Cholera struck with particular force in the wealthy neighborhoods where plumbing and connection to central (polluted) water supplies facilitated its spread. Together with a growing incidence of crime against property, cholera galvanized consternation of the elite regarding the conditions of urban squalor in their midst. According to Ashworth (1954, p. 47):

> Even if he were not his brother's keeper, every man of property was affected by the multiplication of thieves; everyone who valued his life felt it desirable not to have a mass of carriers of virulent diseases too close at hand.

And another pithy account:

> Cholera frightened people. It stirred even the moribund, degraded, unreformed municipal corporations into fits of unwonted sanitary activity. It was the clearest warning of the lethal propensities of the swollen towns of the new industrial era (Flinn, 1965, p. 8).

Cholera frightened the elite, but fear per se is seldom a basis for public action except in such panicky measures as the ostracizing of Jews during the fourteenth-century plague. The translation of fear into rational public response required, not simply emotional rhetoric, but sound scientific investigation and documentation. The period between 1832 and 1860 was to be the "golden age" of sanitary surveys, which would prove to be the advent of the modern public health movement (Rosen, 1958, p. 213).

Besides the cholera scare, two further prerequisites established a foundation sanitary reform. One was the development of the science of statistics and its application to the analysis of social problems. The first British Census was undertaken in 1801 partly to initiate a data base for the calculation of premiums for government-sponsored life insurance. The development of the field of vital statistics was pioneered by William Farr, a dedicated statistician and officer of the Register-General's Office, who foresaw that accurate data on births and deaths was essential to the assessment of the status of health of the population. His statistical reports "provided the ammunition used in the campaigns against disease in the home, in the factory, and in the community as a whole" (Rosen, 1958, p. 227).

The other precondition was the appearance of a political philosophy that advocated legislative intervention in the marketplace to remedy social ills that impeded the economy. The necessary framework was provided by the Philosophic Radicals, a small group of intellectuals headed by the energetic and enigmatic Jeremy Bentham (whose earthly remains dressed in his own clothes today reside in a hallway at University College, London). Among other reforms in the fields of law, education, and birth control, the Philosophic Radicals called for a review of the Poor Laws. Their objective was not to make the poor more comfortable, but rather to centralize relief to the poor and to ensure that the system promoted rather than discouraged working for a living.

That the appointment of a Royal Poor Law Commission in 1832 would eventually lead to the far more lasting contribution of the sanitary survey movement, was in large part due to the driving force of a key individual. Edwin Chadwick, the founder of the urban sanitation movement, exemplified the nineteenth-century tradition of the inspired amateur that included Robert Owen, Frederick Law Olmsted, and Ebenezer Howard. Like Howard, Chadwick would eventually be knighted for distinction in a field in which he had no formal training. Chadwick's introduction to the field that he would transform occurred by chance through his close association with Bentham (for whom he served as secretary before the latter's death in 1832). Through this connection, Chadwick was appointed that year as secretary to the new Poor Law Commission where he remained until 1847. Before 1838, Chadwick's interest in public health was minimal insofar as the commission's attention was largely devoted to factory and Poor Law reform (Flinn, 1965, p. 35). But inevitably his research into the causes of public outlays for "poor rates" (welfare payments) led to the long-ignored physical environment of industrial slums. Together with three physicians, James Kay, Neil Arnot, and Southwood Smith, Chadwick prepared a commission report in 1838 that linked for the first time the incidence of disease fostered by unsanitary living conditions to the economic costs borne by the nation through the payment of poor rates.

This report marked the beginning of a series of sanitary surveys that applied the new science of statistics to the analysis of patterns of illness and death. Essential to this task was a geographic perspective. With reasonably accurate vital statistics supplied by William Farr, the spatial patterns of disease could be mapped and related to environmental factors such as water supply. For example, there were 2,000 deaths from cholera in 1854 in Newcastle-on-Tyne, while in Tynemouth, eight miles away, where new drainage regulations were in force, only 4 deaths occurred (Ashworth, 1954, p. 61).

According to Rosen (1958, p. 35):

> The year 1838, then, was an important turning-point in the history of the public health movement. Although its roots stretch back fifty years, the movement was, before 1838, unorganized, leaderless, and in a legislative sense—the only sense that mattered in the long run—aimless. Essential foundations had been laid, preconditions established, but, important as these were, effective action was missing. This is what Chadwick supplied.

The seminal document of the sanitary survey literature was the 1842 "Report of the Poor Law Commissioners Concerning the Sanitary Condition of the Labouring Population of Great Britain." Written largely by Chadwick, it was based, on the one hand, on a broad empirical investigation of the incidence and causes of disease in poor districts based on statistical surveys and expert accounts. On the other hand, it reflected Chadwick's personal exhaustive research in the British and European literature on epidemiology and urban health. It was graphic and candid in its portrayal of the squalor prevailing in Great Britain's industrial towns. In its effort to relate physical deficiencies to human health and quality of life, Chadwick laid a foundation for future work in urban sociology as well as public health. The report foreshadowed in its style and objectives the "Brandeis brief" and its analogues in twentieth-century American jurisprudence. It also anticipated the Presidential "blue ribbon commission" such as the Kerner Commission on Urban Violence in 1968.

Although the report emerged from a review of the public costs of poor rates, the economic implications are in fact subordinated to issues of human suffering. The first four chapters are concerned respectively with: (1) "General Condition of the Residences," for example, filth and overcrowding; (2) "Public Arrangements External to the Residences," for example, water supply, paving, drainage; (3) "Internal Economy and Domestic Habits," for example, the moral effects of the lack of privacy in the home; and (4) "Comparative Chances of Life in Different Classes." Only after thus reviewing the social and health effects of urban slum conditions does the report turn in Chapter 5 to "Pecuniary Burdens Created by Neglect." Chapter 6 addresses "The Effects of Preventive Measures," including actions by public authorities and employers to improve sanitary conditions.

Chapter 7, "Recognised Principles of Legislation," surveys the state of law and public authority with respect to the problems enumerated earlier in the report. This chapter raises sensitive issues of public intervention in the private building process, and Chadwick is surprisingly indirect and timid. Evidently anxious not to lose his audience by calling too blatantly for an end to laissez faire, this chapter makes no clear-cut proposals for legal or regulatory reform. Instead, Chadwick reviews the state of existing institutions, including fire laws, nuisance, and improvement commissions, noting their respective inadequacies.

The report's final section, "Recapitulation," proposes technological improvements to remedy existing problems rather than legal measures. The public authorities must install "water carriage drainage" and water supply systems to provide for the removal of wastes from the living habitat. With sufficient water flow and U-shaped drains, refuse would be removed automatically through self-flushing, thus eliminating the need for on-site storage of wastes in courts, alleys, and dwellings (Peterson, 1983, p. 17). As to the ultimate disposal of wastes entering public drains, Chadwick was a strong proponent of reviving the medieval custom of "nightsoil," the application of human wastes to agricultural land.

The "Recapitulation" is silent on the need for stronger building laws to require the use of safer construction methods, to reduce overcrowding through occupancy limits, to make adequate provision for ventilation and isolation of wastes, and to limit the number of dwellings per unit of land. Such limitations on the private sector were still too controversial even to be suggested. But the 1842 report performed a unique public service in documenting the nature, extent, causes, and economic implications of urban squalor.

The following year, Parliament established a Royal Commission of the State of Large Towns and Populous Districts (Health of Towns Commission). According to Flinn (1965, p. 67):

> The role of the Health of Towns Commission was thus to substantiate by more systematic and widespread survey the accuracy of Chadwick's axioms, and to point more

precisely to the details of any necessary legislation. In this way the 1843 Commission was a logical and reasonable extension of Chadwick's work.

Although not appointed as a member of the commission (perhaps to avoid the appearance of bias), Chadwick served as unofficial consultant and actually wrote much of its two reports in 1844 and 1845 (Flinn, 1965, p. 67–68). These included legislative proposals that Chadwick felt obliged to postpone in 1842, such as the following (Benevolo, 1963, pp. 91–93):

- delegation of responsibility for sanitary regulation to local health authorities, under royal supervision;
- preparation of detailed sanitary surveys within a district before planning a drainage system;
- coordination of sewer construction with road improvements;
- establishment of minimum sanitary requirements for new dwellings;
- requirements for ventilation and cleaning of existing dwellings;
- construction of new public parks in industrial cities.

In the face of a new outbreak of cholera, Parliament finally adopted England's first comprehensive sanitary legislation: the Public Health Act of 1848. This Act culminated a decade of investigations beginning with the 1838 Poor Law Commission Report. It was squarely based on the 1845 Royal Commission Report which, in turn, drew heavily from the 1842 report, all drafted by Chadwick. Thus a major legislative reform was achieved in response to the perception of environmental threat documented in these and other investigative reports. Like the Act for Rebuilding London after the Fire of 1666, the 1848 Public Health Act demonstrated the capacity of the British legal system to respond (albeit belatedly) to the need for innovation in the face of disaster.

The Act established a General Board of Health and authorized the creation of local district health boards. The former, of which Chadwick was a member, existed only for five years, reflecting perhaps a continued reluctance of Parliament to establish a permanent national role in sanitary reform. Local boards were mandatory only for districts where the annual death rate exceeded 23 per thousand and otherwise were dependent on local initiative.

The powers delegated to the district boards were permissive rather than mandatory. They were authorized to appoint medical officers, and to undertake cleansing, paving, and drainage, and to supply water, but they were not required to do any of these (Flinn, 1965, p. 71).

Local boards, however, were required to "cause to be prepared a map exhibiting a system of sewerage for effectually draining their district . . . " Furthermore, every new dwelling within established health districts must have its own drains and lavatory, a totally unprecedented public requirement. Other provisions dealt with refuse collection, removal of harmful wastes, inspection of slaughterhouses and lodging houses, the paving and upkeep of roads, the establishment of public gardens, water supply, and the burial of the dead (Benevolo, 1967, pp. 96–97).

It is of course one thing to pass a law, and another to bring about the physical changes desired. Parallels with contemporary legislation on clean air and water come to mind. While the 1848 Act did not overnight make working districts habitable, nor even lower the death rate appreciably, it did serve as the beginning of the modern era of public involvement in the growth of cities. According to Flinn (1965, p. 73):

> The Act of 1848 constituted a tentative and uncertain start to govern action in a major field . . . . Nevertheless, it had put a foot through a door which had hitherto defied all attempt at opening, and although the detailed administrative arrangements it laid down were scrapped within half a dozen years, its principle of state responsibility was not discarded. It was this principle which the [1842] *Sanitary Report* had sought to establish.

Chadwick's work and the 1848 Act that codified his recommendations influenced subsequent sanitary reform both in Great Britain and elsewhere. The British Public Health Act of 1875 and Housing of the Working Classes Act of 1890 were direct legacies of the 1848 Act (Benevolo, 1967, p. 100). The English reforms were watched closely in Europe and influenced parallel efforts in the United States where industrialization and immigration were rapidly overcrowding American coastal cities. Between 1800 and 1850, New York City grew nearly tenfold to 660,000, and Boston grew by about 450 percent to 136,000. By mid-century, sanitary investigations inspired by Chadwick's work were underway in both cities, with new public health laws soon to follow. The New York (State) Metropolitan Health Act of 1866 was the first major American law in this field.

The British and American sanitary reforms of the mid-nineteenth century launched not only modern public health but also the field of city planning. Peterson (1983) identifies three results of nineteenth-century sanitary investigations that underlie modern town planning: (1) water-carriage sewerage; (2) sanitary survey planning; and (3) townsite consciousness, as reflected in the urban parks movement. While one may quibble with the exact points of influence, there is no question that sanitary reform gave birth to the urban planning movement of the early twentieth century. The conduct of sanitary surveys helped to develop the methodology of general planning investigations through the assembly, display, and interpretation of diverse data, including physical, demographic, and cultural. Also, the surveys elevated the geographic scale of investigation from selected neighborhoods or problem areas to entire cities, and even metropolitan regions.

Sanitary reforms were the logical prerequisite to the consideration of a broader range of urban planning issues such as transportation, housing, economic development, zoning of land, and environmental protection:

> . . . sanitary legislation would have to develop within the general framework of town-planning legislation, and that, once one problem had been isolated for consideration—that of sanitation—all others would necessarily follow. (Benevolo, 1967, p. 93).

Furthermore, building and public health legislation, primitive though it was, laid a constitutional foundation for the acceptance of broader land-use zoning and environmental regulations in the twentieth century.

REDEVELOPMENT: A CENTURY OF MUNICIPAL IMPROVEMENTS

Advent of Public Responsibility

The rapid growth in urban population during the nineteenth century yielded both intensification and outward sprawl of cities. Existing built-up areas, which were originally tailored to medieval or colonial levels of population, became increasingly overcrowded, leading to the dangerous conditions discussed in the preceding section. Cities also expanded spatially to encroach on adjacent rural lands. Such peripheral expansion was facilitated by the development of means of commuting to work beyond walking distance, such as the horse-drawn streetcar, which opened up the nearby hinterlands of cities such as Boston beginning in the 1850s (Warner, 1978). Electric streetcars and railways (above and below ground) further promoted building booms on the outskirts of major cities in Europe and America. Some of this development occurred within the political limits of the central city, either as already established (as in the case of New York's growth along the length of Manhattan Island) or as expanded through political action (e.g., the merger of New York and Brooklyn in 1898, or the addition of several towns to Boston and Chicago in the 1870s). Increasingly, however, peripheral development occurred outside the political limits of the central city in independent suburbs, thus marking the inception of the modern metropolitan region (Jackson, 1985).

Concomitant to the demographic and geographic growth of nineteenth-century cities was the emergence of government as an active participant in the process of urban development. In 1800, European cities were not only small in size, but also primitive in terms of civic consciousness and public initiative. London, which had possessed a municipal charter of self-government since 1193, nevertheless failed to provide basic urban services such as water supply, drainage, paving, and streetlighting to most of its inhabitants. For those not fortunate enough to live in one of the privately planned estate developments, the haphazard activities of the improvement commissions and a few widely ignored municipal regulations on building materials and encroachment on public ways represented the extent of public interest in their welfare. Paris, which, unlike London, had not burned in recent history, remained medieval in its physical appearance and state of infrastructure.

The impress of public authority on these and other European cities of 1800 was largely of royal origin. London's Hyde Park was a former royal hunting ground that was opened to public recreational use around 1640 (Rasmussen, 1967, p. 92). In Paris also, the Bois de Boulogne and Bois de Vincennes were former royal parks. The Champs Elysees originated in 1670 at the direction of Louis XIV's minister

Colbert. The Place de la Concorde was laid out in the mid-eighteenth century by Louis XV. Napoleon Bonaparte, assuming the trappings of emperor, laid out the baroque Rue de Rivoli. Berlin's Tiergarten was another former royal hunting ground. The great Unter den Linden boulevard, the Champs Elysees of Berlin, was laid out by Frederick William in the mid-seventeenth century (Abercrombie, 1913, p. 222).

American cities in 1800 lacked the royal legacies of their European counterparts. Bridenbaugh (1964) documents the struggle of incipient urban communities to cope with chronic problems of water supply, fires, epidemics, and crime. Public response to these kinds of threats took the form of hastily adopted ordinances often more honored in the breach than the observance. Boston and New York, for instance, passed municipal laws during the seventeenth and eighteenth centuries requiring suitable construction and periodic inspection of chimneys. In the absence of a public fire department, each Boston householder was required to possess a bucket, ladder, and long-handled swab for extinguishing rooftop fires. Lacking a municipal water supply, the standard method for halting the spread of fires in colonial America, as in Stuart London, was to pull down or blow up houses in the path of the conflagration (Bridenbaugh, 1964, Ch. 7).

Necessary public facilities in pre-nineteenth-century American towns were usually provided through private initiative, often with the benefit of a monopoly granted under a license from the colonial legislature or municipal authorities. Thus the Mill Dam constructed in Boston in the 1630s at the present site of Government Center was constructed by private citizens who leased sites for tidal-powered mills (Whitehill, 1968, p. 5). Similarly, the first bridge across the Charles River was constructed privately by entrepreneurs in 1768 whose claim to a monopoly over river crossings was later rejected in a landmark court decision (Kutler, 1971). Boston's first public water supply was established at Jamaica Pond by a private company.

The rapid demographic and spatial expansion of both European and American cities in the early nineteenth century rendered such ad hoc and profit-oriented solutions outmoded as responses to many urban needs. While some services such as urban transportation, mills, and wharves continued to be provided through enfranchised private entrepreneurs, the urgent need to develop larger-scale facilities for common benefit, such as water and sewer systems, parks, highways, and firefighting capabilities demanded that urban governments retool themselves, legally and technologically, to meet modern challenges.

Transition in London: The Nash Improvements

In Europe, the metamorphosis of eighteenth-century aristocratic notions of city planning into broader programs of civic and metropolitan improvements of the middle and late nineteenth century was not the result of a radical overthrow of monarchism. Instead, the major periods of urban restructuring of European capital cities were associated with royal or imperial patronage, not socialist revolution. The

Nash improvements in London between 1812 and 1823 and the vastly more ambitious redevelopment of Paris by Haussmann starting in the 1850s proceeded under the auspices of the Prince of Wales (later George IV) and Napoleon III, respectively. But while these programs were the pet projects of national sovereigns, the nature and extent of authority wielded by those figures was vastly diminished from former times. Regency England and Second-Empire France each denied their monarchs the absolute or "divine" powers asserted by some of their predecessors. The transition from royal privilege to civic initiative as a basis for city planning is perhaps best, and earliest, exemplified in John Nash's distinctive contributions to the face of London.

A hallmark of London, cherished by its admirers and abhorred by traffic engineers, is its long-standing refusal to be a planned city. From the Fire of London to the Nazi blitz, proposals for a sweeping redesign of the metropolis have failed to win popular support. Aside from the neoclassical geometry of the West End squares and their imitators, the city as a whole has grown and evolved organically with a minimum of preconceived planning. The celebrated exception to this history of incrementalism was the work of John Nash. According to Summerson (1962, p. 177):

> Once, and only once, has a great plan for London, affecting the development of the capital as a whole, been projected and carried to completion. This was the plan which constituted the 'metropolitan improvements' of the Regency . . . carried out under the presiding genius of John Nash.

John Nash, born in 1752, was a prominent architect, society habitúe, and close friend of the Prince Regent (who ruled in place of his insane father George III from 1811 until the latter's death in 1820, and then in his own right as George IV until 1830). Nash was to be the leading architectural influence of Regency London whose works included the famous Cumberland Terrace and Buckingham Palace. But it was his role as city planner in the conception and execution of Regent Street and Regent's Park that concerns us here.

The impetus to the projects just mentioned was the reversion to the Crown in 1811 of a sizable tract of rural land at the (then) northern edge of London known as Marylebone Park. The English practice of leasing estates for long terms made the expiration of a leasehold a propitious time for converting land to more profitable uses. The newly installed Prince Regent therefore regarded the reversion of this land as an opportunity to create an aristocratic new suburb for himself and his friends, and also to enrich the royal household.

A major obstacle to this concept was the isolation of Marylebone Park from the financial district in the City, from the Court at St. James, and from the clubs on Pall Mall where upper class gentlemen spent much of their time. In 1809, a preliminary plan for the property prepared by the Royal Surveyor General, John Fordyce, had envisioned the establishment of a new north-south street to connect the Park to the central city. Nash and a partner elaborated on this concept and proposed the construction of an elegant street rivalling Napolean's Rue de Rivoli extending between the future Regent's Park and the Prince Regent's doorstep at Carleton House.

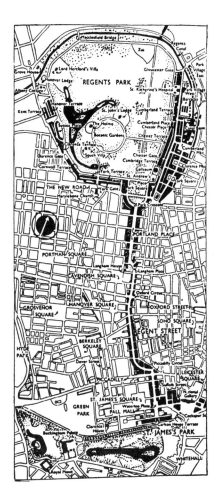

Figure 4.3 London improvements designed by John Nash, 1812–1835. Regent Street connects Regents Park at top to central London at bottom.
Source: Benevolo, 1967. Reproduced by permission of Routledge & Kegan Paul Ltd.

For the park itself, they proposed to convert the site into a vast "picturesque" landscape, punctuated by opulent villas and rows of aristocratic townhouses set in terraces or circular patterns (Figure 4-3).

Nash also proposed the creation of Trafalgar Square, the redesign of several streets in its vicinity, and the construction of Regent's Canal to serve as a navigation artery and water source for the lake and fountains of Regent's Park (Summerson, 1962, p. 177). Altogether, it was for London a program of unprecedented magnitude, aside from the grandiose proposals to rebuild the city after the 1666 fire. And unlike those, it was accomplished.

The importance of this vast undertaking as a transition between royal and public initiative is apparent in the contrast between the ends and the means involved. The Prince Regent made no secret of his wish to eclipse Napoleon as an imperial city builder (Summerson, 1962, p. 181). His objective, unquestioned by

Nash, was to indulge his royal vanity and to make money. But he lacked the financial resources to accomplish the work, particularly the cutting of Regent Street through the labyrinth of Soho. Therefore, an appropriation from Parliament was required. According to Rasmussen (1967, p. 274):

> In order to obtain this, there must have been some social aim in view, and the building of the new street was granted in 1813 as a means of improving the sanitation of the unhealthy quarters.

The alignment of the new street in fact involved an early exercise in "scientific" city planning. The route traversed the western edge of the shabby and cheap districts of Soho, directly adjacent to the more opulent Mayfair neighborhood. Thus Nash, the planner, fulfilled the parliamentary purpose of eliminating some rundown housing while acquiring the right-of-way as cheaply as possible. The choice of alignment simultaneously ensured that the new street would abut and be assimilated to the fashionable West End district, thereby enhancing the value of frontage on both sides of Regent Street for business and residence. In Nash's own words, the new street would provide: ". . . a boundary and complete separation between the streets and squares occupied by the nobility and gentry, and the narrow streets and meaner houses occupied by mechanics and the trading part of the community" (quoted in Davis, 1966, p. 66). (This concept was similar to that of the U.S. urban renewal programs of the 1950s and 1960s.)

Mindful of the need to also satisfy the vanity of his royal patron and potential investors, Nash the architect designed elegant neoclassic facades for his Regent's Park terraces and Regent Street frontage. Park Crescent, where the street meets the park is a fine surviving example of "pure Nash"—a curving, colonnaded exterior of creamy stucco. This sylish facade, reminiscent of Wood's Royal Crescent at Bath, at once provides a satisfying backdrop to the greenery of the park and merges an entire row of townhouses into a continuous, majestic unity.

Finally, Nash the landscape architect achieved brilliance in the design of Regent's Park itself. Between 1796 and 1802, Nash had been a partner with the seminal landscape architect Humphry Repton; the latter's influence on the design of Regent's Park is striking:

> It was from Repton that Nash obtained the idea of 'appropriation,' so that each villa commanded a pleasant prospect whilst not in view from any other villa, the management of the water (except the formal canal) was Reptonian, with its extremities concealed by bridges and by planting; the planting itself, in its blending of wood and lawn, or the contrast between villa and background, owed much to Repton; whilst even the idea of the canal traffic enlivening the scene is akin to Repton's use of cattle to give interest and scale in a park. . . . (Chadwick, 1966, p. 31).

To be sure, there was nothing "public" about the initial plans and execution of Regent's Park. In concept it was simply a grandiose version of the West End squares laid out by landed aristocrats for their peers. The public was not admitted to

Regent's Park until 1838, and even then there was little provided for the enjoyment of the horseless working class (Chadwick, 1966, p. 32). But over time, Regent's Park evolved from a "garden suburb" for the aristocracy to a public pleasure ground bordered by expensive townhouses. Today, Regent's Park is one of London's largest and most heavily used parks and contains, among other amenities, the London Zoo.

Nash was associated with two other park projects in central London, both of which were designed to be public from the outset. One was St. James Park, a Crown property since the reign of Henry VIII, which Nash redesigned in 1828, again with Reptonian flourishes. Chadwick (1966, p. 34) considers St. James Park to be the first English public park, although technically not in public ownership. It thus epitomizes the transition from royal to public in terms of usage if not actual ownership.

Trafalgar Square was, by contrast, public from the outset in both ownership and use. As London's equivalent to Paris's Place de la Concorde, Trafalgar Square has always been the place for crowds to gather to protest governmental policies, to celebrate victories, or to welcome the New Year. This may not have been the intention of Nash, and certainly not of his royal patron. But in 1830, the latter died and the Square opened, symbolizing the passing of royal pretense and the advent of planning for the people.

Haussmann's Transformation of Paris

The Nash "regency program" in London was but an appetizer to the banquet of urban and metropolitan improvements in Paris undertaken by Emperor Napoleon III and his alter ego, Baron Georges-Eugene Haussmann. The rebuilding of Paris between 1853 and 1870, with work continuing until the outbreak of war in 1914, touched every inhabitant of Paris and its suburbs. It ingeniously blended the aesthetic with the functional. It pioneered new methods of finance and public administration. It converted Paris from an overcrowded, unhealthy medieval town into the fabled "City of Light."

The onset of comprehensive redevelopment of Paris coincided with the election of Louis-Napoleon as president of the Second French Republic after the Revolution of 1848. Returning from exile in England, this nephew of Napoleon Bonaparte immediately undertook to revive and continue the program of public works initiated in Paris by his uncle and carried on spasmodically under the Bourbon Restoration. Perhaps influenced by the recent cutting of Regent Street and laying out of Regent's Park in London, Louis-Napoleon turned his attention first to the streets and parks of Paris. Following the coup d'état of 1851 and his assumption of the title of emperor, he selected Haussmann, then a rising lawyer and provincial administrator, to serve as Prefect of the Seine in 1853. According to Chapman (1953, p. 179):

> This was a very special post not because the Seine was the richest and most densely populated Department in France, and Paris the hub of national political life, but also because on grounds of public policy, the Prefect of the Seine acted both as head of the

Department and as municipal head of the city of Paris itself, two posts of immense influence and powers.

Significantly, Haussmann's powers were thus civil and municipal in nature, not delegated directly by the newly installed emperor. While the latter certainly provided much of the stimulus to the redevelopment of Paris, the actual program was a joint undertaking of the state, city, and private sector, utilizing essentially modern forms of legal procedures, financing, and contracting. Regal in scale and inspiration, the rebuilding of Paris was civic and bourgeois in execution.

Louis-Napoleon's motives for undertaking the program are subject to debate. The conventional wisdom ascribes the cutting of the great boulevards and the star-shaped "places" (most notably the Place de l'Etoile) to military considerations, namely the need to protect the government against the socialist rabble who flourished in the labyrinthine streets of Old Paris (Peets, 1927). Certainly the revolutions of 1830 and 1848 may have influenced Louis Napoleon's thinking, but great boulevards are two-way streets, as the Germans demonstrated in 1871 and 1940. A more humanitarian view is suggested by Smith (1907, p. 369), who attributes to Louis-Napoleon an understanding of the "sociological and hygienic condition of modern civilization [which] . . . forced upon him the sympathetic duty of making a suitable home for [his] people." Chapman (1953, p. 182) suggests a threefold motivation: (1) military considerations; (2) economic revitalization through improvement of circulation, markets, and external communications; and (3) "to make Paris into a capital city worthy of France, a capital provided with the light, beauty, and cleanliness essential to human dignity in cities."

In any event, even before Haussmann arrived on the scene, Louis-Napoleon was already engaged in planning new avenues and redesigning Paris's great western park, the Bois de Boulogne. The latter was a former royal hunting ground which the City of Paris took over from the Crown in 1848 on the understanding that the City would improve it as a public park. According to Chadwick (1966, p. 153):

> The Emperor himself was vastly interested in the scheme, which he saw as a future rival to Hyde Park and the other royal parks of London which he had known earlier as a refugee in England.

With the appointment of Haussmann in 1852, the urban rebuilding program began in earnest. Haussmann's first act was to order the preparation of an accurate survey of the city, using temporary timber towers to provide clear sight lines over the tops of buildings. The first phase of new avenue construction was approved in an 1855 law appropriating 60 million francs. This phase involved the construction of new principal north-south and east-west routes to reestablish the "Grande Croisee," which, since Roman times, had intersected at the city's heart (Smith, 1907, p. 372).

The great boulevards laid out under this and subsequent phases of work are the most familiar and cherished elements of post-Haussmann Paris (Figure 4-4). Clearly inspired by classical and baroque precedents, notably the Palace of Versailles, the

Redevelopment: A Century of Municipal Improvements

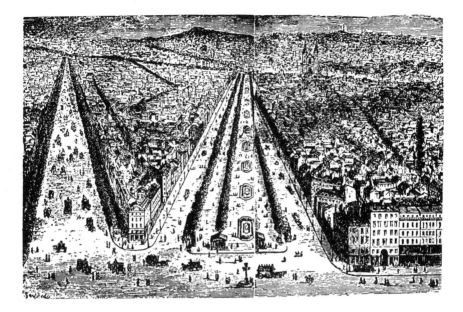

Figure 4.4. Birdseye view of two Haussmann boulevards, Paris, ca. 1870s.
Source: Benevolo. 1967. Reproduced by permission of Routledge & Kegan Paul Ltd.

Haussmann boulevards were widely acclaimed and set the style for ambitious cities around the world. For example, the following ecstatic words were written by an American architectural critic just after the turn of the century:

> The Avenue Napoleon, now de l'Opéra, is a perfect modern French street; not too long, spacious, well-built and furnishing axial vista to a fine symmetrical monument. This is the culmination of the classic scheme of axial symmetry, conceived in the hellenic period, more perfectly suggested in the Roman, carried a little farther in the Renaissance, fully understood by the Bourbon designers in France and brought to an ideal and complete realization by Haussmann in the Avenue de l'Opéra (Smith, 1907, p. 496).

To a greater extent than Regent Street in London, Haussmann's avenues and boulevards extended from point A to point B through whatever lay in their path. Not simply a widening of existing streets, these projects involved the acquisition, demolition, and replacement of the adjoining frontage on both sides of the new street. The result, as visible today along the Boulevards Sebastopol, de Saint-Michel, de l'Opéra, among others, is a broad, tree-lined avenue separated into through-traffic lanes, local-service access lanes, and ample pedestrian sidewalks. The avenues are bordered by substantially uniform facades of stylish, balconied, Second-Empire buildings. Unlike West End London, where residential quarters were isolated from commerce, the new frontage buildings in Paris have always been multipurpose. The ground floor is devoted to shops, cafes, banks, and restaurants. The next three or four principal floors contain elegant apartments for the upper middle class. Above them, the attics beneath the mansard roofs contain artists' studios and garret rooms for the poor (Saalman, 1971, pp. 26–27).

The frontage bordering the new avenues experienced a phenomenal increase in value as Haussmann expected. It was his hope that the city would retain ownership of the frontage and lease or sell it on the open market to capture the increase in value and thereby defray part of the cost of building the streets. This creative use of excess condemnation, however, was opposed by the financial community and was finally prohibited by the Council of State in 1858, which ordered that frontage lots, once cleared, should be returned to their previous owners (Benevolo, 1967, pp. 135–6). Needless to say, this promoted a lively speculation in land expected to be acquired for new streets. Construction of the new frontage buildings proceeded under private auspices but with uniformity of style ensured by a combination of public building restrictions and the dictates of the fashion of the time. In this respect, the rebuilding of Paris resembled the laying out of Back Bay in Boston, which was occurring at this same time.

The architectural critic Sigfried Giedion (1962, pp. 672–4) expresses dismay at the "great length" of Haussmann's boulevards, which he suggests were overly dominated by traffic concerns. But he praises the architectural treatment of the facades:

> Haussmann showed his sagacity in refusing to allow any tricks to be played with facades. Simply and without discussion, he spread a uniform facade over the whole of Paris. It featured high French windows, with accents provided by lines of cast-iron balconies like those used in the Rue de Rivoli under Napoleon I. He employed, unobtrusively, Renaissance shapes of a pleasantly neutral nature. A last touch of the unity which marked baroque architecture can still be felt. The neutral facades and the general uniformity make Haussmann's enormous work of rebuilding better than any other executed in or after the fifties of the nineteenth century (p. 672).

Closely related to the boulevards and avenues is the Napoleon III–Haussmann legacy of parks and open spaces. The smallest examples of these were carved out of the existing medieval clutter at the junctions of major streets. These have since provided oases amid the incessant traffic of foliage, gardens, statuary, playgrounds for small children, and benches for pensioners and the homeless. These small squares were perhaps intended to replicate the atmosphere of the West End squares of London (which themselves were inspired by Henry IV's Place Royale, later Place des Vosges). These squares were supplemented by three larger *parcs-interieures* and the gardens of the Champs Elysees. At the end of the city, the regional parks of Bois de Boulogne in the west and Bois de Vincennes in the east were elaborately redesigned to serve respectively the well-to-do and the working-class populations of Paris on holidays. This three-tiered hierarchy of public open spaces qualifies Haussmann to be honored as "the creator of the first real urban park system" (Chadwick, 1966, p. 152).

Beyond the new avenues and parks, the Haussmann–Napoleon III collaboration contributed innumerable economic and cultural assets to Paris. These notably included the famous covered markets of Les Halles (recently removed to a suburban location), the Opera, the completion of the Place de l'Etoile, several churches and theaters, major additions to the Louvre and the Sorbonne, and various hospitals and other institutions (Chapman, 1953, p. 185). Haussmann also devoted much effort to proposals for regional cemeteries, removed from the city to prevent further contamination of local water supplies. The proposals were rejected, however (Chapman and Chapman, 1957, pp. 126–130).

These projects conspicuously transformed the face of Paris, but, aside from the new parks, did little for the city's health. Industrialization arrived in France a few decades later than in England. But by the 1830s and 1840s, a strong coalition of physicians and social philosophers were calling for the same kinds of sanitary reforms as Chadwick was proposing in England at this time (Rosen, 1958, p. 252). Although these demands arose from socialists, the first significant progress in meeting them was achieved under the bourgeois conservative rule of Napoleon III. Haussmann persuaded his patron of the need to create a regional water supply system for Paris to replace its traditional reliance on local wells, cisterns, and the foul Seine. By 1870, when both the emperor and his chief engineer had been removed from office, work was far advanced on two aqueducts of 114 and 173 kilometers, respectively, to bring fresh water to Paris from the Aisne and Loire river basins. Like Robert Moses, New York's legendary power broker of a later generation (Caro, 1974), Haussmann was adroit at skirting the law to achieve his purpose: "He quietly went ahead with his plans for new aqueducts, his surveys, and the buying of sites, so that when finally permission was obtained he could immediately begin operations" (Chapman, 1953, p. 186).

The other fundamental contribution to public health of the Haussmann era was the Paris sewer system, a network of underground canals varying in width from four to eighteen feet and totalling some 600 miles in length. Principal interceptor sewers and galleries were laid beneath the new avenues as those arteries were constructed, a significant departure from earlier practice of regarding new streets simply in terms of traffic capacity and as decorative amenities. The canals conducted street drainage and raw sewage to an underground reservoir beneath the Place de la Concorde, from which it flowed to the Seine a few miles downstream from the city (Chapman, 1953, p. 187). Haussmann took special pride in the sewer system which was not anticipated in the emperor's program (Chapman and Chapman, 1957, p. 104).

Thus, in terms of public works both seen and unseen, Paris was substantially transformed during the two decades of the Second Empire. Moreover, despite rancor over Haussmann's high-handed financial schemes, work continued on many of his projects under subsequent governments until the First World War.

As important as the physical results of this program, however, were its political and institutional implications. On a much vaster scale than Regency London, the French Second Empire involved the devolution of the authority of the sovereign to the service of the people: The royal or imperial whim was replaced in practical terms by the public interest, or at least the interests of the elite and the bourgeoisie, if not the working classes (Saalman, 1971, p. 23).

In retrospect, the transformation of Paris under Napoleon III and Haussmann was an epic contribution to the evolution of the modern city. In form, it was inherently conservative and imitative. Precedents for baroque city plans were readily available: Wren's unused proposal for rebuilding London; the reconstruction of Lisbon after the 1755 earthquake by Pombal; Peter the Great's new capital, St. Petersburg; L'Enfant's 1792 Plan for Washington, D.C.; and the 1822 plan for Karlsruhe, the capital of Baden—all anticipated to some degree the outward appearance of post-Haussmann Paris.

But in several respects, the rebuilding of Paris was a unique, path-breaking experience that bridged the gap between the cities of the eighteenth and twentieth centuries. Its legacies to modernizing cities elsewhere may be summarized under five headings: (1) aesthetic style; (2) functionalism; (3) metropolitanism; (4) finance; and (5) administration.

Aesthetics. The architecture and plan of Second-Empire Paris were widely celebrated as the ideal translation of classical and Renaissance principles into "modern" city planning (Smith, 1907, p. 406). The broad boulevards terminating in monumental focal points, the small and large parks, the statuary and fountains, the atmosphere of wealth and power—all set the model for aspiring cities around the world. Paris was imitated in late-nineteenth-century redevelopment in Vienna, Berlin, Barcelona, London, Rome, Budapest, and many other European national and regional capitals. In the United States, Haussmann's Paris was enshrined in the City Beautiful movement, whose apotheosis was the White City of the 1893 Chicago World's Fair. Daniel Burnham, principal architect of the fair, drew on Haussmann precedents profusely in his 1902 Plan for Washington, D.C. and his 1909 Plan for Chicago. (See Chapter 7.)

The architectural critic Siegfried Giedion has deplored the setback inflicted on American architecture, particularly the Chicago School, by imitators of Haussmann. Relics of the turn-of-the-century craze for the City Beautiful are found in beaux arts civic buildings across the United States. In the 1920s, however, the City Beautiful was largely displaced by the American Prairie school, and later by the imported influence of the German Bauhaus and the French urban prophet, Le Corbusier (Giedion, 1962, pp. 509–530).

Functionalism. While the appearance of Haussmann's Paris was thus an important, if not totally praised, legacy, a more lasting contribution may be identified in his integration of urban infrastructure into the process of redevelopment. Thus the City Beautiful was also the "city functional." The impressive new avenues were pathways, not only for surface traffic, but also for sewers, as mains, water conduits, and, in the near future, metro tunnels and stations. All of the elements of the "metabolism of cities" were addressed somewhere in Haussmann's plans: housing, communications, food, water, gas, sewerage, commerce, education, culture, recreation, hospitals, cemeteries. Haussmann may be viewed as a founder of modern comprehensive city planning.

Metropolitanism. His planning perspective, however, was not confined strictly to the City of Paris. In two respects, Haussmann contributed to recognition of the metropolitan region as the appropriate geographic unit for management of the urban environment. The first was his successful advocacy of the legal expansion of the city to incorporate surrounding faubourgs or suburbs that had flourished outside the old city walls and therefore outside its taxing jurisdiction. In 1859, 11 communes containing some 400,000 inhabitants (many living in industrial tenements) were legally annexed to Paris, yielding its present area of legal jurisdiction. The annexed areas then had to contribute to the Paris tax revenues but in turn received public improvements costing 352 million francs by 1870 (Chapman and Chapman, 1957, pp. 135–8). Similar enlargements of municipal territories occurred in London in 1888, in New York in 1898, and in Berlin in 1923. In the twentieth century, however, annexations to central cities tapered off as metropolitan systems to provide water and other services to both central cities and suburbs became widespread. The Paris regional water systems, reaching beyond the city limits as necessary to obtain and discharge water, were another Haussmann contribution to incipient metropolitanism.

Finance. A fourth Haussmann legacy was his development of modern fiscal approaches to urban redevelopment. About two-thirds of the total cost of improvements under his direction during the period 1853–1870 was derived from national and municipal grants and the sale of public lands and materials. (The municipal contribution was facilitated by the significant rise in tax revenues attributable to the improvements themselves.) The remaining one-third was financed through borrowing from private banks and other lenders. This "deficit financing," so familiar today, was novel and controversial in Haussmann's time. His optimistic expectations were proven accurate, however, and the loans were repaid (Chapman and Chapman, 1957, pp. 236–7).

Administration. Finally, the Haussmann era marked an administrative revolution: the advent of the modern technocrat. With objectives established by higher authority—emperor, state council, or city—it was Haussmann's role to carry out the will of his superiors. According to Benevolo (1967, p. 134):

> Precisely because he did have the Emperor's support, Haussmann was always able to avoid having to justify his actions politically and could present them as technical and administrative measures deriving from objective necessities. . . . Haussmann set the pattern for the town-planner as a specialist worker who declines all responsibility for initial choice, and therefore in practice for the town-planner who is at the service of the new ruling class.

The transformation of Paris under Haussmann's direction is probably unparalleled in urban history, in the absence of war or natural calamity. Although the program was high handed, expensive, elitist, and unpopular at the time with many Parisians, it created one of the world's most elegant, beloved, and (in the absence of war damage) enduring monuments of late-baroque city planning. Post-Haussmann Paris was a unique blend of the human and the majestic. On the one hand, its alleys, garrets, cafes, and universities nurtured literary and artistic exuberance—the Paris of Renoir, Monet, Gertrude Stein, Fitzgerald, and Hemingway. On the other hand, it served as exemplar of the baroque world capital with its boulevards, parks, museums, and visions of grandeur. And beneath it all lay the sinews of a modern metropolis.

Water Supply for New York and Boston

American urban communities in 1800 were few in number, small in size, and coastal in location. The principal towns and their populations in that year were: Boston, 24,900; New York, 62,900; Philadelphia, 81,000; and Baltimore, 26,500 (Weber, 1899/1963, Table 163). The colonial period had bestowed on each of these settlements the incipient infrastructure of a preindustrial settlement: streets (mostly crooked and unpaved), docks, a town meeting hall, some common open spaces, firefighting implements, a constabulary and jail, and a primitive water supply. As population growth began to soar in the early decades of the nineteenth century, the inadequacy of the latter was perceived to be the chief liability and limitation on urban health and prosperity.

The water problem for New York and Boston was especially acute. Situated respectively on an island (Manhattan) and a peninsula (Shawmut), the cities were bordered by tidal, brackish water with no available fresh water streams. Both therefore depended on private or public wells or rainwater cisterns for their fresh water needs. By 1830, the population of New York had grown to 200,000 and that of Boston to more than 58,000. To compound the problem, in that same year the water closet was invented and thenceforth water consumption per capita would rise rapidly as water-carriage sewerage gradually replaced on-site privies and night soil collection (Weidner, 1974, p. 55).

Although they lacked the ability to foresee the increase in demand for water due to future demographic and technological change, the citizens of New York and Boston were dissatisfied with their existing supplies in 1800. The shallow water table aquifers on which they depended were easily contaminated with wastes from privies. Wells close to the harbor's edge could become brackish due to saltwater intrusion. And with limited surface recharge of local ground water, the overall reliable yield of springs and wells was insufficient for existing demands, let alone future growth. Rainwater cisterns added little to the general supply.

During the early decades of the nineteenth century, the provision of urban water supply was regarded as a private rather than a public function (Blake, 1956, Ch. 4). New York, Boston, Baltimore, and several small towns relied initially on

Redevelopment: A Century of Municipal Improvements

enfranchised private companies in preference to assuming the burden directly. An exception was Philadelphia, where recurrent outbreaks of yellow fever at the turn of the nineteenth century prompted a more aggressive municipal response. In 1801, Philadelphia constructed at public expense a pumping plant on the Schuylkill River powered by two steam engines. This project was designed and promoted by the noted engineer Benjamin Latrobe. It marked both a technological and an institutional breakthrough, namely in the use of the steam engine to pump water, and the use of public taxation to establish a municipal water supply (Blake, 1956, p. 33).

Meanwhile in 1796, the Jamaica Pond Aqueduct Company was chartered by the Massachusetts General Court (legislature) to supply water to Boston. The company laid hollow-log pipelines four miles to Boston from Jamaica Pond in Roxbury, which at that time was an independent town. This was an early example of an extraterritorial water supply. However, the resulting service was deemed inadequate for firefighting or to meet the needs of the growing Boston population (Nesson, 1983, pp. 1–2).

In New York, the Manhattan Water Company was chartered in 1799 with an exclusive franchise to supply the city with water. But contrary to expectations that this company would divert water from the Bronx River in Westchester County (just north of Manhattan), it instead limited its efforts to constructing in lower Manhattan a reservoir that supplied 400 families from local ground water:

> But this water proved both scarce and bad; the company, neglecting the ostensible purpose of its organization, soon turned its attention almost exclusively to banking affairs and thus lost the confidence of the community, and it was not long before the new works were voted a failure (Booth, 1860, pp. 666–7).

In 1811, the first plan for the future expansion of New York was prepared by a special commission established by the state legislature. The "Commissioners' Plan" projected the familiar grid pattern of streets and avenues extending up the length of Manhattan north of the existing limit of development at Houston Street. With future streets marching miles into the countryside of upper Manhattan as far as "155th Street," the plan comprised an accurate, if monotonous, forecast of the spatial expansion of the city (Lymon, 1964, p. 113). The opening of the Erie Canal in 1825, which connected the Hudson River with the Great Lakes, established the city's economic preeminence in the nation and doubled its population to 200,000 in the decade 1820–1830. Local water sources were hopelessly inadequate to serve this rapid rate of growth in terms of quantity, quality, and pressure. Various schemes were debated fruitlessly during the century's first three decades.

Finally, catastrophes in the form of fire (1828 and 1835) and cholera (1832) forced an end to the dithering. Colonel DeWitt Clinton, Jr. was retained by the city's Common Council to assess the water crisis and propose a solution. His report, issued within two months after he was consulted, predicted that Manhattan would reach a population of 1 million by 1890 (he was late by 12 years). His proposal was simple in concept and vast in magnitude, namely, to tap the Croton River 40 miles north of the city to obtain a reliable supply of 20 million gallons per day (gpd) of pure upland water (Clinton assumed per capita demand of 20 gpd, not foreseeing the impact of the flush toilet). Moreover, the elevation of a Croton River reservoir at 200 feet above mean sea level would permit the water to flow by gravity through an aqueduct to be constructed with enough "head" to serve the needs of taller buildings and firefighting in the city (Weidner, 1974, pp. 28–31).

The Croton River project required the construction of storage and conveyance facilities unprecedented since the Roman Empire. With the total cost estimated at several million dollars, the project was too large and too important for private enterprise. Accordingly, the City of New York, under grants of authority from the state legislature, undertook to plan and execute the Croton River project. A water commission was appointed within months after the Clinton Report, financing was approved by the city's voters in 1835, and construction began in 1837.

The project involved five major structural elements: (1) a masonry dam 50 feet high and 270 feet long impounding a reservoir with a surface area of 440 acres and storage capacity of 600 million gallons; (2) a 40-mile covered masonry aqueduct with a cross section of seven-by-eight feet; (3) a 1,450-foot-long "high ridge" to convey the aqueduct across the Harlem River into Manhattan; (4) a 35-acre receiving reservoir located within the future site of Central Park; and (5) a four-acre, masonry-walled distributing reservoir located on the present site of the New York Public Library at Fifth Avenue and 42nd Street. The first Croton River water arrived in Manhattan on July 4, 1842, an event celebrated with church bells, cannon, and a five-mile-long parade (Weidner, 1974, pp. 45–46). The event was both a technological and an institutional threshold: the City of New York had arrived of age.

Only six years later, Boston would hold a similar celebration. In 1845, John Jervis, who directed the Croton River project and was "America's foremost water supply engineer" (Nesson, 1983, p. 4), was hired to study Boston's water dilemma. He recommended adoption of a previously developed plan to create a water supply at Long Pond (later Lake Cochituate) in Natick, a distance of 17 miles from Boston. The Massachusetts General Court adopted a law in March, 1846 authorizing the Long Pond project to be constructed by the City of Boston and providing state backing of municipal bonds to finance it (Nesson, 1983, p. 9). The project was completed in two years and the westward march of Boston's quest for water was underway.

By 1860, Boston's demand had reached 17 million gallons per day (a per capita demand of 100 gpd reflecting the advent of plumbing), which equalled the entire safe yield of Lake Cochituate. With strong public confidence and legislative backing, the city developed a new reservoir in Chestnut Hill and six smaller reservoirs in the Sudbury River watershed to augment the Lake Cochituate supply. Thus by 1878, Boston had tripled its supply to about 63 million gallons per day.

In 1898, Greater New York was formed through the consolidation of Manhattan, the Bronx, Queens, Brooklyn, and Staten Island into a single city of 3.5 million inhabitants. Despite enlargement of the Croton River system with a new aqueduct completed in 1891 and a new, much larger dam in 1906, the city required new sources of water. Between 1907 and 1929 it developed a series of reservoirs and a new aqueduct to draw water from the Catskill Mountains, 100 miles north of the city. The Catskill Aqueduct crossed the Hudson River by means of an "inverted siphon" 3,000 feet long and 1,100 feet below the surface of the river (Weidner, 1974, p. 161).

This feat was repeated in the 1940s when the city reached out to the headwaters of the Delaware River in central New York State. The Delaware River Aqueduct consisted of the world's longest continuous aqueduct tunnel of 105 miles (Weidner, 1974, p. 300). By the 1960s, the combined systems were supplying New York City with over 1 billion gallons per day, of which 41 percent was derived from the Catskill system, 36 percent from the Delaware River, and 18 percent from the Croton River.

Meanwhile, Boston was pursuing a similar course of action but under a different governmental framework. Whereas New York City itself established and continues today to operate its water supply system even in upstate New York, Boston's system in 1895 was sold to a newly created regional authority, the Metropolitan Water District. The district was charged by state legislation to develop new water supplies to serve Boston, and its immediate suburbs. In 1908 the district completed Wachusett Reservoir, in central Massachusetts, connected by an 18-mile aqueduct to the Sudbury-Cochituate system. The 400-billion-gallon Quabbin Reservoir, 75 miles west of Boston, was constructed in the 1930s, completing the present metropolitan Boston water system which serves 2.5 million people in the city and 43 suburbs.

Thus, the water supply systems of both New York and Boston during the nineteenth century evolved from dependence on primitive, privately constructed local sources to large-scale, regional systems constructed and operated by public entities. The transition reflects both the advance of modern technology (for example, the ability to construct underground aqueducts with high-pressure siphons) as well as the evolution of municipal and regional institutions capable of serving the public interest. As in Paris, the emergence of these institutions was characterized by the development of new forms of finance, the application of modern concepts of eminent domain (land taking), and the administrative skills of technical experts.

The New York/Boston model of interbasin diversion from distant, upland sources guided not only the further growth of the water systems for those cities, but also influenced many other American cities as well. Of course, long-distance diversions were not always required: Great Lakes cities such as Chicago, Cleveland, Detroit, and Milwaukee found an ample source at their doorstep. (Chicago, however, is subject to a Supreme Court limit on the amount of water it may divert from the Great Lakes Basin.) Semi-arid western urban regions—Los Angeles, San Francisco, Denver, Phoenix—have aggressively pursued scarce water wherever it could be found, developing whatever technical and institutional measures were necessary to procure it. In some cases, such as the Owens Valley project of Los Angeles, the legal means were shady and the technical execution flawed, as demonstrated in the collapse of the St. Francis Dam in 1905 (Reisner, 1986). But the basic premise behind these western water wars was the same as that which motivated the Croton and Cochituate projects: Employ modern technology and political power to obtain water from wherever it may be found to ensure an unlimited supply of cheap water to the urban populace.

Urban Parks in America: The Olmsted Legacy

Imagine New York without Central Park, Philadelphia without Fairmont Park, San Francisco without Golden Gate Park, Boston without its "Emerald Necklace," and Chicago without its lakefront parks. How impoverished would be the urban habitat of America's older central cities without their distinctive and spatious parks—their "green lungs" as the Victorians called them. The existence of the great city parks is perhaps taken for granted today by the general public. But, like urban water supply systems, parks did not simply happen to fall into place. They were established through deliberate public action, often amid controversy. Their creation during the second half of the nineteenth and early twentieth centuries required vision, money, political power, artistic genius, legal innovation, manpower, and technology. The great urban parks were the embodiment of the maturing city as an instrument for social betterment. They have long been its most visible, accessible, and often most abused artifacts.

The advent of the American parks movement may be traced to 1853, the year that New York's Central Park was authorized by the New York State Legislature. Central Park represented several firsts: (1) the first deliberately planned urban park in the United States (Chadwick, 1966, p. 190); (2) the first park project of Frederick Law Olmsted; (3) the first accomplishment of landscape architecture as a profession in the United States (Fabos, et al., 1968); and (4) perhaps the first use of land-value increment taxation to finance a portion of the costs of a public improvement.

The population of Manhattan increased from 62,000 in 1800 to around 800,000 in 1860. In just two decades of migration from the political turbulence in Europe between 1840 and 1860, it gained about a half-million new inhabitants. In 1845, Dr. John C. Griscom published: "A Brief View of the Sanitary Condition of the City" which, like Chadwick's studies in England, documented the abysmal conditions of overcrowding and incidence of disease in the city's tenement districts (Rosen, 1958, p. 236). One implication of the growing crisis had already been addressed in the development of the new Croton River water supply which reached the city (but certainly not all of its buildings) in 1842. But another need identified in Griscom's report, as in most sanitary reform literature of the time, was for public parks and open spaces where the working class could devote their few hours of leisure time to outdoor recreation and exercise.

In 1853, New York possessed only a handful of small parks—the Battery, City Hall Park, and a few squares provided in the 1811 Commissioners' Plan—all totaling about 117 acres (Olmsted and Kimball, 1928/1973, p. 20). These were supplemented, for the affluent at least, by private "pleasure gardens" modeled on Vauxhall Gardens in London, but by 1850 these were disappearing as their site value for building rose. The only other open spaces available were cemeteries (Olmsted and Kimball, 1928/1973, pp. 21–22).

The proposal to establish a large central park in New York actually arose, not from the sanitary reformers, but from the city's literary and artistic community. During the middle decades of the century, cultivated Americans were aroused by the beauties of nature and wilderness as depicted by artists such as John James Audobon, Thomas Cole, and Frederick Church, writers such as George Perkins Marsh and Henry David Thoreau, and poets such as William Cullen Bryant and Henry Wadsworth Longfellow. In 1844, Bryant, as editor of the New York *Evening Post* wrote that "If the public authorities, who expend so much of our money in laying out the city, would do what is in their power, they might give our vast population an extensive pleasure ground for shade and recreation" (Olmsted and Kimball, 1928/1973, p. 22). Bryant's appeal was reinforced by Andrew Jackson Downing whose journal *The Horticulturist* urged that New York should emulate England in the creation of large public parks.

But unlike London, New York lacked a legacy of royal lands that could be converted into equivalents to Hyde Park, St. James Park, and Regent's Park. In New York, the land for a park had to be purchased by the city from diverse private owners. To obtain a sizable tract of land at a reasonable cost, it was necessary to look beyond the limits of the existing built-up city—about 34th Street in 1850—to the still rural precincts of upper Manhattan. At that time the proposed site of Central Park was a messy landscape of squatters, goats, mud, and rubbish, a 30-minute walk from the existing city. Its advocates, however, correctly anticipated that all of Manhattan would soon be urbanized and thus the park would be central indeed.

In 1853 the New York State Legislature authorized the city to establish Central Park. It was originally to comprise a rectangle of 770 acres, of which 150 acres was occupied by the two Croton reservoirs completed in 1842 and 1858, respectively. The park was subsequently expanded to 110th Street on the north, bringing its total area to 843 acres, about a half mile by 2½ miles (Figure 4-5).

The task of converting this huge and squalid tract of land into one of the world's great urban parks was the masterpiece of Frederick Law Olmsted. Like his contemporaries Chadwick and Howard, Olmsted possessed no particular training for the field that he would dominate. Born in 1822, he studied agricultural science and engineering at Yale and then devoted himself to farming, travel, and writing. He moved from his farm on Staten Island directly to the post of superintendent of the new Central Park project in 1857 (Sutton, 1971, p. 7). In collaboration with Calvert Vaux, he prepared the winning plan in a design competition for the park. In 1858, he was appointed Architect in Chief to execute his own plan.

Olmsted and Vaux's "Greensward plan" for Central Park was influenced by the English picturesque landscape tradition of Nash's Regent's Park and Paxton's Birkenhead Park in Liverpool (Chadwick, 1966, pp. 71–72). The essence of this tradition was deliberate informality, contrast between open meadow, groves of woods and water, and attention to the park's borders with the surrounding city. While as carefully planned and engineered as a formal baroque park such as Ver-

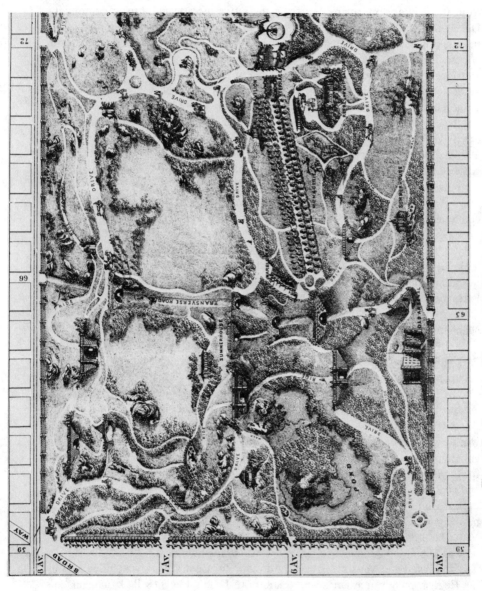

Figure 4.5. Excerpt from "Greensward Plan" for New York's Central Park by Olmsted and Vaux, 1858. *Source:* Fabos, et al., 1968.

sailles, the picturesque style sought to create the illusion of "rus" in "urbs," an artificial bucolic countryside. Olmsted wrote in 1872 that his purpose in designing Central Park was "to supply to the hundreds of thousands of tired workers, who have no opportunity to spend their summers in the country, a specimen of God's handiwork that shall be to them, inexpensively, what a month or two in the White Mountains or the Adirondacks is, at great cost, to those in easier circumstances" (Olmsted and Kimball, 1928/1973, p. 46).

A distinctive feature of Central Park, later widely imitated, was the separation of different forms of circulation. Pedestrians were removed from the path of equestrians and carriages, and internal pleasure routes were isolated from crossroads for through traffic. Where routes serving different purposes met, Olmsted provided under- or overpasses to eliminate stopping points and to enhance the illusion of open countryside.

The building of Central Park was the largest public work yet undertaken in the city of New York, involving thousands of jobs and millions of dollars. Tammany Hall and the "Tweed Ring" hampered Olmsted in the execution of his plans during the 25 years of his official connection with Central Park. He actually resigned from the project five times (Sutton, 1971, p. 9). The politics of Central Park were described by Frederick Law Olmsted, Jr. and Theodora Kimball (1928/1973) and need not be restated here. Central Park has continued to spark civic controversy ever since, as exemplified in the "Tavern on the Green" battle during Robert Moses's term as Parks Commissioner in the 1960s (Caro, 1974).

Despite its political travails, Central Park was a spectacular financial success. The total cost of acquiring its site of 843 acres from private owners was $7.4 million. This amount was reduced by $1.8 million charged to owners of frontage property on the theory of betterment of their property values. Thus, the net cost to the city of acquiring the land was about $5.6 million. The cost of improvements to the site was about $8.9 million, yielding a total cost to the city of $14.5 million (Olmsted and Kimball, 1928/1973, pp. 54 and 95). The project was self-amortizing through increases in property tax collections on surrounding land. In the 1870s, the annual increase in such taxes was estimated to exceed the annual interest on the park project costs by over $4 million (Olmsted and Kimball, 1928/1973, p. 95). Today, condominiums with a view of Central Park cost upwards of $1 million apiece, with city tax revenues accordingly enriched.

From its inception, Central Park was also a functional success. In 1871, usership of the park amounted to some 30,000 visitors per day and over 10 million per year (Olmsted and Kimball, 1928/1973, p. 95), or about 10 visits per capita for the entire population of Manhattan. The park was designed by Olmsted and Vaux to encourage active use in many ways, such as horseback riding, boating, carriage driving, skating, cycling, and strolling. Deliberate informality in the picturesque tradition dominates most of the park plan, with the exception of the formal baroque-style mall. Open meadows, rocky outcrops, wooded areas, and water surfaces encourage spontaneity and the sense of freedom which the "gilded age" associated with a rural landscape.

Olmsted was not reticent in extolling the success of Central Park. In 1880 he wrote:

> ... to enjoy the use of the park, within a few years after it became available, the dinner hour of thousands of families permanently changed, the number of private carriages kept in the city was increased tenfold, the number of saddle horses a hundredfold, the business of livery stables more than doubled, the investment of many millions of private capital in public conveyances made profitable.
>
> It is often said, How could New York have got on without the park? Twelve million visits are made to it every year. The poor and the rich come together in it in larger numbers than anywhere else, and enjoy what they find in it in more complete sympathy than they enjoy anything else together. The movement to and from it is enormous. If there were no park, with what different results in habitat and fashions, customs and manners, would the time spent in it be occupied.
>
> And the Park of Brooklyn [Prospect Park] . . . is sure, as the city grows, to be a matter of the most important moulding consequence—more than the great bridge [Brooklyn Bridge], more so than any single affair with which the local government has had to do in the entire history of the city" (quoted in Sutton, 1971, p. 255).

The use of the term *moulding* [sic] is perhaps deliberately intended to suggest two meanings. First, Central Park and its counterparts in other cities were designed to "mould" the physical form and structure of the surrounding city. Parks were conceived to be oases of open space and bucolic scenery around which the city would grow. Second, parks were intended to "mould" the moral character of the populace. Olmsted frequently pontificated on the benefit of open space and outdoor recreation on the physical and mental health of the city dweller. These views became articles of faith in twentieth-century urban and regional planning.

But Olmsted's most important contribution to modern city planning was the recognition of parks and open space as integral elements of the urban system. According to Sutton (1971, pp. 10–11):

> The success and popularity of Central Park started a trend, and city administrators throughout the country woke up to the advantages of open spaces. The land they were willing to purchase and sacrifice for this purpose, however, was usually some site undesirable for commercial or residential buildings, and in no way integral to the established patterns of city life, for example: the Fens in Boston; the mountain in Montreal; the swamps in Buffalo; the marshlands in Chicago. In general, the officials adopted simplistic notions of a park, separating it in their minds from the activities of the city. *Olmsted's effort was to integrate the two*" [emphasis added].

In this regard, the "emerald necklace" plan for Boston, formulated in the 1880s, was a logical progression from the concept of Central Park (Figure 4-6). The Emerald Necklace comprised a series of major and minor open spaces, some existing and some proposed, to roughly encircle central Boston on its landward side. One anchor for the necklace was the old Boston Common and the Public Garden (laid

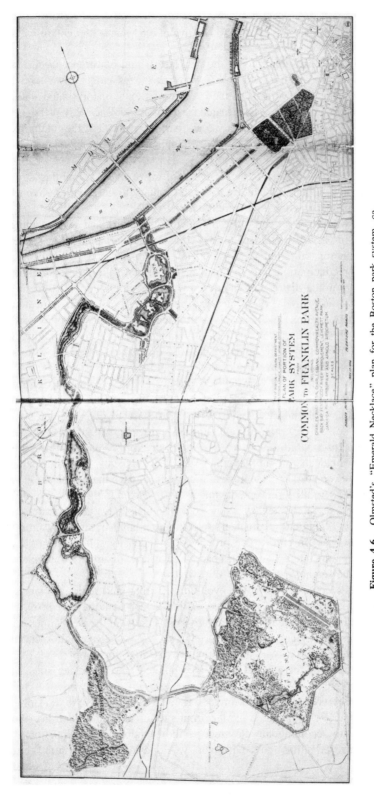

Figure 4.6. Olmsted's "Emerald Necklace" plan for the Boston park system, ca. 1885. *Source:* Fabos, et al., 1968.

out in 1839). The other terminus of the necklace would be the proposed Franklin Park. These were to be connected by a series of parkways and greenways bordering local streams. The achievement of these links was to be incorporated within the ongoing development of new land created by draining the Back Bay marshes in the 1860s. Commonwealth Avenue, a broad parkway, formed the main axis of the fashionable new Back Bay district and simultaneously served as a link in the Emerald Necklace.

The next link was less obvious. Olmsted urged that a remaining stretch of marsh bordering the humble Muddy River, a tributary to the Charles River, be set aside as open space with some modification, rather than be filled. Viewed in isolation, the Muddy River Fens were befouled with rubbish and sewage and generally unpromising as urban park space. But Olmsted enumerated multiple benefits to be achieved in rehabilitating the swamp: (1) abatement of a "complicated nuisance"; (2) "reconciliation of convenient means of general public communication through the adjoining districts of the city; (3) "dressing and embellishment of the banks"; and (4) an element of a "general scheme of sylvan improvement for the city" (quoted in Sutton, 1971, p. 227). Olmsted's "Fenway" is today sadly degraded by an elevated highway and lack of maintenance. It is chiefly known for its namesake, Fenway Park, the home of the Red Sox, and for several cultural and educational institutions in its vicinity.

The two paradigms of Olmsted's legacy were thus reflected in Central Park and the Emerald Necklace. The former involved the use of open space as the "working man's Adirondacks," and the latter, the use of diverse open spaces to interrupt the spread of urbanization. Olmsted enjoyed a wealth of opportunities to apply these two paradigms to various cities of North America. In contrast to Haussmann, who dealt with all planning issues of a single city, Olmsted dealt with a particular issue, open space planning, in more than a dozen major cities. Olmsted is most celebrated for his achievements under the first paradigm, the creation of the great urban parks. But this legacy produced few sequels in twentieth-century America as the nation boarded its streetcars, electric commuter trains, and automobiles and turned its back on cities. The process of low-density suburbanization would ultimately promote interest in open space as a means of urban containment and separation—the second paradigm.

Olmsted's contribution to the evolution of urban and regional planning, as distinct from park design, was furthered in his Chicago activities. In 1869, he and Vaux designed a prototypical garden suburb at Riverside on the DesPlaines River, a few miles west of Chicago (Jackson, 1985, pp. 79–81). Riverside's curvilinear streets, common greenways, and spatious residential lots impressed Ebenezer Howard, who visited there in 1876 and incorporated these characteristics into his influential garden city proposal.

Olmsted returned to Chicago in 1890 as landscape architect to the Columbian Exposition of 1893. He planned the lakefront site for the Fair, which later became Jackson Park, to accommodate the great pavilions and exhibition halls amid open courts, lagoons, reflecting pools, and statuary. Alongside this beaux arts White City was a touch of pure Olmsted, a wooded island "to contrast with the artificial grandeur and sumptuousness of the other parts of the scenery" (Sutton, 1971, p. 194).

Behind the grandiose stage set of the White City, the Fair was celebrated for its unprecedented level of functional organization and planning. Transportation, food, lighting, water supply, waste disposal, and mechanical energy were all incorporated into its design. To highlight its function as a showcase for electricity, President Cleveland activated its lighting and fountains by pushing a button in Washington, D.C. Thus, while the Exposition looked backward to Haussmann, Versailles, and antiquity for its architectural inspiration, it looked forward to the twentieth century in its application of technology to the design of a new community, albeit a temporary one. The contrast between the planned environment and what lay beyond its borders was widely noted according to Mayer and Wade (1969, p. 193):

> The [Exposition] was an artificial city that conflicted with the actual city in almost every important element. Where the American metropolis was chaotic and disorganized, the Exposition was planned and orderly; while the real city was private and commercial, the ideal was public and monumental; where Chicago was sooty and gray, the White City was clean and sparkling.

Olmsted as landscape architect was of course not responsible for the functional arrangements of the fair. But he may be credited for having successfully assimilated the demands of technology and aesthetics in his site design. This was in effect a land-use plan incorporating both the built and the unbuilt elements of a "city." The application of this integrative approach to actual cities and their surrounding regions, most notably in the 1909 plan of Chicago, would be the work of Daniel H. Burnham, a Chicago architect and Director of Works for the fair. In effect, the torch of national preeminence in the art of planning cities passed from Olmsted to Burnham at the 1893 Exposition.

Burnham, whose contributions are considered in Chapter 7, commemorated Olmsted's entire career in a tribute to:

> . . . the genius of him who stands first in the heart and confidence of American artists . . . . He who has been our best adviser and our common mentor. In the highest sense he is the planner of the Exposition. No word of his has fallen to ground since first he joined us . . . . An artist, he paints with lakes and wooded slopes; with lawns and banks and forest-covered hills; with mountain-sides and ocean views. [We honor him] not for his deeds of later years alone, but for what his brain has wrought and his pen has taught for half a century (quoted in Chadwick, 1966, p. 196).

RELOCATION: THE IDEAL COMMUNITIES MOVEMENT

This chapter has so far considered two avenues of public response to urban deterioration during the nineteenth century: public health regulation and municipal redevelopment. These two approaches went hand in hand, the first yielding a variety of building and sanitary codes to be applied by local public authorities and the second fostering the development of paved and lighted streets, water systems, sewers, and parks. The regulatory and the redevelopment approaches addressed the ills of existing cities directly, with gradual, uneven, but generally positive results. The principal European and American cities of 1900, despite their much greater size, were arguably safer, healthier, and more habitable than they had been in the early decades of the century.

In contrast to these efforts to regulate and rebuild existing cities, a handful of influential idealists proposed a completely different approach, namely relocation of workers to new, planned industrial villages in rural settings. Such communities, it was argued, would promote health, happiness, productivity, and morality through the influence of a salubrious environment. Several individuals in Europe and America put their beliefs into practice and created model villages to inspire wider imitation. Although they did not succeed in the latter goal, the experimental communities and the theories of socioeconomic organization that prompted them have deeply influenced twentieth-century planning ideology.

The remainder of this chapter considers the experience of three preeminent nineteenth-century ideal community proponents. To list them is to indicate that this was no tightly circumscribed school or movement, but rather a diverse collection of individualists motivated by quite different goals and assumptions. Those considered here are the Welsh-born utopian Robert Owen, the Chicago sleeping-car magnate George Pullman, and the stenographer/progressive Ebenezer Howard. These and their like-minded contemporaries had little in common except for a repugnance for large cities, an impatience with conventional reforms, and a faith in environmental determinism.

While the industrial model-village schemes emphasized moral values, particularly useful work and temperance, these communities differed from the spiritual utopias that proliferated in the United States during the nineteenth century. The latter included a few long-lasting and economically viable communities such as the Shaker villages of New England and New York, the Oneida settlement in New York, the Amana Colonies in Iowa, and the Mormons in Illinois and later Utah. They also included a variety of more ephemeral utopian experiments, whose religious or philosophical objectives perhaps overshadowed economic and functional practicality, as with Brook Farm in Massachusetts or the 30 Fourierist "phalanxes" established in the United States between 1843 and 1858 (Hayden, 1976, p. 149). Concepts for spiritual utopias floated in the wind of mid-nineteenth-century America like cottonwood seeds. According to Ralph Waldo Emerson in 1840: "Not a reading man but has a draft of a new community in his waistcoat pocket" (quoted in Hayden, 1976, p. 9).

Utopian settlements established for religious or philosophical purposes were by definition limited to adherents to those beliefs, while industrial model towns were intended for a broader cross section of the working class. Also, the spiritual communities valued total isolation from mainstream society, while the industrial communities required access to main transportation routes (obviously a sleeping-car factory had to be connected to mainline railroads). Religious communes, however, undoubtedly influenced the concept and form of industrial model towns and garden cities. Fundamental elements of both types of communities included: (1) centralized control over the use of land and structural development (usually through ownership of the site by sect or corporation; (2) proximity of work and residence; (3) population limits with overflow to be accommodated in new settlements in the vicinity; (4) a rural setting with much open space within and surrounding the community; (5) facilities and programs for social, cultural, and moral betterment.

Owen: From Practice to Theory

Robert Owen (1771–1858) and Ebenezer Howard (1850–1928) opened and closed the nineteenth century symmetrically. Owen moved from practical experience gained in a pre-existing community—New Lanark, Scotland—to articulate a general theory of cooperative socioeconomic organization. Howard first formulated his concept of the "garden city" and then successfully applied it in the establishment of new communities at Letchworth and Welwyn. Both proselytized public opinion but with quite different styles and results. Howard's "peaceful path to reform" promised a humanitarian experiment involving "no direct attack upon vested interests" (Howard, 1902/1965, p. 131). He was rewarded with a knighthood in 1927. Owen's advocacy of labor organization earned him the adulation of subsequent socialists but no knighthood.

The incubator of Owen's far-ranging theories was New Lanark, a village founded in 1783 as a site for cotton-spinning mills on a rapids of the Clyde River in south central Scotland (Figure 4-7). Owen, on marrying the daughter of one of the founders in 1800, assumed the position of manager of the mills, which then employed over 1,100 workers, two-thirds of them children. Owen devoted the next 14 years and much of his personal profits to the improvement of New Lanark, both physically and institutionally. Living conditions first attracted his attention:

> The great bulk of their houses . . . consisted of one single room, and before the door of that room was, as often as not, a dung-heap. One of Owen's first acts was to build another story to each of these houses, thus giving the family two rooms, and to remove the dung-heaps to a less unhealthy and unsightly position (Cole, 1953, p. 54).

Owen also had the streets cleaned and paved and reorganized the provision of food and coal to the inhabitants.

Owen's chief contribution to New Lanark was in the area of education. He espoused the view, remarkable at the time, that children should be in school rather than in the mills, at least until the age of 10. His Institute for the Formation of Character opened in 1816 (by which time he had bought the mills with several fellow investors including Jeremy Bentham). The institute provided child care and instruction beginning when a child could walk. It was designed to provide a balance of classroom teaching, exercise, and training in music and the arts and the "useful arts" (Benevolo, 1967, p. 40).

Figure 4.7. View of Robert Owen's New Lanark, ca. 1818. *Source:* Allen, 1986. Reproduced by permission of New Lanark Conservation Trust.

No more humble than Haussmann or Olmsted, Owen referred to New Lanark as "the most important experiment for the happiness of the human race that has yet been instituted at any time in any part of the world" (Allen, 1986, p. 3). But the scope of his increasingly utopian schemes was rapidly expanding beyond the possibilities for practical implementation in the "experimental cell" of New Lanark (Ashworth, 1954, p. 119). In response to a national inquiry into the problem of unemployment and public unrest after the close of the Napoleonic Wars, Owen articulated his vision for "villages of cooperation." The advent of mechanization, Owen argued in his 1817 "Report to the Committee for the Relief of the Manufacturing Poor," displaced workers from their former jobs, who now were crowding the cities and depending on meager poor-law relief.

Owen proposed the creation of a network of communal, agricultural villages to accommodate this surplus population. His loosely described land-use plan foreshadowed Howard's "green belt" concept and even Frank Lloyd Wright's "Broadacre City" of the 1930s. He envisioned village units of about 1,000 inhabitants, set in a predominantly agricultural landscape. The populace would chiefly be occupied in farming activities although some "manufactories" would also be provided. Like a feudal manor, each village would be largely self-sufficient in food (Cole, 1953, p. 110).

The design of the village nucleus, or "parallelogram" as he called it, was described more precisely. Owen specified the size, use, and arrangement of buildings for communal living, dining, education, and relaxation. Supportive facilities included churches, stables, slaughterhouses, breweries, corn mills, and so on. Missing from this list were courts of law and prisons, which Owen deemed superfluous in a community designed to optimize human happiness and freedom from need.

The villages-of-cooperation concept was restated in Owen's 1820 Report to the County of Lanark. According to Cole (1953, pp. 134–5):

> This Report was very much more than a proposal for dealing with the problem of unemployment in a single Scottish county, or even with the Poor Law as a whole; it is, in fact, the most reasonable and complete statement of Owen's socialism, much clearer and more defined than the wild words of 1817.

Even rational words, however, were insufficient to Owen at this time. According to his son, Robert Dale Owen (1874, p. 211), his father's "one ruling desire was for a vast theater on which to try his plans of social reform." He became convinced that such a theater could be found, not in Britain, but in the hinterland of America. The site of his second practical experiment in social organization was New Harmony, Indiana, on the Wabash River. In 1825, he purchased a communal village already established there by followers of the utopian George Rapp, together with 20,000 acres of alluvial farmland and forest. He and his son moved there to establish a village of cooperation.

The enterprise was a failure. Unlike New Lanark, New Harmony lacked an existing economic base to undergird its social principles. Furthermore, Owen attempted to carry these principles much further than at New Lanark, to make every adult settler an equal partner in the ownership, operation, and economic yield of the land and manufacturing assets of the village: "Liberty, equality, and fraternity, in downright earnest!" (Owen, 1874, p. 254). This goal, to which Owen pledged his personal wealth, was defeated by the unsuitability of the people who had been attracted to settle there " . . . a heterogeneous collection of radicals, enthusiastic devotees to principle, honest latitudinarians, and lazy theorists, with a sprinkling of unprincipled sharpers thrown in" (Owen, 1874). In 1827, Owen declared the project a failure and returned to England, having lost four-fifths of his own fortune.

This was Owen's last attempt to found or restructure a social community himself; subsequent experiments of this kind were conducted by his Owenite disciples (with similarly disastrous results). Owen devoted the remainder of his life to the cause of trade unionism and the advocacy of worker cooperatives.

Pullman: The Perils of Paternalism

George M. Pullman (1831–1897), one of America's most prominent union-busters, was an incongruous successor to Robert Owen in the field of ideal town building. Like the early Owen of New Lanark, Pullman was a capitalist entrepreneur who recognized that a worker is likely to be more productive if he or she is well housed, well fed, healthy, and entertained. But whereas Owen departed from

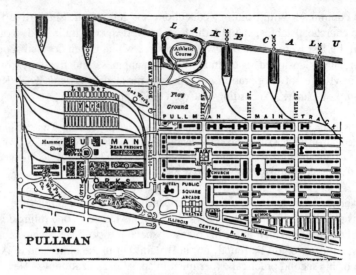

Figure 4.8. Map of Pullman from *Harper's Monthly*, 1885.
Source: Buder, 1967. Reproduced by permission of Oxford University Press.

the profit motive to explore the possibilities of pure socialism at New Harmony, Pullman remained a stalwart industrialist. Ironically, despite the apparent success of his town in terms of bricks and mortar, Pullman's experiment in socioeconomic engineering was ultimately defeated by his obstinate capitalism as surely as Owen's obstinate socialism proved *his* undoing at New Harmony. Pullman, Illinois is probably better known for the great labor strike that occurred there in 1894 than for its physical amenities. Perhaps the underlying similarity of both men was their inability to compromise.

George Pullman invented the railroad sleeping car that bore his name in 1864 and thereafter dominated the construction and operation of such cars on railroads throughout the United States. To establish a new works for his burgeoning Pullman Palace Car Company, Pullman in 1880 purchased 4,000 acres of prairie and boggy wetland adjoining Lake Calumet, 20 miles south of downtown Chicago. This site was certainly not selected as a rural utopia, but for the sound reasons of cheap land and accessibility to mainline railroads. There was, however, nowhere for a work force to live without a long train ride. Making a virtue of necessity, Pullman undertook to build a brand new town as a model of enlightened corporate planning and good employer-employee relations (Figure 4-8).

There were few precedents to draw on: Since New Lanark, only a handful of industrial model towns had actually been constructed in Europe and America (Ashworth, 1954, p. 126). The best known were Titus Salt's woolen-mill town Saltaire in England, completed in 1871, and the Massachusetts towns of Lowell and Holyoke, which produced textiles and paper, respectively. But with the help of a New York architect, Pullman basically designed his town himself.

The basic elements of the town were the car works, the residential district, a commercial and cultural arcade, a covered market, parks, a hotel, a theater, and an interdenominational church. All were built and owned by Pullman through a holding company. Details of the town are provided in the excellent book on Pullman by Buder (1967). In physical terms, it was progressive, humane, and "ideal." Housing of diverse size and rental cost was provided to accommodate both laborers and managers. Even the lowest-cost units consisted of brick row houses lining paved streets at densities of 8–10 per acre, far less crowded than usual at the time. (Even Saltaire had 32 houses per acre). Human wastes were collected and conveyed by pipeline to agricultural land south of the town that produced commodities for sale in the local market. Consistent with Owen, and later, Howard, alcoholic beverages were banned from sale in the town, but schools, a fine library, a theater, and other morally uplifting amenities were provided at Pullman's expense.

The town attracted immediate public attention. According to Buder (1967, pp. 92–93):

> From 1880 to 1893, the town was intensely surveilled. Hundreds of thousands saw this most modern and novel of communities and an overwhelming majority left impressed. . . . Here was an American utopia that people wanted to succeed.

The town was touted at the company's exhibit at the 1893 World's Fair as a place ". . . where all that is ugly, and discordant and demoralizing, is eliminated, and all that inspires to self-respect, to thrift and to cleanliness of thought is generously protected" (Buder, 1967, p. 148).

Unfortunately, similar protection was not extended to the right to form a labor union or to object to company policies on wages, hours, and costs of rent, water, and food in the company town. The paternalism that benefited the 8,000 inhabitants in prosperous years became their scourge when recession forced wage cuts and layoffs in 1894. The hand that fed them could also starve them. The resulting strike lasted three months and provoked the first use of federal troops in U.S. labor history. George Pullman died three years after the strike, vilified by those whom he thought he was helping. The town, severed by legal action from the company in 1904, gradually deteriorated into obscurity until gentrification set in during the 1960s. Pullman's worker row houses are now solid investments for Chicago yuppies.

Howard: From Theory to Practice

It was perhaps inevitable that an Ebenezer Howard should appear at the close of the nineteenth century. England and America were rife with utopian and progressive indignation concerning the state of large cities. Someone had to synthesize the many strands of thought, word, and deed into a practical program. This was Howard's contribution. In his own words: "I have taken a leaf out of the books of each type of reformer and bound them together by a thread of practicability" (Osborn, ed. 1945/1965, p. 131).

In effect, Howard blended Owen's New Lanark cooperative socialism with

Pullman's bricks-and-mortar paternalism (although neither are discussed in his book). He incorporated impressions of landscape design experienced during his visit to Olmsted's Riverside, Illinois. He was influenced by Henry George's theory of a single tax on land rent to recoup undeserved profits of land ownership for the public welfare. He was enthralled with Edward Bellamy's 1889 tract "Looking Backward," which envisioned American society recast on Owenite principles with centralized planning, cooperative enterprise, and equality of income (Fishman, 1977, p. 33). He obtained the idea of a town core and green belt from James Buckingham's utopian plan for Victoria (Osborn, ed. 1945/1965, p. 126). His admiration for small agrarian villages was derived from Peter Kropotkin (Fishman, 1977, p. 36).

There was in fact very little that was original in Howard's garden city proposal. But the assimilation of these and other intellectual "leaves" yielded his seminal tract: *To-morrow: A Peaceful Path to Real Reform* (published in 1898 and reissued in 1902 as *Garden Cities of To-morrow*). According to Lewis Mumford in his preface to the republication of Howard's book (Osborn, ed. 1945/1965, p. 29), Garden Cities ". . . has done more than any other single book to guide the modern town-planning movement and to alter its objectives."

Howard was the last of the great nineteenth-century amateurs in urban reform. By trade, he was a court stenographer and inventor, implying an ability to record faithfully the statements of others and to assemble components into a workable machine. Both skills, his admirers have noted, served him well in formulating his garden city theory, first in assimilating the ideas of the time and second, in visualizing a community as a system or "machine." F. J. Osborn (1945/1965, p. 21), Howard's chief disciple, described him as ". . . not a political theorist, not a dreamer, but an inventor." Another biographer, Dugald MacFadyen (1933/1970, p. 11), identifies a trait of "Americanism" in Howard's personality:

> The special inheritance of the Puritan as we see it philosophically in Emerson, practically in Ford, is a real conviction that mind triumphs over matter, that a clear idea tends to actualize itself by the inherent force that is in it. The mind of old England works from the concrete to the abstract—the New Englander works from the ideal to the real.

While Howard was scarcely a New Englander, the analogy is apt: in contrast to Owen, he effectively moved from theory to the practical.

The garden city idea was represented by Howard's famous magnet metaphor (Figure 4-9) wherein town and country are opposed. The former affords economic opportunity and culture at the expense of health, high prices, and crowding, while the latter provides a healthy environment but also boredom, poverty, and "lack of society." Howard's remedy was represented by a third magnet, "town-country," which incorporated the advantages and minimized the negative features of the other two. This was Howard's fundamental synthesis, which appealed strongly to the Hegelian spirit of the time.

Howard's magnet metaphor had additional significance. He rejected compul-

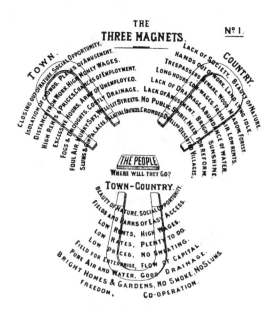

Figure 4.9 Howard's "Three Magnets" diagram. *Source:* Osborn, ed., 1945/1965. Reproduced by permission of The MIT Press.

sion by a central authority as a means of accomplishing resettlement of population to the new garden cities. He viewed migration as a voluntary, individual decision. Thus, his proposed garden city(ies) would offer inducements—social, environmental, and economic—that would draw working class people away from the miserable conditions of the large cities and would intercept rural migrants headed toward the same cities. In Howard's diagram, "'The People' are poised like iron filings between the magnets" (Fishman, 1977, p. 39).

The physical plan of the garden city reflected Howard's "central idea that the size of towns is a proper subject of conscious control" (Osborn, ed., 1945/1965, p. 10). His recommended population size was about 32,000 people (like Buckingham's Victoria), sufficient to attract industry and sustained cultural and social activities but small enough to retain a healthy and uncrowded environment. Such a community was to be situated on a tract of about 6,000 acres, of which the town itself would occupy a central core of 1,000 acres. The remainder would be devoted to a circumferential green belt of agriculture and other rural activities (Figure 4-10). Garden cities were to be located within convenient rail distance of a central metropolis (e.g., London) but should develop a local economic base to discourage long-distance commuting to work.

Internally, the garden city would provide a spectrum of housing opportunities to attract families of different socioeconomic levels (as in Pullman). Dwellings were to be situated along broad sylvan boulevards or local interior streets. The center of the town would be devoted to a community park surrounded by a "crystal palace" or

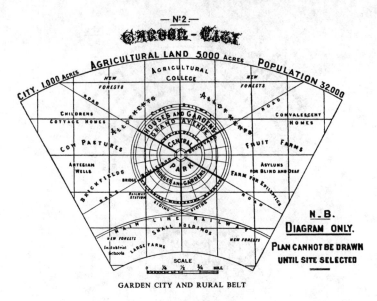

Figure 4.10. Howard's diagram of Garden City and its "Rural Belt." *Source:* Osborn, ed., 1945/1965. Reproduced by permission of The MIT Press.

enclosed shopping arcade together with public and cultural facilities including "town hall, principal concert and lecture hall, theater, library, museum, picture-gallery, and hospital" (Osborn, ed., 1945/1965, p. 53). Privately tended gardens and common open spaces would interlace with the village core and residential areas (the Olmsted factor). An outlying industrial district would accommodate smokeless, "non-sweat" industry.

Crucial to the importance of Howard's proposal was its means of accomplishment. The entire site of the garden city, including its agricultural green belt, was to be acquired, planned, and managed in perpetuity by a limited-dividend charitable corporation or trust. This entity would raise capital from philanthropically inclined private investors in exchange for a modest, fixed rate of return. The trust would derive all of its revenue from rents on land utilized for residence, business, industry, and agriculture. Any revenue accruing to the trust over and above the dividend to investors and operating costs would be returned to the community for beneficial purposes (the Henry George factor). As owner of the land, the trust would strictly control land use and development according to a community master plan.

Limitation of population size would be achieved by limiting the supply of dwelling units and by establishing additional garden cities at a suitable distance. Howard envisioned that the prototype community would lead to a cluster of such towns, separated by their green belts, which in time would comprise a formidable "magnet" drawing the working class populace out of the cities.

Howard and his supporters actually built two garden cities: Letchworth, start-

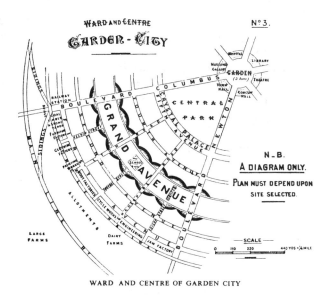

WARD AND CENTRE OF GARDEN CITY

Figure 4.11. Diagram of Garden City "Ward and Centre." *Source:* Osborn, ed. 1945/1965. Reproduced by permission of The MIT Press.

ing in 1903, and Welwyn in 1920. Of the two, Letchworth is the more faithful to Howard's principles and the more widely acclaimed. The author visited Letchworth in September, 1987, and found it thriving. Its 1981 population was 31,560. Its residential district is graced by "cottage picturesque" architecture set amid gardens, parks, and grassy commons. The town plan of 1903 by Barry Parker and Raymond Unwin smoothed the rigid symmetry of Howard's diagrams in favor of a more organic, informal design. It includes prototypical suburban cul-de-sacs, as well as a network of footpaths. The commercial core lacks the central park and crystal palace of Howard's imagination but instead provides retail spaces ranging from a Victorian arcade (reminiscent of Pullman), to a linear shopping street, narrowed into a pedestrian and bus mall, to a contemporary shopping center. Letchworth remains surrounded by an agricultural green belt, albeit comprising only about 2,000 acres now.

Land-use control over the total area of Letchworth is rigid. Permits for land-use conversions must be approved by both the Letchworth corporate authority and by the North Hertfordshire District Council under the national Town and Country Planning Act. In 1963, the original Letchworth corporation, First Garden City, Ltd., was replaced by a public corporation created by an Act of Parliament at the request of the town inhabitants. The new corporation performs the same role as its predecessor as trustee for the local public welfare. In 1982, it turned over 273,000 pounds from excess land rents to some 60 social and cultural organizations in Letchworth to support their activities. Howard's dream is alive and well in Letchworth.

Howard's hope that garden cities would proliferate once their potential was demonstrated was dashed by depression, war, and the rise of the public sector. However, the garden city thesis, as promoted by F. J. Osborn and the Town and Country Planning Association, substantially influenced the postwar New Towns and Greenbelt programs. The British New Towns and their counterparts in France, India, the Soviet Union, Hong Kong, and elsewhere scarcely resemble Letchworth in scale or form of organization. They are predominantly high-rise, publicly constructed communities with populations far exceeding 32,000. Yet the elements of unified ownership and control of land, mixture of functions, and their goal of metropolitan decongestion are all faint images of Howard's third magnet. In the United States, the influence of the movement has been primarily indirect in its impact on the ideology of the zoning movement. (See Chapter 7).

CONCLUSION

This chapter has surveyed three approaches to the problems of rapid urbanization during the nineteenth century: (1) regulation of building and sanitary conditions; (2) redevelopment and expansion of urban infrastructure; and (3) relocation of factory workers to planned model communities in nonurban settings. The first two approaches involved primarily governmental actions. The third approach involved private intiatives taken by well-meaning individuals of widely differing backgrounds and motivations. All three, however, required the development of new legal measures, authorities, and doctrines to modify the existing abusive practices of urban development. Each in turn served as precedent for more expansive intervention into the private land market by government during the twentieth century.

The thread of historical development is picked up in Chapter 7 with the antecedents to the rise of municipal land-use zoning. The next two chapters, respectively, consider the nature of real property ownership and municipal government as a foundation for understanding the respective roles of private and public interests in modern land-use control in the United States.

REFERENCES

ALLEN, N. 1986. *David Dale, Robert Owen and the Story of New Lanark*. Edinburgh: Moubray House Press.

ASHWORTH, W. 1954. *The Genesis of Modern British Town Planning*. London: Routledge & Kegan Paul.

BENEVOLO, L. 1967. *The Origins of Modern Town Planning* (J. Landry, Tr.). Cambridge: The M.I.T. Press.

BLAKE, N. M. 1956. *Water for the Cities*. Syracuse: Syracuse University Press.

BOOTH, M. L. 1860. *History of the City of New York*. New York: W. R. C. Clark and Meeker.

References

BRIDENBAUGH, C. 1964. *Cities in the Wilderness.* New York: Capricorn Books.

BUDER, S. 1967. *Pullman: An Experiment in Industrial Order and Community Planning 1880–1930.* New York: Oxford University Press.

CARO, R. A. 1974. *The Power Broker.* New York: Alfred Knopf.

CHADWICK, G. F. 1966. *The Park and the Town: Public Landscape in the 19th and 20th Centuries.* New York: Praeger.

CHAPMAN, B. 1953. "Baron Haussmann and the Planning of Paris." *The Town Planning Review.* 24: 177–192.

CHAPMAN, J. M. AND B. CHAPMAN. 1957. *The Life and Times of Baron Haussmann: Paris in the Second Empire.* London: Weidenfeld and Nicolson.

COLE, M. 1953. *Robert Owen of New Lanark.* London: The Batchworth Press.

DAVIS, T. 1966. *John Nash: The Prince Regent's Architect.* London: Country Life.

FABOS, J. GY., G. T. MILDE, AND V. M. WEINMAYR. 1968. *F. L. Olmsted, Sr.: Founder of Landscape Architecture in America.* Amherst, MA: University of Massachusetts Press.

FISHMAN, R. 1977. *Urban Utopias in the Twentieth Century.* New York: Basic Books.

FLINN, M. W., ED. 1965. *Edwin Chadwick's Report on the Sanitary Condition of the Labouring Population of Great Britain.* Edinburgh: Edinburgh University Press.

GIEDION, S. 1962. *Space, Time and Architecture.* Cambridge: Harvard University Press.

HAYDEN, D. 1976. *Seven American Utopias: The Architecture of Communitarian Socialism, 1790–1975.* Cambridge: The M.I.T. Press.

JACKSON, K. T. 1985. *Crabgrass Frontier: The Suburbanization of the United States.* New York: Oxford University Press.

KUTLER, S. I. 1971. *Privilege and Creative Destruction: The Charles River Bridge Case.* New York: Norton.

MAYER, H. M. AND R. C. WADE. 1969. *Chicago: Growth of a Metropolis.* Chicago: University of Chicago Press.

MACFADYEN, D. 1933/1970. *Sir Ebenezer Howard and the Town Planning Movement.* Cambridge: The M.I.T. Press.

NESSON, F. L. 1983. *Great Waters: A History of Boston's Water Supply.* Hanover, N.H.: University Press of New England.

OLMSTED, F. L., JR. AND T. KIMBALL. 1928/1973. *Forty Years of Landscape Architecture: Central Park.* Cambridge: The M.I.T. Press.

OSBORN, F. S., ED. 1945/1965. *Ebenezer Howard's Garden Cities of To-Morrow.* Cambridge: The M.I.T. Press.

OWEN, R. D. 1874. *Threading My Way.* London: Turner & Co.

PEETS, E. 1927. "Famous Town Planners: Haussmann." *The Town Planning Review* 12: 181–190.

PETERSON, J. A. 1983. "The Impact of Sanitary Reform Upon American Urban Planning 1840–1890," in *Introduction to Planning History in the United States* (D. A. Krueckeberg, ed.). New Brunswick: Rutgers Center for Urban Policy Research.

RASMUSSEN, S. E. 1967. *London: The Unique City.* Cambridge: The M.I.T. Press.

REISNER, M. 1986. *Cadillac Desert.* New York: Penguin Books.

ROSEN, G. 1958. *A History of Public Health.* New York: MD Publications, Inc.

SAALMAN, H. 1971. *Haussmann: Paris Transformed.* New York: George Braziller.

SMITH, E. R., 1907. "Baron Haussmann and the Topographical Transformation of Paris under Napoleon III." *The Architectural Record.*

SUMMERSON, J. 1962. *Georgian London:* Baltimore: Penguin Books.

SUTTON, S. B. (ED.). 1971. *Civilizing American Cities: A Selection of Frederick Law Olmsted's Writings on City Landscape.* Cambridge: The M.I.T. Press.

WARNER, S. B. 1978. *Streetcar Suburbs.* Cambridge: Harvard University Press.

WEBER, A. F. 1899/1963. *The Growth of Cities in the Nineteenth Century: A Study in Statistics.* Ithaca: Cornell University Press.

WEIDNER, C. H. 1974. *Water for a City.* New Brunswick: Rutgers University Press.

WHITEHALL, W. M. 1968. *Boston: A Topographic History* (2nd ed.). Cambridge: Harvard University Press.

5

PRIVATE OWNERSHIP OF LAND

INTRODUCTION

The property owner is the primary planner of land use in the United States. This statement may appear to conflict with the commonly held notion that planning and resource management are strictly public functions, to be exercised by government officials, boards, and professional staff. But the public role in the United States is essentially reactive to the decisions of the property owner. It is the owner who determines *how* to utilize his or her land in light of geographic, economic, and personal circumstances. It is also the owner who determines *when* a change in existing land use should occur. It is the owner's decision to change the use of land that triggers the public reactive role.

This chapter is devoted to a summary of the institution of land ownership and the role that the owner plays in the unfolding pattern of land usage. This will set the stage for examination of the various levels of public involvement in the land-use decision process in subsequent chapters.

WHAT IS REAL PROPERTY?

The legal concept of property distinguishes between two classes of property: real and personal. *Real property* consists of land, improvements such as buildings, vegetation, subsurface minerals, and, depending on state law, sometimes water. *Personal property* consists of chattels or movable objects owned by an individual. These may include motor vehicles, livestock, furniture, equipment or inventory, and

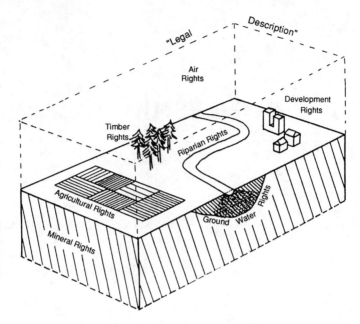

Figure 5.1 The Geography of Legal Rights in Land: Elements of Real Property Ownership.

personal possessions. Crops are often treated as personal property separate from the land on which they grow. The distinction between real and personal property is important in several ways: (1) they are taxed differently; (2) they are bought and sold with different procedures; and (3) a buyer of real property is not necessarily entitled to personal property located thereon.

The concept of land as real property is three-dimensional. The classic Latin motto states: *Cujus est solum, ejus est usque ad coelum ad infernos* (roughly translated: "whomsoever owns the soil, owns also to heaven and to hell). This means that the ownership of real property includes not only the horizontal land surface within the parcel of land in question, but also vertically downward and upward to the limits of practical or legal feasibility. Thus, real property ownership extends to subsurface rights below, and air rights above the land in question. (Figure 5-1).

It is a useful legal cliche to describe land ownership as a "bundle of red and green sticks." The green sticks represent rights that an owner enjoys in making profitable or pleasurable use of his or her land. Green sticks include the right to build on or above land, the right to till the soil and reap the benefits of its fertility, to cut its timber, to exploit its subsurface minerals, and in many states to use its water resources. The entirety of such rights is known legally as the *fee simple absolute* (as discussed later in this chapter).

The red sticks represent the duties of real property ownership that are necessary to enjoy the green sticks. These include principally the duty to pay property taxes, to refrain from creating a "nuisance," and to conform to applicable public laws, such as zoning, building codes, subdivision laws, and sanitary codes.

GREEN STICKS: RIGHTS OF REAL PROPERTY OWNERSHIP

The principal rights that bestow value on real property are of two types: material and spatial. These reflect the dual nature of land as a wedge of the physical material of the earth and alternatively as a volume that may be enclosed by buildings.

Material Rights

The material elements of land value are derived from the physical circumstances of the site: its soil fertility, its surface and ground water resources, its elevation, slope and solar exposure, its climate, its minerals and energy resources, its suitability as habitat for flora and fauna, and its scenic amenities and potential for outdoor recreation. The location of the site influences the economic feasibility of exploiting any of these characteristics, although remoteness may be overcome where the economic value of the resource is high, for example, oil and gas fields or high-quality ski slopes.

The value of land for agriculture is a complex function of soil type and quality, slope, drainage, location, and rainfall or irrigability (without considering exogeneous factors such as world price levels and economic conditions). Mineral development, where economically feasible, normally preempts agricultural use where surface disturbance due to strip mining is necessary. Where mineral or energy resources may be derived with minimal surface disturbance by means of tunnels or gas and oil wells, both farming and mineral extraction may coexist within the same tract of land.

Mineral rights are usually leased or sold to mining companies and thus are severed from ownership of the surface. This may lead to future problems of surface subsidence and collapse due to undermining. (See discussion of the *Mahon* and *Keystone Bituminous Coal* cases in Chapter 8). Acid mine drainage, which pollutes downstream water, is another harmful externality of surface coal extraction.

Water rights are defined by diverse legal doctrines that apply in different states and regions of the United States. East of the Mississippi and in certain western states, the doctrine of riparian rights prevails. *Riparian* refers to land adjoining a stream, lake, or tidal body of water. Owners of riparian land are vested under long-standing common law doctrine with special rights to use such water. Riparian owners may withdraw water for reasonable use, impound streams for water or hydroelectric power, build a wharf, discharge wastes, collect shellfish, and in other ways make economic use of the water as a resource.

The rights of each riparian owner, however, are constrained by an obligation not to substantially interfere with the enjoyment of similar rights of other riparians upstream and downstream. Riparian law was the customary principle for allocating water rights in areas of ample supply until the early twentieth century. With the growth of urban water supply systems and large-scale sources of industrial and urban pollution, riparian law was inadequate for water resource management. Various public programs and policies have largely superseded the riparian doctrine to manage the quantity and quality of water resources. (Trelease, 1979).

In certain arid western states such as Colorado and Wyoming, water rights have been severed completely from the ownership of land and no riparian rights are recognized. Historically, this resulted from the need to assemble large quantities of water for agriculture and mining in areas that provided inadequate average stream flow and severe annual variations. Thus, land owners in the West must separately purchase the right to water and have no vested interest in streams that happen to flow across or past their land. This doctrine is known as *prior appropriation*. The contrast between eastern and western water rights law has long been recognized as a classic case of law being shaped by geographical context (Whittlesey, 1935).

Other valuable material attributes of real property include a variety of natural resources such as timber, fruits, berries, herbs, nuts, maple sap, shellfish, finfish, wildlife, and so on. Hunting and fishing are however limited by state and local laws regarding species, size, season and size of catch.

Spatial Rights

Alternatively, the fee owner may choose to build on the land, that is, to enclose a portion of the air column above the site for usable interior space. Where geology and economics allow, such enclosure may proceed downward as well as upward from the land surface in the form of excavation for a basement, a parking garage, a shopping concourse, an art gallery. The above-surface air column may be enclosed by a structure built on pilings or stilts so as to retain most of the surface or subsurface for a different purpose such as railroad tracks or a highway. Structures so elevated are referred to as *air rights* development.

The owner may choose to subdivide a structure into dwelling or commercial units that may be sold individually in fee simple as *condominiums*. In this way, the overall air column may be divided legally as well as physically. Condominiums above the first floor are units of real property that "float" in mid-air supported by the framework of the building. Ownership of a condominium unit ("condo") also involves a proportional ownership of the common elements of the overall building—its basement, exterior walls, roof, stairs, elevators, and the like. Those facilities are maintained by a condominium association whose costs are paid from monthly assessment fees charged to each condo owner.

The exploitation of real property *qua* space is thus substantially different from the use of real property qua matter. Of course, the material character of a site is not totally irrelevant to the type and cost of spatial development to which it is suited. Buildings must be physically situated on or above the land. Many sites are physically hazardous for development, for example, floodplains, coastal hazard areas, seismic faults, wetlands, and unstable slopes. But the private market tends to ignore such limitations, or tries to overcome them through engineering solutions in profitable locations. Thus, Boston's Back Bay, a former marsh, is home to the Prudential and John Hancock office towers, which stand on footings extending to bedrock. Canyons of the San Gabriel Mountains bordering the Los Angeles basin are filled with expensive homes that occasionally are destroyed in chapparal fires or debris flows from upslope (McPhee, 1989).The willingness of Californians to risk earthquake hazards in their locational decisions is well known (Palm, 1986).

Real property *qua* space is thus an investor's abstraction: physical site limitations are the concern of the engineer and architect. Potential adverse effects on future site occupants, neighbors, community, region, or nation tend to be overlooked by the private market. There are, however, the red sticks.

RED STICKS: DUTIES OF REAL PROPERTY OWNERSHIP

Nuisance

The most long-standing duty of real property ownership in countries sharing the British common law tradition is to refrain from creating a nuisance that interferes with the rights of adjoining property owners and/or the general public. Nuisance may be either private or public. A *private nuisance* is a harmful externality inflicted primarily on the occupants of neighboring property. Traditional forms of private nuisance include sickening odors, cutting off light and air, overflow of water, accumulation of wastes that harbor vermin and insects, or allowing trash or dust to blow onto neighboring property. Courts in individual lawsuits have recognized certain new forms of nuisance such as recurrent noise, raucous parties, floodlighting, windmill-driven generators, race tracks, and animal feedlots. In 1990, a trial court in Massachusetts held that ringing of a church bell every hour all night long was a nuisance to adjoining homeowners (who had moved there when the clock was inoperative).

A *public nuisance* is one that affects a wider area or the public at large, as in the emission of air or water pollution, conducting an immoral or antisocial business, such as an AIDS-infested bath house, or dumping hazardous wastes. The borderline between a private or public nuisance is not always clear cut. However, public nuisances may be defined by state or local law and thus comprise criminal offenses (Prosser, 1971, Ch. 15). A public nuisance may also be a violation of an environmental statute, e.g., on hazardous waste disposal.

Private nuisances may be remedied (enjoined) by a court upon suit of the injured party or parties. Civil nuisance suits, however, involve wide latitude for judicial discretion. The court attempts to "balance the equities" to determine the relative costs to the defendant and society of abating the activity in comparison with the harm to the plaintiff if the activity continues. If the plaintiff moved to the area after the harmful activity was underway, the court may deny relief because the plaintiff "came to the nuisance." Relief may also be denied if the plaintiff waited unreasonably long to file suit against the objectional activity ("laches") or is also creating a nuisance ("unclean hands").

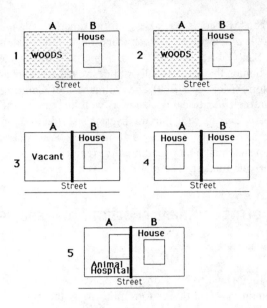

A inflicts externalities upon B ---
When may B validly claim judicial protection
under the doctrine of nuisance?

Figure 5.2 Externalities and Nuisance: A Hypothetical Scenario.

Not every harmful externality in a geographical sense comprises a legal nuisance. Consider the following hypothetical scenario as illustrated in Figure 5-2. Assuming no public zoning or other public regulations are in effect, when should a court reasonably grant relief against A?

1. Homeowner B enjoys the proximity of A's woods next door. B benefits from the shade, the privacy, the birdlife, and also the opportunity for his children and pets to cavort in A's woods, and to gather firewood there.
2. A, after discovering B's unauthorized use of his property, builds a fence that deprives B's family of their access to the woods. B is thus "harmed" but can he complain legally of being denied the right to trespass?
3. A cuts down his woods, thus additionally depriving B of the external benefits of shade and natural surroundings. Should the law prevent this externality?
4. A constructs a house similar to that of B, who now loses his privacy and whose property declines in value. Should A be prevented from doing so?
5. Instead of a house, A builds an animal hospital that will involve 24-hour floodlighting, incessant barking, vehicular traffic, and the stench of the furnace in which decreased pets are cremated. Does this comprise an unreasonable externality that a court might find to be an actionable nuisance?

Very likely! The earlier actions would probably be considered non-actionable harms—no legal relief would be available to B.

Property Taxes

Another unpleasant duty of a property owner is to pay all taxes and special assessments levied against the real estate. These include *ad valorem* taxes levied annually by the local municipality, county, and special districts within whose jurisdiction the property is situated. One-time charges in the form of special assessments are also the responsibility of the property owner. Special assessments are often charged for public improvements that benefit the real property in question, as in the case of sidewalk or street reconstruction, new sewers, water lines, or other facilities of localized, as distinct from citywide, benefits. (Taxes are discussed further in Chapter 6).

Public Regulations

Beyond nuisance and tax obligations, the landowner is confronted by a bewildering array of further public constraints in the form of land-use and building regulations imposed at the local, state, and federal levels. These include, for example, building codes, zoning, subdivision regulations, wetland restrictions, seismic design requirements, floodplain regulations, and public nuisance laws. The rest of the book is largely concerned with such public limitations of private property development.

TYPES OF OWNERS AND LEGAL INTERESTS

Land may be owned by many kinds and combinations of owners. These may include individuals, spouses, partnerships, corporations (business or nonprofit); or public entities, including federal, state, county, municipal, or special district agencies. Ownership may be divided among multiple parties each holding a fractional interest in the real property. For example, one-third of a parcel of land might be owned by a family partnership, and the other two-thirds by two spouses as *joint tenants* (when one dies, the other succeeds to his or her share). Real property is also subject to various claimants who hold interests less than fee, for example, tenants, lessees of specific rights, and even squatters.

To better appreciate the complexity of ownership, which may plague the best-intentioned plans for the future use of particular land (e.g. as public open space), the following summarizes the types of legal interests that may be encountered.

Fee Simple Absolute

The highest and most complete interest in real property is known as the *fee simple absolute,* also known as the fee simple or just the "fee." This concept involves both a geographical and a temporal dimension. As described in the foregoing, a fee simple interest in real property is three-dimensional, extending within the bounds of the parcel perimeter upward and downward, and indefinitely into the future. Land owned in fee may be sold, rented, given away, or devised to heirs by means of a will. According to Moynihan (1962, p. 29) the fee simple absolute "denotes the maximum of legal ownership, the greatest possible aggregate of rights, powers, privileges and immunities which a person may have in land."

The owner may divide the fee simple in several ways: (1) physical partition, (2) severance of specific rights, and (3) division in time.

Physical partition of the land into two or more pacels allows a portion to be sold or given away, and the rest retained. Farmers sometimes split off roadside building lots to raise cash, or may give them to offspring as homesites or for resale. This lowers the eventual inheritance taxes on the overall farm.

Second, the fee may be divided in terms of usage rights. For instance, the right to extract minerals such as oil or coal or to cut timber may be sold or leased to another party. In such cases, the owner is divested of the right to exploit these resources in exchange for a one-time payment or a series of royalty or "stumpage" payments over time.

Third, the fee simple may be split into time periods, creating both present and future interests held by different persons. Some examples follow.

Life Estates and Future Interests

Students of property law are exposed to the complexities of future interests and life estates. Such interests in real property result from the splitting of a fee simple into various time periods or "estates" by means of a will or trust. Typically, a will may specify that, upon the death of the owner, his or her property will pass to the spouse for that individual's lifetime and thereafter descend to surviving children. The surviving spouse thus receives a *life estate* consisting of the right to beneficial use of the real property for his or her remaining lifetime. Children living at that time are vested with a *future interest* in that they will inherit the property upon the death of the second parent. When that event occurs, the children's future interests automatically become present possessory interests in fee simple. Where more than one child is involved, each then owns an undivided fraction of the total fee simple, which is awkward to sell if cash is needed. Alternatively, the property may be physically partitioned and allocated by mutual agreement, or it may be sold and the proceeds divided among the heirs.

Spouses often own real property as *joint tenants with rights of survivorship*. When either spouse dies the survivor automatically is vested with a fee simple ownership of the property (subject to any existing leases, liens, or other limitations).

The esoteric complexities of life tenancies and future interests need not detain us further. It is enough to say that ownership of property may be divided among multiple parties, some with present and some with future interests, some possessory and some contingent. All of these must be ascertained through a title search when the property is sold.

Leaseholds and Tenancies

Owners of real property may obtain income from their land by renting or leasing it for various purposes. The words *rental* and *lease* are interchangeable, although they respectively suggest terms of less than, or more than, a year in duration. In either case, a written agreement between the landlord on the one hand and tenant or lessee on the other specifies the duration of the arrangement and the rights and duties of each party. In the case of an apartment rental, the landlord normally must provide clean, livable space while the tenant must promise to preserve the premises in good order and to pay rent on a regular schedule. Provision for utilities and heat should be specified in the rental or lease agreement. The owner remains liable for the payment of property taxes and conformity to local zoning and other public regulations. Either party may withdraw upon notice to the other party according to the terms of the agreement.

Farmland is frequently rented on a year-to-year basis where the owner does not wish to farm the land but wants to keep it available for sale when needed. Rental also helps the tenant farmer who needs extra land but cannot afford to buy it. Longer-term leases are used for mineral extraction, recreational facilities, and forestry where the lessee needs greater security to protect a major investment in equipment and labor.

A less formal, temporary arrangement is the license. Under a *license,* the owner may allow someone to occupy or use the land indefinitely by virtue of friendship, family relationship, or other reason. Normally a written document is not used and rental payments may or may not be involved. The distinguishing feature of a license as compared with a formal rental or leasehold is that the occupant may be required to leave at any time by the owner. This situation is sometimes called a *tenancy at will.*

Easements

Easements are limited interests in real property held by individuals other than the principal owner, or by the general public. Easements are most easily described in terms of the purposes they commonly serve. Typical kinds of easements include: (1) *utility easements*—corridors for utility lines, pipelines, or other purposes may be established permanently across private land through purchase of easements providing rights of access along a defined narrow strip or right of way. (2) *easements of access*—building lots lacking frontage on a public way may be accessible by means of an easement granting the right to cross intervening property owned by another person. (Zoning laws, however, may prohibit construction on a lot with inadequate frontage, regardless of easements.) (3) *recreation easements*—public access to rivers, coastal shorelines, ponds, or upland areas may require easements of access across intervening private land.

Each of these types of easements involves a formal conveyance of a right of access to some party, for example, a power company, a neighbor, or a public recreation agency. In the absence of such an easement, the desired access could occur only by illegal trespass or by license, which could be revoked at any time.

In order to obtain legal access on a continuing basis, the party desiring such benefit may offer to purchase an easement from the property owner. As with a leasehold, an easement is expressed in a formal written document that sets forth the terms of the agreement, its duration, the obligations of each party, and provisions for its termination. An easement cannot be revoked by a subsequent buyer of the property. The buyer thus receives title to the land subject to the easement unless the holder of the latter is willing to release it.

The property over which the easement extends is referred to as the *burdened estate*, while adjoining land that is served by the easement is the *benefited estate*. In the case of a public recreation easement, no specific land is benefited, and the easement is said to extend to the public *in gross*.

A different kind of easement of considerable interest to students of land-use control is the *scenic easement*, otherwise known as an *easement of development rights*, or a *conservation easement*. This device does not necessarily involve access to the property burdened by such an easement. It does create rights in parties other than the owner to whom the easement runs, to prevent the owner from altering the character of the premises through development, subdivision, cutting of timber, or other means, as provided in the legal instrument creating the easement. This is sometimes referred to as a *negative easement* in that the owner of the land subject to the easement is prevented from doing certain things. Such easements are commonly used to preserve the natural or open character of particular land while leaving legal title in the original owner. In this manner, a community or private conservation organization that seeks to preserve open land may pay owners to convey a scenic easement that binds them; existing uses such as a residence, farming, or woodlot management may be continued and modest changes or additions to existing buildings may be permitted under the terms of the easement. Basically, such an easement provides a flexible tool for preserving open land by agreement with the owner without transferring ownership of the entire property to a public agency. Usually the property tax valuation of the land is reduced in light of the scenic easement. If the owner is willing to donate a scenic easement to a charitable or public entity, a federal tax deduction may be claimed by the donor.

Covenants

Covenants are commitments or promises contained in a deed or contract affecting the ownership of real property. Developers usually include various restrictive covenants in deeds transferring the ownership of residential building lots. By accepting covenants, the lot buyer promises to use the lot in a specified manner, for example, to construct a single-family home, to install and maintain landscaping, to refrain from nuisance-type activities, and in general to use the land in an orderly and considerate manner. Since each lot buyer in a subdivision must agree to an identical set of covenants, each is considered to acquire rights against other lot buyers to enforce the covenants, even after the developer has disappeared from the scene. The lot buyers are then said to be "third-party beneficiaries" of the original agreement between each buyer and the developer. Covenants are said to "run with the land" and are enforceable against subsequent purchasers of the property to which they apply. (The buyer is presumed to have notice of any covenants, easements, or other restrictions that have been properly recorded with the registry of deeds.)

Covenants are a powerful tool of land-use control since they are created by voluntary (but irrevocable) agreement between a buyer and a seller. They may reinforce public zoning and other land-use regulations and indeed may deal with subjects beyond the scope of zoning, such as landscaping and architectural design.

Enforcement of restrictive covenants is justified by the fact that they are contractual in nature and accepted voluntarily. However, once a covenant is in effect, it may be eventually subject to termination by a court on the ground that conditions in the area have so changed that the covenant is rendered meaningless. In this way it differs from an easement, which normally is irrevocable for the duration of its stated term of effectiveness.

Liens

Liens are basically claims against real property asserted by parties to whom the owner owes money. If property taxes are unpaid, the local or county government will file a tax lien against the property and eventually seize it for sale. A painter, carpenter, or other contractor who is owed money for work done upon a premises may file a *mechanic's lien* against the property. Like a tax lien, the mechanic's lien may eventually be "perfected" by a court order requiring the property to be sold to satisfy the debt. Anyone purchasing real property must be careful that unsatisfied liens are cleared by the seller before transfer of the property so that the buyer will not acquire the outstanding debts.

Options

An option is a contract between a landowner and a prospective buyer under which the latter has a right to buy the property at an agreed price within a stated time period. An option allows a potential buyer to prevent the land from being sold to someone else pending decisions regarding such matters as financing and zoning. The cost of an option is a small fraction of the total price, perhaps 10 percent. But it is not refunded if the full price is not paid within the agreed time period. The option price is usually deducted from the sales price.

Options are useful to public and private agencies seeking to preserve open land. If a parcel of land is about to be subdivided or developed, the agency may seek to purchase an option to maintain the status quo for a few weeks or months to see if funding can be obtained to buy the tract. There is, however, no obligation on the part of the owner to sell an option and thus tie up the marketability of the property. If no option can be obtained, a public agency may use its eminent domain power to take the property upon payment of a price determined in court. A private agency or individual, however, has no such power.

ACQUISITION AND DISPOSITION

Real property is *alienable,* that is, it may be transferred from one party to another. The usual ways in which real property may be alienated, either voluntarily or involuntarily, are: (1) purchase or sale; (2) gift; (3) inheritance; (4) involuntary forfeiture; and (5) eminent domain (condemnation).

The marketability of real property through purchase or sale is a key element of the value of property as an economic asset. The law of property seeks to promote marketability by providing an orderly and secure process by which title may be transferred from one party to another. Normally the conveyance of ownership of real property is a two-stage process. The first stage involves the execution of a *contract* between the buyer and seller specifying the property in question, the price to be paid by the buyer, and any covenants or commitments to be performed by either party.

The second stage is the *closing,* when the legal deed to the property is transferred to the buyer in exchange for the purchase price or promissory notes (if the seller is financing the buyer). Normally, the buyer borrows a significant part of the purchase price from a bank or other lending institution. To secure its loan, the lending institution requires a *mortgage* or promise by the borrower to repay the loan in regular monthly installments. Failure to do so will result in foreclosure by the lender and sale of the property to recover the outstanding balance of the loan plus interest. Thus the mortgage agreement between the buyer and bank must occur prior to the closing between the buyer and seller. Immediately after the closing, the deed and mortgage instrument are recorded so that the respective interests of the lender and buyer are protected.

Real property may be given to anyone whom the owner selects. Frequently, portions of property belonging to a parent may be conveyed as a gift to children or other relatives. Real property may be donated to nonprofit organizations for charitable or conservation purposes. It may also be donated to a governmental agency. In the case of a gift to a public agency or a qualified private organization (for example, Massachusetts Audubon Society, The Nature Conservancy, and so on), the value of the gift may be deducted from the donor's taxable income. To be valid, a gift of land must be legally accepted by the recipient. Occasionally, land may be offered to a public entity that is unwilling to accept it due to future maintenance costs and loss of tax revenue. No gift occurs in that case.

Real property, like personal property, may be devised by the owner through a will which is properly signed and witnessed. Where no valid will can be found, property of a decedent is inherited by "next of kin" under state law. If no relatives can be found, property escheats to the state.

The value of inherited property is reduced by death taxes levied by federal, state, and sometimes local governments. A branch of property law involves the establishment of trusts and other devices for conserving the value of the decedent's estate by minimizing taxes.

One form of involuntary forfeiture is *foreclosure* by a lender when a borrower fails to make mortgage payments on time. This is, unfortunately, common during periods of economic recession, as in the oil-producing states of Texas, Louisiana, and Oklahoma during the mid-1980s. Similarly, the county or municipal government may foreclose on property that is subject to unpaid taxes.

A different kind of forefeiture occurs through the process of *adverse possession*. This refers to the occupancy of land by a squatter for a lengthy period of time. If the occupance is "open and notorious," and without consent of the owner, the occupant may claim legal ownership after a period of years specified by state law. The purpose of adverse possession is to protect *squatter's rights,* where the owner takes no interest in the land and someone else works hard to make productive use of the land. This is chiefly a doctrine affecting rural land. Adverse possession seldom arises in the high-priced metropolitan land market.

A *prescriptive easement* is a form of adverse possession asserted by the general public rather than a private occupant. It arises where the public uses a footpath, vehicular right of way, beach, or other private land without permission of the private owner. If continued for the statutory period, the owner may be precluded by statute or common law from preventing the continuation of the public trespass. But the public right is limited to the particular right-of-way or tract itself and does not extend to the rest of the owner's property.

Eminent domain or *condemnation* is the taking of title to private land by a public agency or publicly regulated utility company. This topic is discussed in connection with public acquisition in Chapter 9.

LAND-USE CONVERSION

The conversion of land from open to developed status (that is, from material to spatial form of use) is a crucial and essentially irreversible decision. Once occupied by buildings and pavement, a parcel of land is unlikely to revert to agriculture or other material uses (except possibly subsurface mining). Economically, land that is ripe for development or is in fact developed is orders of magnitude more valuable than agricultural or other rural land. While an acre of average cropland on the urban fringe might be worth up to $10,000 for farm purposes, the same acre might be worth $100,000 or more if suitable for three or four residential building lots. If the site could be developed with a multistory office building, the same acre might be worth over a million dollars.

In an idealized private land market, the decision as to how land should be utilized and particularly when it should be converted from material to spatial usage, is theoretically up to the private fee simple owner. As described in textbooks on urban and economic geography, the owner is motivated by considerations of "economic rent" or the desire to maximize the economic return from the use of the land, which in turn is determined in part by the location of the land with respect to some focal point.

Absent public intervention, the private market allocates land to that use which optimizes the economic rent that can be realized at the location in question. Location in today's complex land market is not necessarily with reference to a downtown central business district (CBD). Distance (or driving time) to an interstate highway interchange, satellite regional "mini-city," major airport, or recreational amenity such as a seashore or ski resort certainly influences the timing and substance of land investment decisions. According to a cliche of the real estate business, the three most important determinants of land value are "location, location, and location."

But location can no longer be regarded as a question only of distance to anything. The political or institutional context within which the landowner and other participants in the development process make their respective decisions may outweigh the importance of distance per se. Kaiser and Weiss (1970) identify three sets of factors that underlie the timing and form of development: (1) *Contextual factors*—the political and socioeconomic characteristics of the community and region; (2) *Property characteristics*—the physical and institutional aspects of the site itself; and (3) *Decision agent characteristics*—for example, willingness to take risks or respect for the natural qualities of the land.

Among these sets of factors, the public sector looms large, both at the community/regional scale and at the site level. Distance to employment is not irrelevant, but zoning, taxation, development incentives, natural hazard, and environmental regulations may matter even more.

Before turning to the public sector, there is one more topic to be considered in this review of real property, namely, the legal geography of property and political boundaries.

LEGAL BOUNDARY DESCRIPTIONS

A legal description is a means of defining the precise location and extent of parcels of real property and of units of political jurisdiction. A legal description permits the exact boundaries and area of legal and political units to be determined, frequently within a margin of error of only inches. This subject provides a refreshing dose of precision in constrast to the somewhat vague concepts that pervade discussions of land-use policy. While routine use of legal descriptions is the province of the lawyer more than of land use planners and geographers, anyone concerned with the use of land must understand why and how it is differentiated legally.

Uses of Legal Descriptions

Legal descriptions of property and political boundaries serve several purposes. First, legal descriptions permit *precise definition* of the boundaries and extent of private property. This is of the utmost importance to anyone investing in real property who must know exactly what land they own. At many dollars per square foot on the urban fringe and much more in the CBD, buyers must be assured they are receiving all the land that they bargained for. Similarly, mortgage lenders must be assured that the real property that is security for a loan is exactly the same land that the borrower is buying. Where doubt exists, a survey must be made to delineate the boundaries of the parcel in question.

A second use of legal descriptions is to *avoid disputes* between adjoining property owners regarding their common boundaries. Legal descriptions sometimes refer directly to adjoining parcels by name of the owner so as to ensure that no overlap exists. Thus, one boundary of a parcel may be described as "the south boundary of property owned by one C. W. Miller." Although Miller may subsequently transfer his land to someone else, it is clear that the boundary of the parcel at the time it was owned by Miller is to serve as the boundary of the adjoining property. However, a dispute may arise as to where Miller's boundary really was.

Certainty as to the location of boundaries also *promotes efficiency* in the use of land. Adjoining property owners may use their entire land area to the parcel boundaries without fear of trespassing on neighboring land.

A fourth use of legal descriptions is to identify the ownership of land for purposes of *property taxation*. Without precise boundary descriptions, some land might be taxed to more than one person and other land might escape taxation altogether.

Types of Legal Descriptions

Three types of legal descriptions are prevalent in the United States: (1) metes and bounds; (2) descriptions based on the Federal Land Survey; and (3) lots numbered on recorded subdivision plats. All three types are used today in various locations and circumstances.

Metes and bounds define the perimeter of a parcel of land through a series of straight line segments and references to physical features (Figure 5-3). Typically, the survey directions begin at one corner of the parcel, which is marked at the time of survey by a stone monument, iron pin, or other device. From this starting point, each straight line segment is defined by a precise compass direction and distance (usually expressed in surveyor's measurement units—chains, rods, and links). Corners or points where the boundary changes direction may be marked by further monuments or pins.

Physical features such as streams, roads, or railroads may be mentioned in a metes and bounds description. Questions may arise concerning changes in stream channel, shorelines, or road alignments.

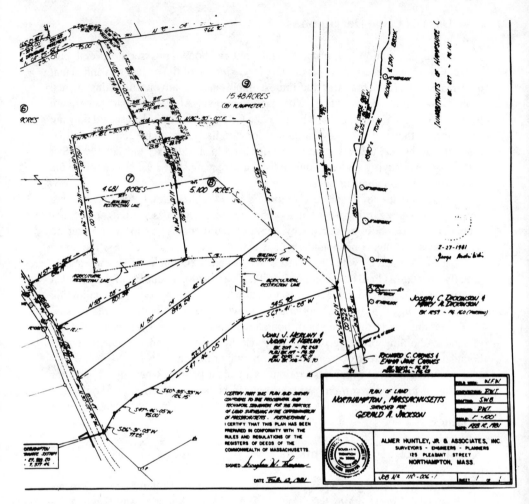

Figure 5.3 Excerpt from a "Metes and Bounds" Property Survey.

Two further difficulties with early metes and bounds surveys appear in this description of the western boundary of Massachusetts, prepared in 1787:

> Beginning at a monument erected in 1731 by commissioners from Connecticut and New York, distant from the Hudson River twenty miles, and running north 15 degrees, 12' 9" east 15 miles 41 chains and 79 links to a red or black oak tree marked by said commissioners, which said line was run as the magnetic needle pointed in 1787." (quoted in U.S. Geological Survey, 1976, p. 13).

First, the reference to a "red or black oak tree," besides being botanically uncertain, leaves the boundary unmarked with the eventual disappearance of said tree. Second, the use of magnetic compass directions, which was universal at this time, left uncertainty as to the degree of compass error or divergence between magnetic and true north, which varies from one place to another on the earth's surface. Normally a survey using a magnetic compass would note the *deviation* or departure from true north for the area in question.

Metes and bounds legal descriptions thus present many problems when old boundaries are resurveyed. Error was introduced from various sources including inaccuracy of instruments, careless procedures, change of physical conditions, and illegibility of early field notes. Yet property and political boundaries laid out in early times must be followed as faithfully as possible. The modern surveyor is armed with an array of new technology, including astronomical triangulation, satellite imagery, inertial guidance vehicles, and laser beams. Despite all this paraphernalia, where an ancient boundary must be redrawn, the modern surveyor must attempt to follow exactly in the footsteps of his or her colonial predecessor.

Metes and bounds descriptions are cumbersome and susceptible to mistakes. A small error in direction, "west" instead of "east," for instance, will send the boundary off into the blue, and no area is thereby enclosed. A major source of error is simply the copying by hand or typing of a metes and bounds description. Metes and bounds was too laborious to serve the needs of settlers of the trans-Appalachian public domain. Accordingly, at the dawn of its existence, the nation quickly devised a new and simpler system: the Federal Land Survey.

The *Federal Land Survey* system was probably the most lasting and influential achievement of the pre-Constitutional Confederation. The basic principles for the Federal Land Survey were established in the Land Ordinance of 1785 and the survey itself commenced in that year. From its "point of beginning" where the Ohio River crosses the western border of Pennsylvania, the survey extends westward to the Aleutian Islands of Alaska and southward to Key West, Florida. The survey ultimately has covered all of the continental United States with the exception of the original thirteen colonies, West Virginia, Kentucky, Tennessee, Texas, and parts of Ohio (U.S. Geological Survey, 1976).

The basic geographic unit of the federal survey is the *township*—a rectangle (actually trapezoid) of approximately six miles on each side containing 36 "sections" of one square mile each. Townships are identified with respect to "principal meridians" and "baselines," which are designated for each state in which the survey is used. Townships are located with reference to their positions east or west of a principal meridian and north or south of the designated baseline (Figure 5-4).

Principal meridians and baselines may continue across several states. The 40th parallel is the baseline for part of Illinois and also for Kansas and Nebraska, which it divides. Further west, the same parallel serves as the baseline for Colorado (Boulder's Baseline Street lies on this parallel). In some areas local principal meridians and baselines were established to serve the needs of early mining or other settlements, for example, at Grand Junction, Colorado and the Mormon region bordering Great Salt Lake in Utah.

The rectangular federal survey has been described as "a striking example of

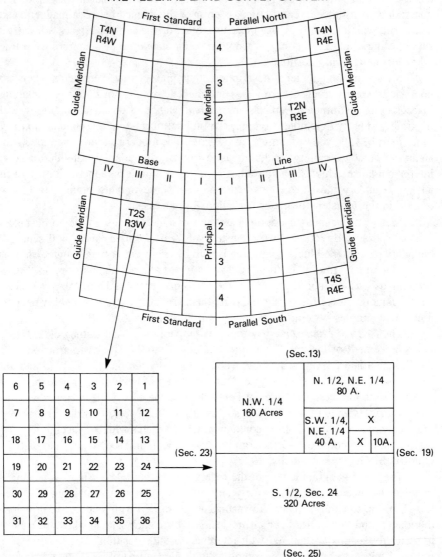

Figure 5.4 *A hypothetical excerpt from the Federal Survey System.* For states or portions of states surveyed under this system, one or more east–west base lines and north–south principal meridians have been designated. Units in the upper diagram represent individual survey townships of 36 square miles each, designated with reference to the applicable base line and principal meridian. At lower left, one township is broken out—in the second tier south of the base line and the third range west of the principal meridian. At lower right, one square mile "section" of this township, "Sec. 24," is further subdivided into parcels designated as fractions or fractions of fractions of Sec. 24. What are the survey designations and acreages of the two parcels marked X? See text

geometry triumphant over physical geography" (Pattison, 1957/1970, p. 1). Simply, the curvature of the earth dictates that townships will be trapezoidal not square, since their northern boundaries must be shorter than their southern borders. This problem is resolved by offsetting the north-south boundaries of ranges of townships at periodic intervals north and south of the baseline (See Figure 5-4). Other problems arise from natural obstacles such as large water bodies and mountain ranges.

Despite these complications, the Federal Land Survey proved indispensable to the disposition and management of the federal lands. (See Chapter 10). Altogether some 700 million acres have been transferred out of the public domain, most of it in areas covered by the federal survey. The survey permits precise legal descriptions of parcels of land to be expressed in terms of township, section, and fractions of sections. The 640-acre section is an ideal unit for land transfer and subdivision into fractional parcels.

By way of illustration, let us determine the legal description of Parcel X in Figure 5-4. First, the fractional portion of section 24 is expressed as:

The north half of the southeast quarter of the northeast quarter of section 24 *plus* the southwest quarter of the southeast quarter of the northeast quarter of section 24.

Next, the township containing section 24 must be located with reference to the overall survey grid: "Township 2 south, Range 3 east of the principal meridian." The full legal description abbreviated as per custom is:

N. ½ of the S.E. ¼ of the N.E. ¼ of sec. 24 plus the S.W. ¼ of the S.E. ¼ of the N.E. ¼ of sec. 24 of Twp. 2S, Range 3W of the ＿＿ Principal Meridian in ＿＿ County, State of ＿＿, an area of 30 acres.

Obviously this system is far superior to metes and bounds for describing large tracts of land in sparsely settled areas. The federal survey has been most useful in the settlement of level or moderately hilly lands between the Appalachians and the Rocky Mountains. Anyone flying over the nation's heartland on a clear day witnesses the physical legacy of the federal survey in the infinite pattern of rectangular farms, fields, road systems, and urban devopment, all conforming to the pattern of townships and sections laid out by the federal survey.

Naturally, local variations require departures from the strict use of federal survey directions. Legal descriptions for land bordering rivers, lakes, or tidal waters may refer to that feature as a boundary. Metes and bounds may be used where necessary to describe irregular tracts that do not fall precisely into survey fractions.

Urbanization of land covered by the federal survey is shaped by the preexisting grid system. The older street system of cities as farflung as Chicago, St. Louis, Los Angeles, and Seattle follows section and half-section boundaries. Development after the 1950s, however, was more oriented to the new interstate highway system, which cut through cities without regard to the survey grid. The monotonous results of grid-plan development yielded to a new look in suburban growth after

World War II. As influenced by the Garden City movement, this new look emphasized curving streets, irregular lot shapes, and abundant landscaping and greenery. The spread of the residential subdivision brought with it a new system for describing the precise location of parcels of land.

A third system of legal description that is today employed throughout the United States is the *subdivision plat or plan*. A subdivision plat is literally a map of a tract of land that is to be divided into smaller lots for sale to individual buyers (see Figure 9-1). Subdivisions usually anticipate residential development, although land is sometimes subdivided for commercial or industrial development. An example of the latter is the typical industrial park. Buildings may be constructed by the subdivider (who then serves as developer) prior to sale of the improved lots to the ultimate buyer. Altenatively, land may be subdivided for sale as a vacant lot with the buyer responsible for constructing a suitable building. (See Chapter Nine.)

In either event the proposed plat or plan of subdivision must be submitted for approval to the local planning authorities in accordance with state law. The subdivider must conform with local subdivision regulations concerning street layout, width, and paving; utility placement; drainage; and other details of project development. Once the plan is approved, it is filed with the county recorder of deeds, like any other instrument affecting land title.

The value of subdivision plat as a convenient means of legal description then becomes obvious. As individual lots within the approved subdivision are sold they may be simply described as lot no. ____ shown on a subdivision plat recorded on ____ at book no. ____, page no. ____ at the ____ County Registry of Deeds. The plat on file provides exact survey dimensions of each lot. There is no need for a lengthy metes and bounds or federal survey legal description in the deed for each lot (although the lot boundaries must be precisely described in the original plat).

Recording of Legal Interests

Precise legal descriptions are necessary, but not sufficient, to ensure that land may be efficiently bought, sold, and utilized. Equally important is a public record of land ownership and all claims against each parcel of land. Public knowledge of the precise legal status of title to real property is vital. With vast sums of money at stake, the buyer and mortgage lender must be able to ascertain whether they are receiving "clear title" from the seller. Defects that cloud the title include potential claims under wills of former owners, tax and creditor liens, easements, unexpired options, and other irregularities. Once such defects are identified, the seller is required to eliminate them or adjust the sale price to gain the buyer's consent to accept them.

Such claims are discovered through a *title search*. All deeds, probated wills, subdivision plans, liens, options, and other documents affecting title to real property must be filed with the registry of deeds for the county in which the land is located. Claims based on nonrecorded documents are usually not enforceable, but the buyer is presumed to have notice of all claims on record (*caveat emptor*).

All recorded documents are listed under the name of the owner then in possession of the premises. Past claims may be located by tracing the chain of title backward starting with the current owner to see what has been recorded under each owner's name concerning the property in question. The seller is expected to provide an abstract of title which is checked by an attorney for the buyer and mortgage lender. Many states also require a borrower to purchase a title insurance policy to protect the lender against undiscovered claims.

CONCLUSION

This chapter provided a brief survey of the institution of real property in the United States, including the rights and duties of the property owner and the purpose and forms of legal descriptions for specifying property boundaries. This topic is traditionally the subject of an entire first-year course in law schools; clearly, this chapter is not meant to provide a definitive knowledge of the field of property law. Instead, it is intended to introduce the reader to the kinds of complexity and some of the terminology pertaining to the ownership of property.

Why is such a background necessary in a book concerned primarily with public land-use control? The first reason is that the common law of real property is itself a system of land-use control. The long-established rights and duties of ownership, the procedures for acquiring and alienating land, and the recording of interests in real property using exact legal descriptions, all comprise a de facto system for putting land to use. Before public land-use controls were introduced in the late nineteenth century, private property rights were basically the "only game in town." The abuses fostered by that system in terms of overcrowding, unsanitary conditions, and a variety of harmful externalities on neighboring areas led to the assertion of public powers to constrain the excesses of the private land-use system. But public controls remain reactive, negative, and supplementary; the private property owner and investor still retains most of the initiative in land-use conversion.

A second reason is that an understanding of the legal nature of land is as important to those involved in land planning, acquisition, and management as for a doctor to understand the working of the human body. Land is more frequently understood in terms of its physical parameters: geology, geomorphology, hydrology, climate, and vegetation. But overlaying the physical terrain is a "legal terrain" of diverse interests in land. Public land-use control programs must cope with the latter as with the former. While detailed knowledge of ownership of land is not requisite for "broad-brush" zoning, the administration of land-use controls at the level of individual parcels and developments requires recognition and understanding of the nuts and bolts of property ownership. This chapter therefore serves as a foundation for the discussion of public roles in land-use control which follows.

REFERENCES

KAISER, E. J. AND S. F. WEISS. 1970. "Public Policy and the Residential Development Process." *Journal of the American Institute of Planners* 36(1): 30–37.

McPhee, J. 1989. *The Control of Nature*. New York: Farrar, Straus & Giroux.

Moynahan, C. J. 1962. *Introduction to the Law of Real Property*. St. Paul: West Publishing Co.

Palm, R. 1986. "Coming Home." *Annals of the American Association of Geographers*. 76(4): 469–479.

Pattison, W. 1957/1970. *Beginnings of the American Rectangular Land Survey System: 1784–1800*. Columbus: Ohio Historical Society.

Prosser, W. L. 1971. *Handbook of the Law of Torts* (4th ed.). St. Paul: West Publishing Co.

Trelease, F. J. 1979. *Cases and Materials on Water Law*. St. Paul: West Publishing Co.

U.S. Geological Survey. 1976. *Boundaries of the United States and the Several States*. Professional Paper No. 909. Washington: U.S. Government Printing Office.

Whittlesey, D. W. 1935. "The Impress of Effective Central Authority on the Landscape." *Annals of the Association of American Geographers* 25(1): 25–97.

6

AMERICAN MUNICIPAL GEOGRAPHY AND LAW

OVERVIEW

Modern municipal and county governments in the United States are descendants of the medieval municipal corporation discussed in Chapter 3. That institution was described as a "legal person" that could own land, make and enforce local laws, sue or be sued, and exist indefinitely until terminated by process of law. The same characteristics, with a few embellishments, apply as well to contemporary municipal governments, which will be the subject of this chapter.

It is perhaps surprising that everyday life in the United States is so pervasively influenced by a medieval institution. Like the character in Moliere's play *Le Bourgeois Gentilhomme* who discovered that he had been talking prose all his life without realizing it, we are governed by an intricate web of municipal corporations without recognizing the fact. Only the annual property tax bill may indicate the various units that overlie our real property and charge us for the services they provide, for example:

- the local municipality—an incorporated city, town, village or equivalent unit;
- county;
- school districts; and
- other special districts.

Municipalities are the local governments for most of urban America. Municipalities are authorized by state laws and home rule doctrines to plan and regulate the use of private land within their respective jurisdictions in diverse ways. Some of these powers, which will be discussed more fully later, are the following:

- preparation of master or comprehensive plans, including land-use plans;
- adoption and amendment of local land-use zoning laws;
- enforcement of building, zoning, and other public land-use regulations through building permits;
- review of proposed plans for land subdivisions;
- regulation of floodplains, shorelands, wetlands, and other sensitive areas; and
- acquisition and management of public open space.

These functions differ as to the relative balance of state and local authority. Zoning, for instance, allows considerable local discretion, while building codes and wetland regulations are often legislated by the state and administered locally. Different classes of municipalities such as villages, townships, or cities of different population size may vary in the powers delegated to them by the state legislature. And states differ widely as to the balance of state versus local authority. But in general, the local municipality may be viewed as the "workhorse" of public land use control.

Counties, the second major class of municipal corporation, are nearly ubiquitous in the United States. Each state is divided into counties or equivalent units (e.g., parishes in Louisiana). A total of 3,041 county-type units virtually blanket the nation's land area. A few cities are independent of any county, notably Baltimore, Maryland; St. Louis, Missouri; and some 40 cities in Virginia. Denver, Colorado is a combined city/county unit. There are also several examples of metropolitan governments involving complete or partial merger of city and county functions: for example, Miami-Dade County, Florida; Indianapolis-Marion County, Indiana; and Nashville-Davidson County, Tennessee.

Many counties display a split personality. For some purposes they are agents of the state, as in the management of the county courthouse, prison, roads, and welfare programs. In another sense, they are units of local government that exercise plenary power, including land-use planning and zoning, over unincorporated areas (i.e., areas not yet absorbed into an incorporated municipality).

Counties differ from state to state in their functions and importance to land-use control. In southern New England, where all land is incorporated, counties in Massachusetts are comparatively impotent, and those in Connecticut and Rhode Island have been abolished as governmental units. In Maryland and Virginia, however, counties serve as the basic units of local government, except in freestanding cities. Elsewhere in the nation, counties generally provide local government services to rural areas and unincorporated urban settlements.

Certain metropolitan counties such as Cook (containing Chicago, Illinois), Nassau and Suffolk (on Long Island, New York), and Los Angeles serve as major regional governments (Advisory Council on Intergovernmental Relations (ACIR)

1974; Duncombe, 1977). Los Angeles County, for instance, with 7.9 million people in 1984 has a public works budget of tens of millions of dollars. Municipalities within the county, including the City of Los Angeles and some 76 suburban jurisdictions, retain their local zoning prerogatives. However, the county or its surrogate special districts influence land use within municipalities indirectly through the location, timing, and capacity of regional facilities such as flood control, storm water drainage, sewage treatment, or water supply.

Special districts, the third major class of municipal corporations, comprise a very numerous and diverse potpourri of governmental units established for many purposes. Special districts numbered 28,588 in 1982, an increase of 2,626 or nearly 10 percent in just five years (U.S. Bureau of the Census, 1986, Table 470). Functions performed by individual special districts range from aviation to zoo administration. In some metropolitan areas, they provide such critical services as mass transportation, water supply, sewage treatment, solid waste disposal, parks and recreation, and air and water pollution control. Most districts, however, are created to provide only a single function. Districts serving different purposes may overlay each other, as well as general-purpose units of government, but districts of the same type are geographically mutually exclusive.

Special districts almost never engage in land-use planning and zoning; these remain municipal and county prerogatives. But special districts, especially those operating at a regional scale, clearly influence land usage through the spatial distribution of their services and facilities. The location and capacity of sewer and water lines, for instance, is a crucial factor in land-development patterns. Bollens (1957, p. 30) refers to a large number of special districts as "phantom governments" unknown to the people served by them.

With this overview in mind, we now consider the historical evolution of substate governments in the United States and their geographical and legal characteristics today.

ORIGINS AND DIFFUSION

The evolution of municipal institutions in America, while based in part upon precedents transplanted from England and the Continent, was strongly influenced by the physical and cultural circumstances of colonial settlement. Thus, differences among the colonies in terms of factors such as geomorphology, climate, religion, and economic organization yielded contrasting forms of local governments. Although the original reasons for these variations have long since disappeared, historical differences in form, function, and terminology survive to the present time.

The New England Town

The New England town was the earliest form of local government to appear in the American colonies. It has been justly celebrated as ". . . not only the most original but also the most democratic and, perhaps for that reason, . . . [it has displayed] remarkable power of survival" (Wager, 1950, p. 46). Despite its origin as a refuge for religious dissidents, colonial New England was noted for its intolerance of individualism, as portrayed in Arthur Miller's play *The Crucible*. Nevertheless, paradoxically, the New England town was to approach the ideal of participatory democracy as closely as any governmental institution devised in this nation. And its continuity may be attested by anyone who ventures to the north country in late March to attend an annual town meeting. The officers, procedures, customs, and even some of the issues debated date back to the seventeenth century.

Circumstances of early settlement which shaped the evolution of the New England town included: (1) a common religious bond—Puritanism; (2) a strong sense of independence and defiance of higher authority (other than God); (3) a harsh climate; (4) an intractable terrain requiring cooperative effort for productive utilization; and (5) fear of Indian attack. These factors jointly influenced the establishment of small, compact settlements often widely separated from each other. "The Puritan concept of community presupposed a clustering of people, a physical grouping that would enhance interaction and social cohesion" (Meinig, 1986, p. 104).

During the seventeenth century, individual settlements originated with groups of families holding land as common proprietors under grant from the colonial assembly of Massachusetts Bay (Boston), Plymouth, New Haven, Hartford, or Providence. Upon reaching a sufficient state of permanence, indicated by the ability of a settlement to support a minister, the community would be officially incorporated as a town by the colonial legislature or assembly (McManis, 1975).

These early settlements were closely knit through ties of family, religion, and common purpose in converting the surrounding land to pasture and cropland. Dwellings were clustered in villages, close to the meeting house, which served both civil and ecclesiastical purposes. Today, in many old New England towns the "first church" and town hall stand side by side (Separation of church and state precludes sharing the same structure today as they once did.) (Figure 6-1).

The original motive for emigrating from the Old World, the quest for religious and economic independence, carried over to the process of settlement and town formation. Each town regarded itself as a self-governing institution, beholden to neither the colonial assembly nor the crown. In some cases, these claims were supported by formal grants or charters; in others, they were simply uncontested. In either case, the towns generally went their own ways.

From earliest settlement, the New England town served not only as a unit of political, religious, and social organization, but as a resource-management entity as well. Morrison (1965, p. 70) describes the allocation of land and water resources within the typical town as follows:

> A committee was appointed to satisfy Indian claimants, to settle on a village site, and lay out lots. Home lots and the meetinghouse, which served both as church and town hall, were laid out around a village green, with a surrounding belt of planting lots for growing crops. Salt meadows on the coast, or river meads in the interior, valuable for the wild grass which could be cut and stored for winter forage, were laid out in long

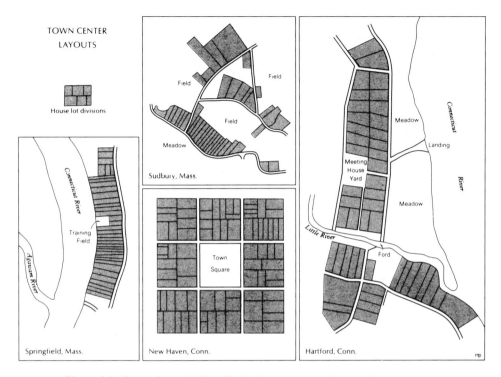

Figure 6.1 Types of colonial New England town centers. *Source:* McManis, 1975. Reproduced by permission of Oxford University Press.

strips and usually cultivated in common. The rest of the township for many years remained the property of the community, where anyone could cut firewood and timber, or pasture cattle.

Over time, the original pattern of land allocation was drastically transformed as family holdings were split by conveyance or inheritance or augmented by purchase (either from the town's own reserve lands or from other households). Furthermore, certain individuals, such as William Pyncheon, the founder of Springfield, Massachusetts, amassed sizable personal holdings through direct grants from the colonial assembly. These lands were conveyed to purchasers through the private market, as is the case today, yielding great wealth for the initial proprietor. Most New England towns settled in the eighteenth century were founded by individual proprietors or speculators who laid out roads, lots, and sites for churches and schools and then sold farm units to the public.

As in Thomas More's *Utopia*, early New England settlements maintained strict control of their population growth. As an early form of "growth management," towns limited the right of strangers or even relatives of residents to settle within their corporate limits (Porter, 1922, p. 31). Sometimes unwanted residents were "warned out" and forced to leave the community. Another means of limiting population while exiling dissenters was to encourage exodus by small groups to found new settlements. According to Morrison (1965, p. 70): "When members of a village community felt crowded for space, they petitioned the colonial assembly for a new township, the ideal size being six square miles."

Many New England towns were spawned through splitting large territories into smaller ones (McManis, 1975, p. 62). Over 100 of the 351 towns and cities in Massachusetts originated by vote of a larger, pre-existing town to subdivide itself and set off a newly settled portion as a separate town (Grant, 1890). In this manner, the closely linked institutions of church and town could replicate themselves in each new settlement.

Whatever its origins, the New England town assumed governmental status upon formal incorporation by the colonial (or later, state) legislative body. From that time on, the towns theoretically served as governments of, by, and for their inhabitants.

Tidewater Counties

The Tidewater region of coastal Maryland and Virginia fostered an entirely different system of local government administration based on the county. Many of the factors that contributed to the hegemony of the town in New England were missing in the Tidewater. The climate was mild. Soils were level, easily worked, and relatively free of glacial rubble, which figures so prominently in the landscape and character of New England. The lengthy shoreline of Chesapeake Bay, deeply incised by navigable estuaries, afforded convenient maritime routes connecting the region's hinterland with England and other overseas destinations. The native populations of the area were relatively friendly, obviating the need for compact, protective communities.

Most important, the primary impetus toward settlement of the Tidewater was economic rather than religious. The close affinity between church and town which prevailed in New England never emerged here. Indeed, towns per se scarcely appeared, as cultivation of the tobacco staple was conducted on extensive plantations, widely separated, and largely self-sufficient as economic and social units. Thomas Jefferson (1782/1944, p. 227), a geographer in addition to his many other talents, explained the absence of towns in Virginia in 1782:

> We have no townships. Our country being much intersected with navigable waters, and trade brought generally to our doors, instead of our being obliged to go in quest of it, has probably been one of the causes why we have no towns of any consequence. Williamsburg, which till the year 1780 was the seat of our government, never contained above 1,800 inhabitants; and Norfolk, the most populous town we ever had, contained but 6,000. . . .

This lack of municipalities was deplored in some quarters. According to a letter of 1688 by the Reverend John Clayton (quoted in Morrison, 1965, p. 87):

> This conveniency [of maritime access] was an impediment to the advance of the country. . . . The country is thinly inhabited; the living solitary and unsociable; trading confused and dispersed; besides other inconveniences.

John Reps (1972) recounts the efforts by the colonial assemblies of Maryland and Virginia to overcome this lack of towns. Acts passed in both colonies around 1680 designated a number of sites for the establishment of new port and market settlements. These acts authorized trustees appointed in each county to acquire through eminent domain the sites, approximately 100 acres in size, from their plantation owners and to lay out towns thereon. The width of streets and the size of lots were specified in the authorizing act, an extraordinary early example of town planning legislation. However, despite the fact that these ports were to be granted certain monopoly privileges with regard to external trade, most in fact were never established, and only one—Norfolk, Virginia—thrived. In Jefferson's pithy words:

> . . . the *laws* have said there shall be towns; but nature has said there shall not, and they remain unworthy of enumeration. (1782/1944, p. 227)

Maryland and Virginia have never developed a strong municipal level of government. Except for freestanding, self-governing cities such as Baltimore and Richmond, local government functions—including land-use control—have been largely exercised by counties.

The county as an institution of government was directly transplanted from England to the Tidewater colonies. The English county was (and remains) a fundamental unit for the administration of justice, the collection of taxes, and the maintenance of roads and other public facilities. County business was dominated by landed gentry, and it was from this social class that the plantation owners of the Tidewater colonies originated.

The county thus characterized the Plantation South, as the town represented Yankee New England. Evolving from distinct geographical conditions, each institution reflected and in turn helped to shape the culture of the region that it served. In the words of the perceptive de Tocqueville (1832/1969, p. 81):

> We have seen that in Massachusetts the township is the mainspring of public administration. It is the center of men's interests and of their affections. But this ceases to be so as one travels down to those states in which good education is not universally spread and where, as a result, there are fewer potential administrators and less assurance that the township will be wisely governed. Hence, the farther one goes from New England, the more the county tends to take the place of the township in communal life. The county becomes the great administrative center and the intermediary between the government and the plain citizen.

The Prairie Synthesis

The contrasting traditions of the New England town and the Tidewater county were to be synthesized in the settlement of the Northwest Territories, the region bounded by New York State, Canada, the Ohio River, and the Mississippi. The clash and eventual merger of the two traditions was most dramatic in Illinois. The earliest pioneer settlement of Illinois emanated from the middle Atlantic and southern colonies. In 1818, Illinois' first census reported that 75 percent of its population was from Kentucky, Tennessee, Virginia, and Maryland (Billington, 1970, p. 91). County government was thus early established in the Midwest states:

> Then the tide turned. The Erie Canal was responsible. The opening in 1825 of that all-water route between the Hudson River and Lake Erie shifted the center of migration northward as New Englanders and men from the Middle Atlantic States found the gateway to the West open before them. Now the Great Lakes, not the Ohio River, formed the pathway toward the setting sun. . . . In 1834, 80,000 people followed this route westward (Billington, 1970, p. 92).

The northeastern settlers brought their towns with them, which quickly spread across the northern plains in modified form as the middle western township or *civil township*. In Illinois, this institution was soon adopted by most but not all of the pre-existing counties. Thus, the New England and southern traditions of local government merged in Illinois and its neighboring states.

The middle western township, however, only remotely resembled its ancestor, the New England town. By the time of the opening of the Erie Canal in 1825, the concept of the "pure" New England town had already been modified—some might say debased—in New York, New Jersey, and Pennsylvania (Porter, 1922, Ch. 3). The township (or town) lost its roots in the town meeting as it moved westward and became merely a county administrative unit. Civil townships were eventually established in 16 states outside New England: New York, New Jersey, Pennsylvania, Ohio, Indiana, Michigan, Wisconsin, and Iowa, and in portions of Illinois, Missouri, Kansas, Nebraska, Minnesota, North and South Dakota, and Washington State (Wager, 1950, p. 35). These are essentially relics of pioneer sentiment which sought to transplant the form, if not the substance, of town government to the prairie sod.

Civil or governmental townships must not be confused with "Congressional" or "survey" townships established under the Federal Land Survey described in the previous chapter. The federal survey township is usually a rectangle six miles square. It is numbered rather than named and has no governmental status as such.

But to confuse the issue further, *civil* townships or other governmental units in states covered by the federal survey were sometimes established according to *survey* township boundaries. For example, several Illinois civil townships coincide with survey townships and are thus 36 square miles in size.

The county has fared better than has the town outside its region of origin in this country. From the Southeast, counties diffused eventually to all 50 states, in one form or another ranging from modern metropolitan governments to the rural counties of American folklore. Of the latter, Wager wrote nostalgically in 1950:

> Across the face of America are county seats, and at the heart of each a courthouse in a courthouse square. These courthouse squares have a peculiar American flavor; they

suggest spaciousness and leisure, there are benches where old men gather to reminisce and almost invariably a monument to honor the county's military heroes. The courthouse, if old, has a certain museum quality; it is cold and austere with a slightly musty smell. The newer ones are brighter and have more of the air of office buildings. But nearly everywhere there is an atmosphere of secretiveness in county offices, a certain reluctance to reveal with frankness the details of public business. Politics is played with more finesse than in more robust days, but the professional spirit has not yet been fully adopted. On the other hand, there is a friendliness and warmth—a certain "down to earth" quality about county government that is as American as pumpkin pie (pp. 4–5).

Growth of Central Cities

Towns and counties, as we have seen, originated in the agrarian societies of New England and the Tidewater. Urban communities were tiny before the Revolution. Urban settlements accounted for only about 5 percent of the national population in 1790. This was to change rapidly with national independence and the industrial revolution. Contemporary American cities, however, developed spatially and legally quite differently. Three examples, Boston, Chicago, and Los Angeles, are illustrative.

Boston. The town of Boston grew from 18,320 inhabitants in 1790 to 33,787 in 1810 and to 43,298 in 1820 (Whitehill, 1959, pp. 73–74). This rapid expansion was accompanied by urban problems, for example, fires, water shortages, overcrowding of land, disposal of human and animal wastes, street paving, lighting, and crime. In 1822, Boston adopted a city charter, which replaced the town meeting with a mayor and city council. This conversion of Boston from a town to a city symbolized the advent of the modern urban municipality, which was to become the principal instrument of local government in the newly emerging nation of cities and suburbs.

Boston's city charter of 1822 clothed it with the formal attire of urban government but left many problems unresolved. Among these was the question of what geographical area the city should encompass. Like its counterparts elsewhere:

> . . . Boston desired more land for residential and economic development. There was also an interest in increasing the city's size, reflecting not only America's passion for bigness but also Boston's need to present a prestigious statistical position to national and international commercial interests. Furthermore, Boston felt that if it were a large political unit, it would have greater political influence in the state legislature and in Congress (Wakstein, 1972, p. 287).

The results of this quest for bigness in terms of annexation and boundary changes are depicted in Figure 6–2. Two early defectors from the original territory of the town of Boston were Brookline in 1705 (which today remains legally independent of Boston although surrounded by it on three sides) and Chelsea in 1739

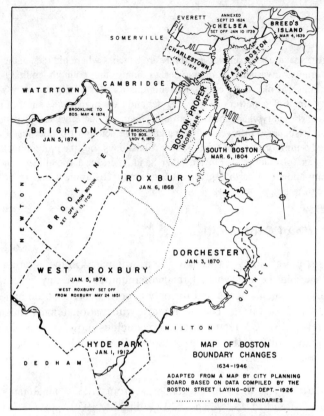

Figure 6.2 Territorial evolution of Boston, 1634–1946. *Source:* Wakstein, 1972. Reproduced by permission of the American Planning Association.

(which Boston voters refused to take back in 1856). These "setoffs" were consistent with the process of subdivision of towns in Massachusetts noted earlier.

The reverse process of territorial *expansion* commenced with the annexation of South Boston in 1804 by act of the Massachusetts General Court (legislature). Although this addition was not submitted to the voters of either community, subsequent annexations (or, properly, consolidations where both communities were previously incorporated) involved mutual consent by local referendum. During the mid-nineteenth century, the consolidation issue dominated local politics.

The primary motive for outlying communities to unite with Boston was to ensure a dependable water supply (Teaford, 1979, pp. 54–55). Between 1867 and 1874, the towns of Roxbury, Dorchester, Charlestown, and West Roxbury voted to consolidate with Boston. These additions increased the population of Boston by 116 percent and its territory by 441 percent (Wakstein, 1972, p. 280). But this explosive rate of growth was short-lived. With the annexation of Hyde Park in 1912, the annexation/consolidation movement ended in Boston. The establishment of "metropolitan districts" for the provision of water, sewage collection, and parks in the late nineteenth century deprived the central city of much of its attraction. Also, the construction of horse-drawn and later electric streetcar lines facilitated residential growth on the urban fringe within easy commuting distance of central city jobs (Warner, 1978).

Table 6–1 Boston, Chicago, and Los Angeles: Population and area of central city and MSA.

	Boston	Chicago	Los Angeles
Population			
1980, central city (1,000s)	562	3,005	2,996
1980, MSA* (1,000s)	2,759	7,057	7,445
% central city	20%	42%	40%
Land Area			
central city (sq. mi.)	46	222.6	463.7
MSA (sq. mi.)	1,769	3,720	4,069
% central city	3%	6%	11%

*MSA = Metropolitan Statistical Area as designated by the U.S. Bureau of the Census

The modern suburban ideal of "having it both ways" was thus well established by the turn of the century. Suburban residents could enjoy the jobs and amenities of the city while living in a quasi-rural, small-town environment. With increasing ethnic, racial, and economic polarity between central city and suburbs, the latter elected to preserve their municipal independence. Today the city of Boston accounts for about 20 percent of its metropolitan-area population, and its land area is just three percent of the total metropolitan region (Table 6–1).

Chicago. During the mid- to late nineteenth century, most American cities shared Boston's quest for territorial and demographic expansion. West of the Appalachians, this process typically involved annexation of adjoining unincorporated land on the verge of development whose owner(s) sought the benefits of municipal services. Consolidations of existing municipalities with central cities were also common.

Chicago was originally incorporated in 1833 with a territory of only three-eighths of a square mile on the southwest shore of Lake Michigan, as designated by the Commissioners of the Illinois and Michigan Canal (Mayer and Wade, 1969, p. 14). The city was to be the northern terminus of the canal that opened in 1848 to connect Lake Michigan with the Illinois-Mississippi river system.

Annexations to Chicago occurred in two phases (Teaford, 1979, pp. 44–45). Before 1870, the city expanded to about 26 square miles through a series of annexations enacted through acts of the state legislature without local referenda. In 1872 and 1887, the legislature established procedures for further annexations, including local referenda as in Boston. Between 1889 and 1893 several entire townships and villages were added to Chicago despite court battles waged by opponents (Figure 6–3). A severe problem for Chicago and its suburbs alike was the contamination of the lake as a drinking water source by raw sewage discharged into it. As in Boston, a solution was ultimately found through a regional special district, the Chicago Sanitary District (later renamed the Metropolitan Sanitary District of Greater Chicago). In 1899, the sanitary district completed its Ship and Sanitary Canal to convey Chicago's wastes away from Lake Michigan and toward the Illinois-Mississippi

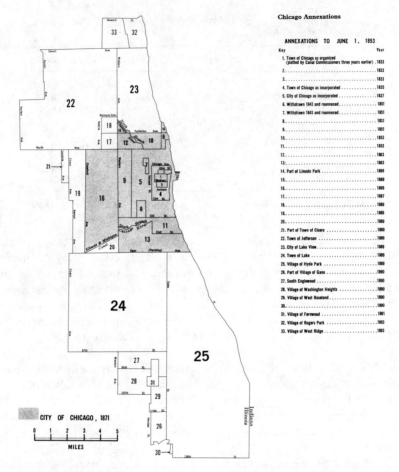

Figure 6.3 Territorial evolution of Chicago, 1833–1893. *Source:* Mayer and Wade, 1969. Reproduced by permission of The University of Chicago Press.

river system. By the turn of the century, Chicago had reached about 90 percent of its present area of 222 square miles and annexation had virtually stopped as municipalities chose to retain their autonomy. In 1980, Chicago accounted for about 42 percent of its metropolitan-area population, twice the proportion of Boston. Its land area of 222 square miles is five times that of the city of Boston (Table 1).

Los Angeles. Los Angeles provides a western example of municipal accretion, largely occurring during this century. The city was originally incorporated in 1850 with a legal territory of 25 square miles centered on the Spanish plaza. Prior to 1900, urban growth was largely limited to this municipal territory. Thereafter, annexation added vast new areas to the city—extending southward to San Pedro Harbor, which serves as the port of Los Angeles, westward to the Santa Monica Mountains, and northward into the San Fernando Valley (Figure 6–4).

Origins and Diffusion

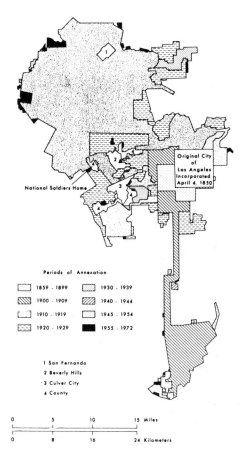

Figure 6.4 Territorial evolution of Los Angeles, 1850–1972.
Source: Reynolds, 1976.

The expansion of the corporate jurisdiction of Los Angeles into the San Fernando Valley in 1905 was inextricably associated with the city's quest for water. According to Marc Reisner (1986, p. 77) in his book *Cadillac Desert,* the annexation of the valley was prompted by a desire to increase the city's assessed tax value so that more bonds could be issued to pay for a 250-mile aqueduct from the Owens Valley in the Sierra Nevada to Los Angeles:

> If the assessed valuation of Los Angeles could be rapidly increased, its debt ceiling would be that much higher. And what better way was there to accomplish this than to *add to the city?* Instead of bringing more people to Los Angeles—which was happening anyway—*the city would go to them.* It would just loosen its borders as [William] Mullholland [Chairman of the Los Angeles Department of Water and Power] loosened his silk cravat and wrap itself around the San Fernando Valley. Then it would have a new tax base, a natural underground storage reservoir, and a legitimate use of its surplus water in one fell swoop [emphasis is original].

Table 6-2 Number of substate units by type, 1977–1982

	1977	1982
Counties	3,042	3,041
Municipalities	18,862	19,076
Townships	16,822	16,734
Special districts	25,962	28,588
School districts	15,174	14,851
Total	64,688	82,290

SOURCES: *Statistical Abstract of the U.S., 1980*, p. 309 and *Statistical Abstract of the U.S., 1986*, Table 470.

As with Boston and Chicago, the process of growth eventually ran its course as all remaining territory became incorporated in communities determined to preserve their independence. Some of these, notably Beverly Hills and San Fernando, are entirely surrounded by Los Angeles. Others ring its perimeter except where it borders public lands or the Pacific Ocean. Annexations since 1955 have been limited to modest-sized tracts scattered along the city's periphery, as well as some inholdings that could not conveniently be incorporated as separate municipalities. As of 1976, there were 76 incorporated cities within Los Angeles County in addition to the City of Los Angeles. Conversely, the city itself contained 20 postal community names, such as Hollywood and San Pedro, which once were independent communities (Nelson and Clark, 1976, pp. 21 and 28).

Los Angeles now occupies a land area of 463.7 square miles, more than twice the size of Chicago and 10 times the size of Boston. Despite this impressive size, Los Angeles occupies only 11 percent of the land area of its Metropolitan Statistical Area. And despite its enormous expansion earlier in this century, Los Angeles resembles other central cities in exercising jurisdiction over the land and resources

Table 6-3 U.S. cities by population size, 1980

	No.	Pop. (mill.)	% U.S. Pop.
1,000,000+	6	17.5	7.7
500,000–1,000,000	16	10.8	4.8
250,000–500,000	34	12.1	5.4
100,000–250,000	117	17.0	8.7
50,000–100,000	290	19.7	10.3
25,000–50,000	675	23.4	12.2
10,000–25,000	1,765	27.6	6.8
Total	2,897	128.1	55.9

SOURCE: *Statistical Abstract of the U.S., 1986*, Table No. 17.

of only a modest and irregular fragment of the total urbanized region of which it is the heart.

Suburban Proliferation: Hardy American Weed

The evolution of American metropolitan areas during the "Gilded Age" of 1870–1895 may be viewed as a race between new municipal incorporations on the one hand and the enlargement of existing municipalities through annexation and consolidation on the other. The former process involved the piecemeal fragmentation of urbanizing regions among hundreds of small units of municipal authority. The latter process attempted to integrate small fragments into larger urban communities. As we have seen, the annexation movement thrived briefly in various cities only to subside as central cities met resistance from incorporated suburban jurisdictions that wished to remain autonomous from both the central city and from each other. Municipal fragmentation has since prevailed as the determinant of the political geography of most American metropolitan areas, although small-scale annexation to existing municipalities continues in many areas.

A major factor behind the proliferation of incorporated municipalities in this country was the adoption in most states of very permissive incorporation laws beginning in the early 1800s. As surveyed by Teaford (1979, pp. 6–7), state laws typically involve very low minimum population requirements, petition by a majority of the inhabitants or property owners of an area proposed for incorporation, and a pro forma review by a court, the county commissioners, or other authority. Basically, little hindrance was imposed on the desire of a community to incorporate itself. This permissiveness was consistent with American traditions of self-government and home rule. It further facilitated the establishment of new "cities" in remote mining areas, ports along the inland river system, and elsewhere that the path of pioneer settlement led (Reps, 1965). But as practiced in burgeoning metropolitan areas, permissive incorporation has fostered a spirit of rivalry, and sometimes hostility, which has obstructed efforts to promote regional solutions to metropolitan problems.

Despite the stereotypical images of suburbia offered by fiction writers such as John Cheever and John Updike, suburbs differ greatly in size, economic function, and growth characteristics. The origins of "streetcar suburbs" in the late 1800s are examined in Warner's (1978) classic study. By World War II, suburban growth had proliferated and diversified. Harris (1943/1959) found that the degree of suburbanization differed from one functional type of central city to another, and that suburbs themselves appeared to be either "bedroom" or "manufacturing." By the mid-1980s Jackson (1985, p. 5) observed that "American suburbs come in every type, shape, and size: rich and poor, industrial and residential, new and old."

Today, we may distinguish five types of suburban jurisdictions which have little in common except for their legal autonomy within a larger metropolitan area. First, many older suburban municipalities such as Evanston, Illinois; Boulder, Colorado; Cambridge, Massachusetts; and White Plains, New York have sizable territories and populations and display a broad range of urban functions and land uses. These would be viewed as *full-fledged cities* in their own right but for their proximity to larger urban complexes. Some, such as Stamford, Connecticut, have experienced spectacular growth and prominence as a haven to major corporations seeking lower-cost commercial space and more amenable surroundings than older central cities. Others, such as Newark, New Jersey, have experienced the opposite phenomenon: disinvestment, declining economic base, pervasive social and ethnic problems, and a crumbling infrastructure.

Second, white collar *bedroom suburbs,* many very affluent such as Grosse Pointe, Michigan; Lake Forest, Illinois; or Lincoln, Massachusetts represent the archetypal suburb caricatured by fiction writers O'Hara, Cheever, and Updike. Such communities often are characterized by large lots, huge homes, manicured lawns, and an absence of manufacturing, multi-family housing (except luxury condominiums), and any other unwanted land uses. Municipal autonomy has permitted these municipalities to insulate themselves from the rest of the metropolitan region from which their wealth arises.

A third type of suburban municipality is the *industrial* blue collar community such as Chelsea, Massachusetts; Elizabeth, New Jersey, or Hamtramck, Michigan. These may have originated as industrial enclaves or as speculative working-class land subdivisions. They are likely to be ethnically and racially diverse in comparison with their affluent neighbors. They are often afflicted by poverty, unemployment brought about by plant closings, and inadequate schools and other public facilities. While sharing many of the social, economic, and environmental ills of the central cities, they may lack the resources and technical sophistication to qualify for federal and state assistance. Nor do they have the population size or notoriety of a Newark to attract state and national attention to their plight.

A fourth type is the "splinter" municipality of miniscule size and sometimes negligible population, which are scattered here and there in many metropolitan areas. According to the Advisory Commission on Intergovernmental Relations (1969, p. 75):

> The overwhelming majority of metropolitan local governments are relatively small in population and geographic size: . . . about half of the nearly 5,000 municipalities in SMSAs have less than a single square mile of land area, and only one in 5 is as large as 4 square miles. Two-thirds of them have fewer than 5,000 residents; one-third fewer than 1,000.

Historically, splinter municipalities often were incorporated to promote or avoid moral concerns. Local laws differ widely on liquor, gambling, and other vices, as does local enforcement of state laws on matters such as prostitution. Teaford (1979, p. 20) found that the emerging fragmented metropolis of the early twentieth-century "resembled a moral checkerboard with alternating squares of state law." In Minnesota, for instance, a law allowing municipalities to own and support themselves from municipal liquor stores gave rise to a number of incorporations involving only a city block and a few hundred persons. Avoidance of taxing and zoning policies of neighboring jurisdictions is another frequent motive, shared alike by affluent, exclusive suburbs and industrial enclaves. Burns Harbor, Indiana, for example, was incorporated in 1968 by Bethlehem Steel Corporation. Total assessed valuation of Burns Harbor in 1970 was listed at $39 million despite the presence of a substantial portion of a new quarter-billion-dollar steel plant (Platt, 1972, p. 160).

A fifth type of suburb may be described as the "mushroom" variety. Mushroom communities begin either as small, rural towns on the metropolitan fringe, or simply as agricultural or wooded land. At some point, development interest focuses on the site, perhaps due to construction of a new highway or proximity of the advancing frontier of urbanization. Given the necessary financing and legal approvals, the site erupts from its bucolic state into a late-twentieth century "urban village," (Leinberger and Lockwood, 1986) complete with expensive apartments, high- and low-rise condominiums, a super-regional shopping mall, restaurants, office parks, and one or more major hotels.

A good example of a mushroom suburb is Schaumburg, Illinois, which in the late 1950s was a sleepy farm village 20 miles northwest of Chicago's Loop. Schaumburg erupted in the late 1960s as part of the general development boom triggered by the construction of several interstate highways and the rapid expansion of O'Hare Airport. By 1970, Schaumburg had grown to about 18,500 people and already boasted the world's largest indoor shopping mall. Its spatial area grew rapidly with annexation of surrounding farmlands. Although the pace of development in the 1970s lagged behind earlier predictions due to the national recession, the "village" had reached a population of 52,000 by 1980.

Although the details of local political geography differ from one state to another, Schaumburg clones may be identified in many fast-growing U.S. metropolitan areas, for example: Tysons Corner, Virginia; King of Prussia, Pennsylvania; Framingham, Massachusetts; Walnut Creek, California. Mushroom communities pose distinct and intractable issues for public policy. These include lack of affordable housing for persons employed in the community, lack of educational, cultural, and aesthetic amenities due to absence of civic traditions and involvement; a lack of rootedness and sense of community; impossible traffic and parking problems; and a general lack of integration of planning and design among the various developments that comprise the mushroom suburb (Muller, 1975; Leinberger and Lockwood, 1986).

The typical American metropolitan region thus comprises one or more central cities surrounded by a mosaic of incorporated suburbs of the various types listed above, interspersed (except in New England) with pockets of unincorporated land under county jurisdiction. Figure 6-5 depicts the chaotic municipal geography of Cook County, Illinois. And this does not even include special districts. Figure 6-6 depicts the corporate area of a regional special district, the Metropolitan Sanitary

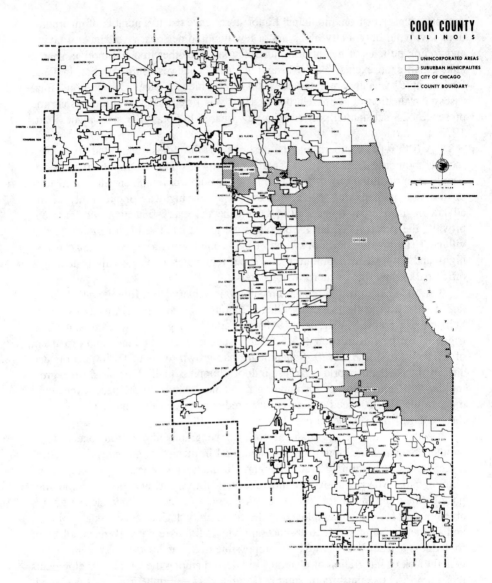

Figure 6.5 Municipal geography of Cook County, Illinois.

District of Greater Chicago, which provides most of Cook County with sewage treatment. But Cook County is also overlain by an average of 4.8 special districts at any given location, hundreds in all, which contribute to the overall political fragmentation of the county and region (Stetzer, 1975).

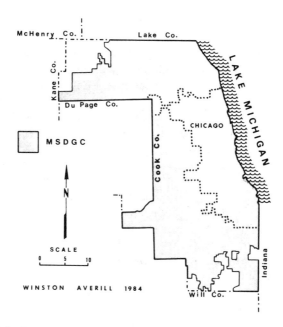

Figure 6.6 Corporate jurisdiction of the Metropolitan Sanitary District of Greater Chicago.

MUNICIPAL JURISDICTION AND POWERS

Big or small, modern municipal corporations, like their medieval ancestors, are both legal and geographical creatures. They have a spatial territory and certain legal powers which are exercised within that territory. This duality of the municipal institution is implied by the term *jurisdiction,* defined by *Webster's New Collegiate Dictionary* as:

- the limits of territory within which authority may be exercised;
- the power, right, or authority to interpret and apply the law . . .

Geographical Jurisdiction

Territory is a fundamental attribute of any municipal corporation. Without a definite, formally delineated spatial area, a municipal government is a nullity. Furthermore, according to American custom, municipal territories must be contiguous, that is, must consist of a single, connected spatial unit rather than scattered, disconnected tracts of land. And like parcels of real estate, the boundaries of a municipal or other political territory must be precisely delineated through the use of a formal legal description, which is filed in the appropriate state office.

Municipal territories are highly variable in size and shape. Municipalities range in size from Oklahoma City with 604 square miles in 1984 (three-fifths the size of Rhode Island and more than twice the area of New York City) to many places that occupy less than one square mile each. Larger municipal territories are most common in the south central and southwestern states. Tulsa, Oklahoma City, Houston, Dallas, and Phoenix have expanded rapidly under state laws allowing unilateral annexation of adjoining unincorporated land by central cities. Thus, Tulsa in 1966 more than tripled its size from 50 to 160 square miles. Houston expanded from 433.9 square miles in 1970 to 560 in 1980 (mostly involving undeveloped rangeland). Such dramatic increases in the territories of central cities are unusual. Most older cities such as Boston, Chicago, and Los Angeles have long been static within their "corsets" of incorporated suburbs.

Some municipal boundaries coincide with tangible physical features such as rivers, lakes, or tidal coastlines. Water reference features are simple to recognize but may shift over time due to river channel meandering and coastal shoreline change. Where a river or stream is a political boundary, the bordering floodplains on either side are in different political jurisdictions, which complicates public response to flood hazards (Platt et al., 1980). Some state and county boundaries follow upland topographic features such as drainage divides or crestlines of mountain ranges (e.g., the eastern boundary of Idaho). The regions that such boundaries traverse are usually remote and unsettled and therefore not incorporated as municipalities.

Other municipal boundaries follow well-defined cultural features such as railroad rights of way, highways, or canals. But many municipal boundaries are arbitrary, consisting of zigzag segments, panhandles, outliers, and indentations. The reasons for each jog in a municipal boundary lie in past political and personal decisions to add or withhold territory from a particular municipal unit. The aggregate result in most U.S. metropolitan areas is a political geography of byzantine complexity (Figure 6–6).

Despite their irregularity, corporate boundaries are important determinants of urban morphology. The spatial juxtaposition of county, municipality, and special district influences the permissible uses of land and thereby land values. It determines the availability of many public services. It affects property tax liability. It also defines who has authority to plan and manage future growth.

Municipal geography, however, is difficult to map. Boundaries of communities that are annexing territory change frequently. Even communities with stable boundaries are sometimes too small to depict accurately on a regional-scale map. Special districts are usually not indicated on metropolitan maps due to the problem of displaying several kinds of districts overlaying each other (Stetzer, 1975).

But complexity should not imply vagueness. At the microscale, the boundaries of a municipal unit may be ascertained precisely. Where land investors are in doubt, the exact jurisdictional status of individual parcels may be determined by referring to legal records in the appropriate municipal or state offices. For broader planning needs, such as the management of watersheds, aquifers, agricultural land, shorelines, or floodplains, the location of all corporate jurisdictions overlying the resource may be identified precisely, albeit laboriously. Computerized geographic information systems may be used to plot the "legal stratigraphy" of public authority over land.

Both incorporation and enlargement (annexation) of municipal governments are governed by state laws and court decisions. These are largely procedural in nature and provide relatively little substantive direction regarding the desired size, shape, and symmetry of municipal units. Approval of annexations differs from one state to another but typically involves one or more of the following actions: (1) direct state legislative approval; (2) approval by the annexing municipality (either by popular referendum or action of the elected governing body); (3) approval by a stated percentage of voters and/or property owners in the area proposed to be annexed; (4) impartial determination by a judicial or administrative agency (Sands and Libonati, 1981. sec. 8.30).

While state policies differ, some general principles governing municipal boundary changes may be briefly stated. First, municipalities are always *mutually exclusive* in their geographic area. No area of land is incorporated in more than one general-purpose unit of local government. Nor may one municipality "pirate" territory from another incorporated area. Land must be unincorporated to be eligible for annexation. (Consolidation of incorporated units is rare today).

Second, annexed land must normally be *contiguous,* or physically connected to the annexing municipal territory. In suits challenging proposed annexations (for example, by neighboring municipalities competing for lucrative development sites), courts generally interpret contiguity to mean a substantial common border reflecting a mutuality of interest in the sharing of municipal responsibilities. Contiguity is thus related to the feasibility of extending municipal services to adjoining land. There are, however, examples of annexations of areas tangent only at a corner or connected by a "ribbon." An extreme example is illustrated in Figure 6-7 wherein the City of Broken Arrow, Oklahoma resembles the result of a shrapnel explosion. Isolated fragments of incorporated area are connected to the rest of Broken Arrow by 10-foot-wide strips. These strips have been annexed unilaterally by Broken Arrow to provide contiguity to a new subdivision or shopping center that seeks to be included in Broken Arrow. Oklahoma, however, allows municipalities to establish legal "fencelines" encompassing unincorporated areas within a specified distance beyond the present corporate limits. Since other municipalities cannot "invade" this area, Broken Arrow in due course will incorporate most or all of the territory within its legal "fenceline."

A third constraint is that annexations must usually be mutually agreeable to both the annexing municipality and the annexee. Often a majority of property owners or the owners of a majority of the land in question are sufficient to bind dissenting owners whose inclusion is necessary to achieve contiguity. Agreement by the municipality may take the form of a city council resolution, a popular referendum, or some other means. Many states permit "pre-annexation" agreements be-

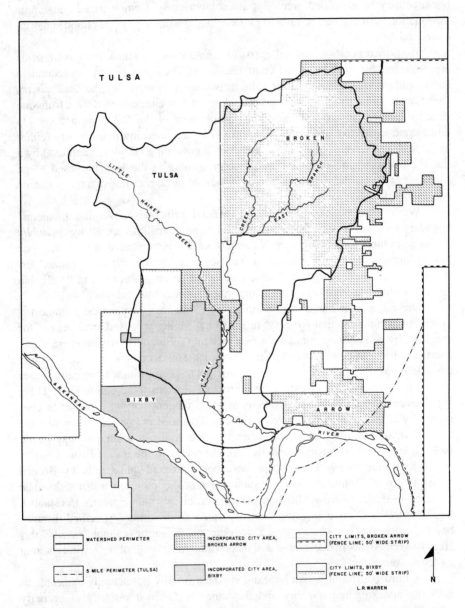

Figure 6.7 Municipal annexation: political geography of the Haikey Creek Watershed in Oklahoma.

tween annexee property owner(s) and annexing municipality regarding the provisions of public services, zoning treatment, and other matters. Such agreements are common where the annexation is made in anticipation of impending development of the land.

Municipalities are usually limited in the exercise of their allotted powers to their corporate limits. But certain functions require actions beyond those boundaries, such as urban water supply. Since most communities cannot satisfy their water needs strictly from internal sources, the quest for water often leads to external sources. Many states allow municipalities to acquire land and water rights outside their boundaries to develop and protect public water supplies. (This authority may not, however, include the extraterritorial use of eminent domain.) Extraterritorial land also may be acquired in some states for nature refuges, parks, sewage treatment plants, and airports.

Municipal services are often sold extraterritorially. For example, Boulder, Colorado sells water to surrounding unincorporated areas under contract. Chicago likewise supplies water from its Lake Michigan treatment works to suburban communities, at higher rates than its own residents pay.

Counties and special districts rarely operate outside their legal territories. Special district boundaries, however, may be modified to annex additional areas not yet served by an equivalent district.

Legal Jurisdiction

Scope of municipal powers. The fundamental rights and powers of municipalities have changed little since medieval times:

- to sue and be sued as a "body politic;"
- to enter into contracts;
- to acquire, hold, and dispose of land;
- to establish ordinances or bylaws; and
- to possess a seal.

To these must be added a modern necessity:

- to obtain revenue, to borrow, and to spend.

Recitation of these broad powers, however, does not suggest the permissible scope of their applicability. For what purposes may municipalities enter into contracts, buy land, make ordinances, or impose taxes? How much autonomy and immunity from state review may local governments claim? (Clark, 1985).

> In 1868, the legal authority Thomas Cooley observed that 'the American legal system is one of complete decentralization, the primary and vital ideal of which is, that local

affairs shall be managed by local authorities.' During the following four decades Americans seemed dedicated to realizing the ideal as Cooley had stated it. Local self-determination was the rallying cry of Americans, and this meant that each fragment of the metropolis would enjoy the right to govern itself and to decide its destiny. Local government was a sacred element of the American civil religion, and the nation's lawmakers were devout in their adherence to the faith (Teaford, 1979, pp. 5–6).

But, as with other religions, the hopes of the faithful were sometimes badly served. At the turn of the century, American cities were dominated by bosses and political machines that turned the ancient municipal prerogatives to personal gain. Abuses were (and still are) rampant with respect to municipal contracts, land deals, and patronage in public employment. In the early decades of the twentieth century, the corruption of municipal government prompted a reform movement that reasserted tighter control by states over their wayward municipal "children."

Contrary to the myth of local autonomy, municipalities in fact are not sovereign entities. They are incorporated under state law and thereafter are controlled by the state in many respects such as selection of officials, contracts, bonded indebtedness, taxation, and land-use control. The history of municipal government in the United States has been a continuing struggle between states and local governments regarding the relative degree of control imposed by the former on the latter (Clark, 1985).

One outgrowth of this struggle was the *home rule movement* under which certain states, beginning with Missouri in 1875, adopted state constitutional amendments expanding the scope of municipal autonomy. Under this doctrine, a community could perform functions if they were not forbidden by the state legislature or otherwise in conflict with constitution or statute. The home rule movement reached only about 15 states, however, and has been characterized as ". . . an uncertain privilege, for it depends entirely upon the whim of the legislature and may at any time be repealed or modified" (Zink, 1939, p. 121).

Regardless of home rule status, questions arise frequently as to the validity of a municipal action in light of state delegation of authority. Clearly, latitude is needed to go beyond the literal provisions of state law, or municipal governments would be stifled in responding to local needs and circumstances. The fundamental question therefore is: How may a municipality ascertain the limits of its available powers?

The classic formulation in response to this question is *Dillon's Rule,* which was first expressed in 1911 and remains solid law today:

> . . . a municipal corporation possesses and can exercise the following powers and no others. First, those granted in express words; second, those necessary or fairly implied in or incident to the powers expressly granted; third, those essential to the accomplishment of the declared objects and purposes of the corporation—not simply convenient but indispensable (Sands and Libonati, 1982, sec. 13.04).

Dillon's Rule is a long-settled principle that is cited by courts in resolving challenges to municipal innovation. It is an elastic test, however, and affords much judicial discretion in applying it to actual controversies.

Revenue. An important issue in the long-standing conflict between states and local governments is that of *revenue*. Municipalities chafe under state restrictions such as tax limitation measures adopted in the 1980s in California, Massachusetts, and elsewhere. Revenue may be obtained from a variety of sources, which local governments constantly seek to augment.

The fundamental revenue source for most local governments is the *ad valorem tax*. This comprises an annual tax imposed on all taxable real property based on a specified percentage of the latest assessed value of each parcel (including land and buildings). The percentage of tax levied (tax rate) is uniform throughout the taxing jurisdiction, although tax rates often differ substantially from one governmental unit to another. Even different classes of private real property must be taxed at the same rate. Tax rates are usually expressed in terms of "mills per dollar of assessed value" or "dollars per $1,000 of assessed value." The ad valorem tax levy is not applied to tax-exempt property, which includes all publicly owned land and buildings plus property owned by charitable, religious, educational, or other exempt organizations as defined by state law.

Computation of the municipal tax rate and individual tax payment is simple in concept. First, the governmental unit determines its budgetary needs for the next fiscal year. It deducts revenue from nontax sources such as federal and state transfer payments to determine the net amount that must be obtained from ad valorem property taxes. It then divides the needed revenue figure by the total taxable assessed property value within the jurisdiction to yield the tax rate. Thus, if a town needs $750,000 from property taxes and has $50 million total assessed property valuation, the computation is:

$$\frac{\text{Needed Revenue}}{\text{Total Assessed Value}} = \text{Tax Rate}$$

$$\frac{\$750,000}{\$50,000,000} = 1.5\% = (\$15 \text{ per } \$1,000 \text{ or } 15 \text{ mills})$$

For an individual owning property assessed at $80,000, the tax payable (tax levy) is $1,200.

Certain kinds of local public improvements may benefit particular property owners more than the public at large, for instance, the repaving of sidewalks or replacement of local sewers. Such improvements may be financed through *special assessments* instead of the general ad valorem tax levy. A special assessment is a tax imposed on property benefited by a local public improvement. It may be one lump sum or payable over several years.

Large-scale public projects such as new bridges, police and fire stations, public garages, schools, and parks require a sizable outlay of funds at one time. The normal mechanism to assemble such funds is the *general obligation bond*. Such bonds are sold to investors through national and international bond markets. The rate of interest payable over the term of the bond is a function of the creditworthiness of the issuing governmental unit, as well as the term of the bond and the general economic climate. Some states impose constitutional limits to the amount of bonded indebtedness that a municipality or county may incur in relation to its total assessed value. Such limits, however, do not usually apply to special districts. Thus, states such as Illinois, with strict debt ceilings, are awash in special districts created to circumvent debt limits on general-purpose local governments.

General obligation bonds issued by any governmental unit are repaid out of the ad valorem tax revenue and other monetary resources of that unit over time. If that proves insufficient, such bonds are backed by the "full faith and credit" of the issuer, who must raise taxes, sell property, or otherwise raise the funds to repay the bondholders.

Certain public facilities such as swimming pools, golf courses, and solid waste disposal facilities, however, are expected to pay for themselves over time through the imposition of user fees. Such facilities may be funded through *revenue bonds,* which are exempt from ceilings on general obligation bonds. Revenue bonds are repaid only from fees generated from the facility in question and are not backed by the issuer's "full faith and credit."

Fees for permits and services comprise a growing source of local government revenue. In Massachusetts, for instance, when a referendum (Proposition 2½) limited ad valorem tax revenue in many communities, fees for sewer, water, sanitary landfill, and other services were raised to offset the shortfall.

Finally, *intergovernmental transfers* comprise a significant proportion of revenue to municipalities, counties, and some special districts. During the 1970s, the federal government consolidated many of its individual grant programs to local governments under the new programs of "general revenue sharing" and Community Development Block Grants. State governments also make transfer payments (mostly from sales and income taxes) to local governments. Such allocations are usually earmarked for specific purposes such as education, welfare, housing, and parks.

Beyond these traditional revenue sources, certain municipal governments are authorized by state law to tax personal and corporate income and even to impose a local sales tax. Taxes on hotel rooms, meals, liquor, and cigarettes yield additional revenue in some cities. As state and federal aid to local governments declines in the 1990s, the latter may be expected to redouble their efforts to devise new sources of revenue. (See Chapter 9 regarding "impact fees".)

Land-use powers. Municipalities are the front line of public land-use control. Authority to engage in planning, zoning, and related functions is conferred on local governments (including both municipalities and counties) through state enabling acts, which in turn are subordinate to the state and federal constitutions. The principal powers that local governments exercise are: (1) the *regulatory power* (also called the *police power*) and (2) the *spending power,* including particularly the power of *eminent domain.*

Figure 6–8 depicts the central position of the local elected body in the land-use control process. The name of the body differs from one community to another, for example, city council, town meeting (New England), village board of overseers,

Municipal Jurisdiction and Powers

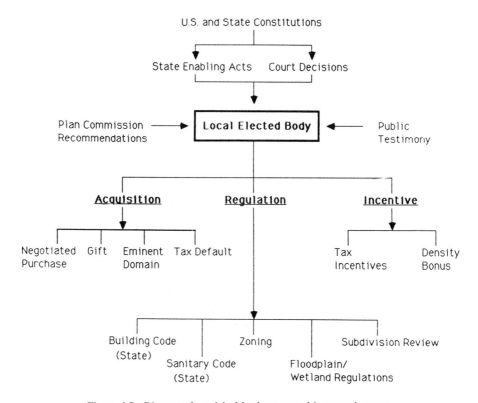

Figure 6.8 Diagram of municipal land use control inputs and outputs.

county commissioners. Whatever its name, it serves as the primary decision maker for the local government. Subsidiary roles may also be performed by appointive administrative boards such as the planning commission, board of zoning appeals, board of health, and board of public works.

The local elected body is essentially a "black box" that processes inputs and produces outputs. The inputs include: (1) constitutional and statutory constraints (as interpreted by the municipality's legal counsel); (2) expert advice from planning staff or consultants; and (3) public opinion as expressed in verbal and/or written testimony at public hearings and through informal means. Outputs include the various forms of public intervention available to the municipal government including: (1) regulation, (2) acquisition, and (3) incentives and persuasion.

The "black box" is actually semi-opaque since all meetings of the elected body and local boards are required to be publicized and open to the general public. A public meeting differs from a public hearing. At the former, the public may attend but has no right to testify; the latter is a forum for receiving testimony. Both must be publicized in accordance with statutory requirements or actions taken at such occasions may be nullified by a court if challenged.

The "black box" is also a "Pandora's box" which we open in the next three chapters.

CONCLUSION

Once the proud legal expression of the autonomous medieval city, the municipal corporation in America is now an ironic metaphor for governmental inadequacy in the face of external economic, political, and environmental forces. It is a victim of its own success, having been replicated in such vast numbers that each individual municipality retains only a fragmentary role in the management of the overall metropolitan area. The final word on this subject is from political scientist Robert Wood (1961, p. 1) referring to the New York Region:

> On the eastern seaboard of the United States, where the state of New York wedges itself between New Jersey and Connecticut, explorers of political affairs can observe one of the great unnatural wonders of the world: this is a governmental arrangement perhaps more complicated than any other that mankind has yet contrived or allowed to happen. A vigorous metropolitan area, the economic capital of the nation, governs itself by means of 1467 distinct political entities (at latest count), each having its own power to raise and spend the public treasure, and each operating in a jurisdiction determined more by chance than by design. The whole 22-county area that we know as the New York Metropolitan Region provides beds for about 15 million people and gainful employment for about 7 million of them. Its growth, which is rapid, takes place almost entirely in its outer, less crowded parts, and this means that the Region is becoming more alike in the density of its population and jobs, more alike in community problems. But the responsibility to maintain law and order, educate the young, dig the sewers, and plan the future environment remains gloriously or ridiculously fragmented.

REFERENCES

ADVISORY COMMISSION ON INTERGOVERNMENTAL RELATIONS. 1969. *Urban America and the Federal System.* Washington, D.C.: ACIR.

———. 1974. *The Challenge of Local Government Reorganization.* Washington, D.C.: ACIR.

BILLINGTON, R. A. 1970. "The Frontier in Illinois History," in *An Illinois Reader* (Clyde C. Walton, ed.). Dekalb, IL: Northern Illinois University Press.

BOLLENS, J. C. 1957. *Special District Governments in the United States.* Berkeley: University of California Press.

DE TOCQUEVILLE, A. 1832/1969. *Democracy in America* (J. P. Mayer, ed.; George Lawrence, trans.). Garden City: Doubleday Anchor.

DUNCOMBE, H. S. 1977. *Modern County Government*. Washington, D.C.: National Association of Counties.

GRANT, R. D. 1890. *The Policy and Practice of Massachusetts in the Division of Towns* (Testimony before the Legislative Committee on Towns).

HARRIS, C. D. 1943/1959. "Suburbs," in *Readings in Urban Geography* (H. M. Mayer and C. F. Kohn, eds.). Chicago: University of Chicago Press.

JACKSON, K. T. 1985. *Crabgrass Frontier: The Suburbanization of the United States*. New York: Oxford University Press.

JEFFERSON, T. 1782/1944. "Notes on Virginia," in *The Life and Selected Writings of Thomas Jefferson* (A. Koch and W. Peder, eds.). New York: Random House Modern Library.

LEINBERGER, C. B. AND C. LOCKWOOD. 1986. "How Business is Reshaping America." *The Atlantic* 258(4): 43–52.

MAYER, H. M. AND R. C. WADE. 1969. *Chicago: Growth of a Metropolis*. Chicago: University of Chicago Press.

MCMANIS, D. R. 1975. *Colonial New England: A Historical Geography*. New York: Oxford University Press.

MORRISON, S. E. 1965. *The Oxford History of the American People*. New York: Oxford University Press.

MULLER, P. O. 1975. *The Outer City: Geographical Consequences of the Urbanization of the Suburbs*. Resource Paper No. 75-2. Washington, D.C.: Association of American Geographers.

NELSON, H. J. AND W. A. V. CLARK. 1976. *Los Angeles: The Metropolitan Experience*. Cambridge: Ballinger.

PLATT, R. H. 1972. *The Open Space Decision Process*. Paper no. 142. Chicago: University of Chicago Geography Research Series.

———, et al. 1980. *Intergovernmental Management of Floodplains*. Monograph no. 30. Boulder: University of Colorado Natural Hazards Center.

PORTER, K. H. 1922. *County and Township Government in the U.S.* New York: The MacMillan Co.

REISNER, M. 1986. *Cadillac Desert: The American West and Its Disappearing Water*. New York: Penguin Books.

REPS, J. W. 1965. *The Making of Urban American: A History of City Planning*. Princeton: Princeton University Press.

———. 1972. *Tidewater Towns: City Planning in Colonial Virginia and Maryland*. Charlottesville: University of Virginia Press.

REYNOLDS, D. R. 1976. "Progress Toward Achieving Efficient and Responsive Spatial-Political Systems in Urban America," in *Urban Policymaking and Metropolitan Dynamics* (J. S. Adams, ed.). Cambridge: Ballinger.

SANDS, C. D. AND M. E. LIBONATI. 1981. *Local Government Law*. Chicago: Callaghan.

STETZER, D. F. 1975. *Special Districts in Cook County: Toward a Geography of Local Government*. Research Paper no. 169. Chicago: University of Chicago Department of Geography.

TEAFORD, J. C. 1979. *City and Suburb: The Political Fragmentation of Metropolitan America 1850–1970*. Baltimore: The Johns Hopkins University Press.

WAGER, P. W. 1950. *County Government Across the Nation*. Chapel Hill: University of Carolina Press.

WAKSTEIN, A. M. 1972. "Boston's Search for a Metropolitan Solution." *Journal of the American Institute of Planners* (Sept.), pp. 285–295.

WARNER, S. B., JR. 1978. *Streetcar Suburbs: The Process of Growth in Boston 1870–1900*. Cambridge: Harvard University Press.

WHITEHILL, P. 1959. *Boston: A Topographic History*. Cambridge: Harvard University Press.

WOOD, R. 1961. *1400 Governments*. Cambridge: Harvard University Press.

ZINK, H. 1939. *Government of Cities in the United States*. New York: Macmillan.

7

EUCLIDEAN ZONING: ORIGINS AND PRACTICE

INTRODUCTION

Local municipal governments since the 1920s have been the front line of public response to private land-use initiatives. And the most prevalent instrument wielded by local governments in performing that function is the two-edged blunt sword of land-use zoning. Although it originated in Germany in the late nineteenth century, zoning is a quintessentially American institution with the blend of idealism and greed that that implies (Toll, 1969). It is also a twentieth-century phenomenon, especially associated with the decade of 1916–1926, marking, respectively, the years of its inception in the New York City Zoning Ordinance and of its constitutional approval by the U.S. Supreme Court. That decision, *Village of Euclid* v. *Ambler Realty Co.* 272 U.S. 365 (1926), lent the nickname "Euclidean zoning" to the institution that dominated local land-use control until the 1970s and survives in thousands of communities today.

In the words of planning lawyer Richard F. Babcock (1966, p. 3): "Zoning reached puberty in company with the Stutz Bearcat and the speakeasy. F. Scott Fitzgerald and the Lindy hop were products of the same generation." In a more statistical vein, the arrival of zoning coincided roughly with the 1920 Census, which reported that urban Americans outnumbered rural inhabitants for the first time.

Zoning is certainly not the only land-use regulatory measure available to local governments. They also possess important powers to regulate land subdivision, to enforce building code requirements, to regulate the use of floodplains, wetlands, or seismic risk areas, to designate historical districts, to control signs, and a variety of other functions under state enabling legislation and/or home rule authority. Furthermore, like other levels of government, local governments may acquire property for public purposes, either through negotiated sale or through eminent domain (condemnation). (See Chapter 9.) But land-use zoning is the most widespread local land-use control tool, in use in every major U.S. city (except the famous holdout, Houston, Texas) and many thousand smaller communities and counties. It is the broadest of land-use control techniques, applying to virtually any private use of land and many public uses within zoned jurisdictions. And it certainly has been the most contentious of land-use institutions, generating passionate advocacy during the 1920s and 30s, equally vehement denunciation and proposals for reform during the 1960s and 70s, and weary resignation in the 1980s.

This chapter and the next focus on the origins, nature, practice, and policy implications of zoning, with brief attention given to related local land-use tools. Some of the latter are also considered in later chapters.

PRECURSORS TO ZONING

Three strands of long-term response to urban deterioration in the nineteenth century—regulation, redevelopment, and relocation—were traced in Chapter 4. Each of these approaches fostered corresponding lines of action on a much larger scale in response to urban and metropolitan growth during the twentieth century. In the United States, the three approaches have respectively yielded (1) land-use and environmental regulations; (2) urban redevelopment and revitalization; and (3) suburbanization, as encouraged by federal home ownership incentives and the interstate highway system. The advent of zoning in 1916 may be viewed as a further enlargement of the public regulation approach that originated in the British Public Health Act of 1848, the New York Metropolitan Health Act of 1866, and their counterparts elsewhere.

In the shorter term, 1900–1916, there were several immediate precursors to the appearance of zoning. These included: (1) the City Beautiful Movement; (2) the Garden City Movement; (3) public legislation regulating nuisances and building heights under the "police power"; (4) the advent and proliferation of the skyscraper; and (5) the Progressive Movement. Collectively, these engendered an atmosphere of public receptivity to governmental intervention in the private land market, to a modest degree at least, as reflected in the rapid diffusion of zoning and its eventual approval by the U.S. Supreme Court.

The City Beautiful Movement

The twentieth century opened with the commissioning of the nation's first city and regional plan, which was prepared for Washington, D.C. in commemoration of its centennial as the federal seat of government. A special blue ribbon panel, subsequently known as the McMillan Commission after its Senate sponsor James McMillan, was appointed in 1900 to report on "the development and improvement of the entire park system of the District of Columbia" (U.S. Congress, 1902, p. 7). Reflecting the continuing influence of the 1893 Chicago World's Fair, the panel

Figure 7.1 The MacMillan Commission Plan for the Mall in Washington, D.C., 1902. *Source:* Grutheim, 1976.

consisted of four principals of that enterprise: architects Daniel H. Burnham and Charles F. McKim, landscape architect Frederick Law Olmsted, Jr. (who had assumed his ailing father's practice), and sculptor Augustus Saint-Gaudens.

The McMillan Commission report of 1902 proposed to superimpose on the basic framework of the 1792 L'Enfant Plan a monumental redevelopment program which Gutheim (1976, p. 38) describes as "an inspiring set piece of the City Beautiful Movement that was to sweep the nation." The mall, as the city's major axis, was proposed to be redesigned, replanted, and extended to include the future reflecting pool and Lincoln Memorial. It was to be cleared of encroachments and embellished with new public buildings, fountains, gardens, and statuary—a democratic Versailles by way of Haussmann (Figure 7-1). Most of the Commission's Mall improvements were subsequently carried out, including the replacement of an encroaching rail terminal by the new Beaux-Arts Union Station northwest of the Capitol designed by Burnham in 1903.

In contrast with the monumental scale and design of the city's core, the commission was influenced by F. L. Olmsted, Jr. to recommend a more functional system of parks and open spaces along the Potomac and Anacostia rivers and Rock Creek, and the laying out of scenic routes such as the Mount Vernon Parkway. Thus the work and concepts of Olmsted, Sr., who died in 1903, were pursued by his son.

The McMillan Commission plan established Daniel Burnham as the high priest of City Beautiful architecture and city planning. The District of Columbia plan was rapidly followed by plans for Cleveland (1903), San Francisco (1905), and Manila (1905) (see Wilson, 1990). So successful was Burnham's architectural practice that he donated his services in preparing these city plans (Hines, 1979, p. 217).

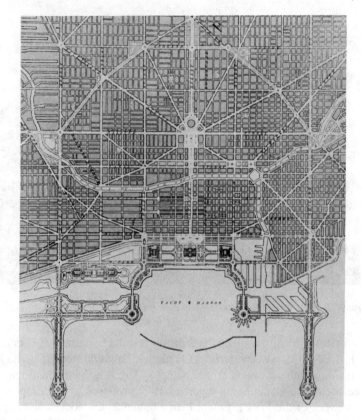

Figure 7.2 Burnham and Bennett's 1909 Plan of Chicago: downtown and harbor improvements. *Source:* Mayer and Wade, 1969. Reproduced by permission of The University of Chicago Press.

But his own city of Chicago most intrigued and challenged Burnham. His 1909 *Plan of Chicago,* prepared in collaboration with Edward Bennett for the Commercial Club of Chicago, was his masterpiece. It was in 1907 in the course of preparing that plan that he uttered his famous adage: "Make no little plans; they have no magic to stir men's blood and probably will not be realized" (Wrigley, 1983, p. 71).

The Plan of Chicago was a major conceptual leap beyond the earlier City Beautiful plans. Ironically, its most City Beautiful element—a baroque civic center at the confluence of radiating boulevards flanked by Haussmann-style building facades—was never built: It evidently did not "stir men's blood" (Figure 7–2). By contrast, it was the functional elements of the plan that changed the face of the city over succeeding decades. The Plan itself, surely one of the most magnificent documents in the history of city planning, addressed a variety of public needs including parks and forest preserves, streets and boulevards, bridges, and rail and port facilities. It also included a chapter on legal aspects of achieving its proposals. As popularized in a manual and textbook, the Plan influenced a generation of Chicago taxpayers to support, among other projects, the completion of the city's lakefront park system; the double-decking of Wacker Drive and a traffic bridge across the rail yards south of the Loop; the consolidation of rail terminals; and, at a regional scale, the establishment of the Cook County Forest Preserve system (Figure 7-3). Aside from the formal and elegant Grant Park—the city's "front yard"—most of the Chicago plan improvements were practical in nature, serving the objectives of economic efficiency, convenience, and public well-being. *The Plan of Chicago* thus marked a transition from the earlier City Beautiful plans dominated by pompous aesthetics to the utilitarian world of the postzoning comprehensive plan.

Like its predecessors, however, the Chicago Plan largely disregarded needs other than public improvements, such as housing and neighborhood planning. Goodman (1968, p. 22) notes as a serious defect in all City Beautiful plans ". . . the lack of legitimation of any public control over the private actions that were decisive in setting the quality of the urban environment. The early planners merely avoided the issue when they made 'planning' coterminous with parks, boulevards, and civic centers." A more acerbic appraisal of the City Beautiful movement has been offered by Lewis Mumford (1955, p. 147): "Our imperial architecture is an architecture of compensation: It provides grandiloquent stones for people who have been deprived of bread and sunlight."

The Garden City Movement

As discussed in Chapter 4, a very different vision of the planned community was provided by Ebenezer Howard's book *Garden Cities of To-morrow* and the founding of Letchworth in 1903. American admirers of his theories established the Garden Cities Association of America (GCAA) in 1906, which attracted the ". . . same kinds of civic and political leaders who had supported Howard's ideas in England" (Schaffer, 1982, p. 32). This organization proposed to house 375,000 families in a series of garden communities to be constructed in several eastern states, but without result. F. L. Olmsted, Jr.'s 1912 design for Forest Hills, New York, sponsored by the Russell Sage Foundation, however, reflected garden city design (a reciprocal tribute to Howard's admiration for Olmsted, Sr.'s design for Riverside, Illinois).

The GCAA was dissolved in 1921 and replaced two years later by the much more influential Regional Planning Association of America (RPAA), of which the young Lewis Mumford was a charter member. This group successfully promoted several garden city–style projects, most notably Radburn, New Jersey (1927). The influence of Howard's concept therefore was felt more strongly in America at the time of the 1926 *Euclid* decision that upheld zoning than at the inception of zoning a decade earlier.

But while the *process* of garden city establishment was relatively unfamiliar before the 1920s, the *form* of the garden city—particularly its emphasis on the sylvan, low-density residential neighborhood and the segregation of homes and commerce—accorded perfectly with the image of turn-of-the-century American

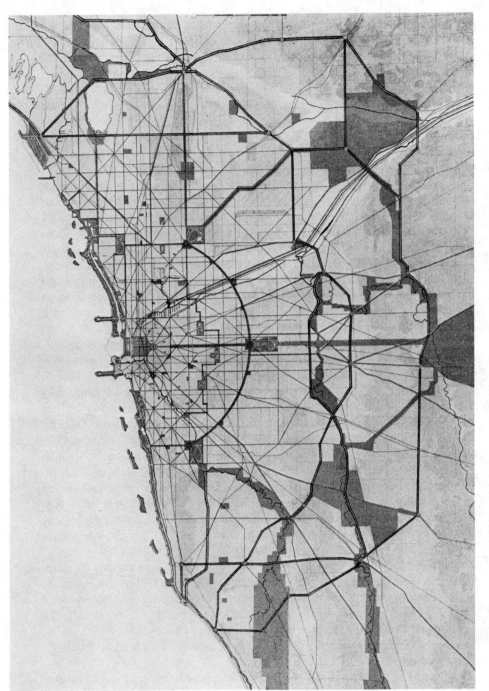

Figure 7.3 Burnham and Bennett's 1909 Plan of Chicago: proposed parks and forest preserves. Reproduced from the original plan.

suburbia. Lacking the control provided by a unified land ownership trust of the Letchworth type, it was natural for the new suburbanities to seek a legal means of perpetuating the bucolic environment that they had fled the cities to attain. Zoning was therefore welcomed as a means to achieve this objective.

Nuisance and Height Regulations

In 1915, a year before New York's first zoning ordinance, the U.S. Supreme Court upheld an unusual ordinance adopted by the City of Los Angeles. This law prohibited the operation of brick kilns within a specified three-square-mile district in an area recently annexed to the city. The measure was challenged by the owner of a clay pit and brick kiln that predated both the ordinance and the annexation of the site by the city. The Court upheld the regulation against the charge that it substantially reduced the value of the property and did not apply to similar businesses elsewhere in Los Angeles (*Hadacheck v. Sebastian* 239 U.S. 394).

The *Hadacheck* decision has been criticized by legal scholars for its harsh and arbitrary impact on a preexisting business (Williams, 1975, sec. 116.03). But whatever its value as a constitutional precedent, the case reflects the permissive state of public nuisance law on the eve of the appearance of zoning. *Hadachek* was consistent with a line of Supreme Court decisions extending back to the 1870s that upheld state and local regulation of particular uses of private property without compensation. These cases contrasted with the prevailing ethos of laissez faire and social Darwinism, which opposed even minimal public intervention in the economy, as reflected in the Court's disapproval of a New York statute regulating child labor in *Lochner v. New York* (198 U.S. 45, 1905) (Toll, 1969, pp. 16–18).

Where a private use of land or economic activity threatened the public interest, as opposed to the interest of individual workers, the Court was remarkably progressive. As early as 1872, a New Orleans ordinance that vested a monopoly in livestock slaughtering in one enterprise and banned all competitors without compensation was upheld by the Supreme Court on the ground that:

> Unwholesome trades, slaughter-houses, operations offensive to the senses, the deposit of powder, the application of steam power to propel cars, the building with combustible materials, and the burial of the dead, may all . . . be interdicted by law, in the midst of dense masses of population, on the general and rational principle that every person ought to use his property as not to injure his neighbors; and that private interests must be made subservient to the general interests of the community. This is called the police power . . . (*Slaughterhouse Cases*, 16 Wall 36, 1872).

In 1887, the Court upheld a Kansas law that prohibited the manufacture and sale of alcoholic beverages and closed existing breweries (*Mugler v. Kansas*, 123 U.S. 663), stating:

> A prohibition simply upon the use of property for purposes that are declared by valid legislation to be injurious to the health, morals, or safety of the community, cannot, in

any just sense be deemed a taking or an appropriation of property for the public benefit [in violation of the Fifth Amendment to the U.S. Constitution]. Such legislation does not disturb the owner in the control or use of his property for lawful purposes nor restrict his right to dispose of it . . . (123 U.S., at 667–8).

In a more tenuous case, the Court upheld a ban on the production and sale of oleomargarine in *Powell* v. *Pennsylvania* 127 U.S. 678 (1888). And it supported a state ban on the use of fish nets in certain waters in *Lawton* v. *Steele* 152 U.S. 133 (1894) declaring:

> The extent and limits of what is known as the police power have been a fruitful subject of discussion in the appellate courts of nearly every state in the Union. It is universally conceded to include everything essential to the public safety, health, and morals, and to justify the destruction or abatement . . . of whatever may be regarded as a public nuisance. . . . A large discretion is vested in the legislature to determine, not only what the interests of the public require, but what measures are necessary for the protection of such interests (152 U.S., at 135).

A key element of the developing doctrine of the police power at the turn of the century was the recognition of the need for prospective regulation of private activity rather than simply after-the-fact mitigation. As stated by Ernst Freund in his 1904 treatise *The Police Power:*

> The common law of nuisance deals with nearly all the more serious and flagrant violations of the interests which the police power protects, but it deals with evils only after they have come into existence, and it leaves the determination of what is evil very largely to the particular circumstance of each case. The police power endeavors to prevent evil by checking the tendency toward it and it seeks to place a margin of safety between that which is permitted and that which is sure to lead to injury or loss. This can be accomplished to some extent by establishing positive standards and limitations which must be observed, although to step beyond them would not necessarily create a nuisance at common law.

The proposition that the police power might address conditions that would not necessarily constitute a traditional common law nuisance opened the door wide to the introduction of public regulation of the use of land and the dimensions of buildings. Furthermore, there was growing precedent for the establishment of different regulations for different geographical districts of a community, that is, incipient zoning. *Hadacheck,* for instance, involved restriction on land use applicable to a particular area of Los Angeles.

The leading prezoning decision on building dimensions that utilized a districting approach was the 1909 Supreme Court opinion in *Welch* v. *Swasey* 214 U.S. 91. *Welch* involved a Massachusetts law that imposed a limit of 125 feet for new buildings in designated commercial districts of Boston and either 80 or 100 feet in residential districts. The Court upheld the measure, assuming that the legislature had good reasons for making such a distinction. (It suggested that women and children are more likely to be at risk from fire in residential areas and thus buildings should be smaller).

The *Welch* decision reflected an oft-expressed judicial policy to presume that a legislative act is valid in the absence of overwhelming proof that it is arbitrary or capricious. While justified in the interest of judicial efficiency, deference given by courts to legislative determinations even of tiny suburban enclaves has often insulated questionable zoning practices from searching judicial review.

These decisions reflect the degree of judicial tolerance of police power measures at the eve of comprehensive zoning. Supreme Court opinions, it must be remembered, are only the tip of the iceberg of what is happening in the society at large. For each disputed measure reviewed by the highest court, dozens were resolved in lower courts and hundreds went unchallenged. Height limitations, for instance, were quite widespread by the time of the *Welch* decision. As early as 1889 a height limit was imposed in Washington, D.C. to enhance views of the Washington Monument and the Capitol. (Due to that measure, Washington remains strikingly horizontal today in contrast to the vertical profiles of other American central cities). By 1913, 22 U.S. cities had some form of height controls, of which four (including Boston and Washington) applied such limits on the basis of districts (Delafons, 1969, p. 20).

Skyscrapers

The proliferation of prezoning height limitations reflects the evolution of legal institutions in response to perceived changes in technology and the urban environment. In this case, the physical change involved the unprecedented concentration of workers and economic activities in high-rise office buildings of virtually unlimited height and bulk. Two technological innovations underlay this architectural revolution: the elevator and steel frame construction. The former was invented by Elisha Graves Otis and introduced to the world at the 1853 New York Crystal Palace Exposition (Toll, 1969, p. 47). The latter involved the support of a tall building by a steel skeleton in place of weight-bearing walls. Burnham and Root's 16-story Monadnock Building, completed in Chicago in 1891, was the tallest and last product of the era of weight-bearing-wall construction. The steel frame was inaugurated in the 10-story Home Insurance Company Building in Chicago (1885) and the 11-story Tower Building in New York (1888). According to Toll (1969, p. 51):

> Until the steel frame appeared, seemingly immutable structural principles curbed the economic appetite. The steel frame, in effect, repealed this restraint and the law offered no substitute. Quite literally the sky became the limit.

The skyscraper thrived particularly in Manhattan. In 1905, Burnham's Flatiron Building reached 20 stories and 300 feet. In 1907, the Singer Building on Lower Broadway reached 612 feet. In 1919, the Metropolitan Life Insurance Building rose to 700 feet, only to be topped by the Woolworth Building in 1913, at nearly

800 feet. By that year there were over 50 buildings in Manhattan over 20 stories and nine above 30 stories (Delafons, 1969, p. 21).

The Singer, Metropolitan, and Woolworth towers were exemplars of the skyscraper style—ornate, slender, tapering to a pyramid or cupola—they fairly represented the mood of exuberance and prosperity of prewar America. But at street level, these and their bulkier neighbors cut off light and air from business districts and flooded the sidewalks with office workers. Furthermore, fire equipment could reach only 100 feet above ground (Toll, 1969, p. 153). The tragic Triangle Shirtwaist Company fire on March 25, 1911, which gutted a 10-story building and killed 146 young women, prompted investigation into the need to control the size and design of commercial structures (Toll, 1969, p. 26). Soon thereafter, the loss of 1503 lives in the sinking of the "Titanic" on April 15, 1912, due to an insufficient number of lifeboats, further aroused the public to the perils of laissez faire in the face of technological change and corporate arrogance.

The Progressive Movement

The precursors to zoning just discussed—the City Beautiful and Garden City movements, judicial sanction of police power regulations, and public alarm concerning tall buildings—reflected the spirit of reform known as the Progressive movement that swept the nation in the first decade of this century. The essence of the movement in the words of Richard Hofstadter (1955, p. 5) was:

> . . . that broader impulse toward criticism and change that was everywhere so conspicuous after 1900. . . . While Progressivism would have been impossible without the impetus given by certain social grievances, it was not nearly so much the movement of any social class, or coalition of classes, against a particular class or group as it was a rather widespread and remarkably good-natured effort of the great part of society to achieve some not-very-clearly-specified self-reformation. Its general theme was the effort to restore a type of economic individualism and political democracy that was widely believed to have existed earlier in America and to have been destroyed by the great corporation and the corrupt political machine; and with that restoration to bring back a kind of morality and civic purity that was also believed to have been lost.

With respect to cities, progressives were greatly concerned with the two interrelated evils of overcrowding and political corruption. Both of these conditions were products of the continuing surge of immigration to eastern seaboard cities from Europe. The highest-ever annual level of immigration to the United States was recorded in 1907 when 1.3 million foreigners arrived. In 1910, 13.3 million foreign-born persons were living in the United States, comprising one-seventh of the nation's total population (Hofstadter, 1955, p. 176). These destitute and largely non-English-speaking refugees readily supported the big city political machines, which offered them jobs and food in exchange for votes.

Reform proposals were first articulated through a deluge of muckraking writings. Jacob Riis' *How the Other Half Lives* (1890/1972) documented through prose and photography the deplorable state of New York's tenement districts. In 1904 another journalist, Lincoln Steffens, scrutinized big city bossism and corruption, in *The Shame of the Cities.* The moral implications of urban overcrowding were lamented by the Reverend Josiah Strong in his 1898 tract *The Twentieth Century City.* Upton Sinclair in *The Jungle* (1906) exposed the abuses of the meatpacking industry. These and other classics of muckraking literature stimulated an outpouring of more than 2,000 articles on social conditions in American magazines and newspapers between 1903 and 1912 (Ciucci et al., 1979, p. 188).

Apart from the issue of corruption in city governments, the most politicized urban concern of progressives was congestion. Like "environmental deterioration" in the 1970s, *congestion* was used by reformers in the century's first decade to cover a variety of ills: disease-ridden tenements, mobbed streets in office districts, loss of light and air, and inadequate open space for recreation. Congestion was the rubric under which progressive ideology was translated into practical city planning measures, including zoning. One of the first progressives to advocate planning to achieve social reform was Frederick C. Howe, whose 1905 book, *The City: The Hope of Democracy,* urged the adoption of German city planning practices.

Howe's book strongly influenced Benjamin C. Marsh, a young social activist and political organizer. In 1907, at the age of 28, Marsh was appointed executive secretary to the newly formed Committee on Congestion of Population (CCP) in New York City. Like James Chadwick, who launched the British sanitary reform movement in his capacity as secretary to prestigious committees whose reports he wrote, Marsh made good use of his position. In 1908, he organized an exhibition on the evils of congestion, which was displayed at the American Museum of Natural History. In the following year, after travels in Europe, Marsh privately published a short book: *An Introduction to City Planning,* which opened with the heading: "A City Without a Plan is like a Ship Without a Rudder." Like Howe, Marsh advocated adoption by American cities of German planning techniques including public control over street location and design, advance acquisition of land prior to development, taxation of increased value of land (to discourage speculation), and the districting or zoning of urban land to regulate building height and volume. In contrast to Burnham's *Plan of Chicago,* published later the same year, which largely addressed public improvements, Marsh emphasized the need for public regulation of *private* land development.

Marsh then organized on behalf of CCP the First National Conference on City Planning and Congestion to elaborate and publicize the themes of his book. This event, which is often considered the birthplace of the modern planning movement in the United States, was held in Washington, D.C. on May 21–22, 1909. According to Kantor (1983, pp. 69–70):

> The forty-three conferees met in an air of excitement and hope. Many of the nation's leaders in urban affairs attended, including Frederick Howe, Jane Addams, . . . John Nolen, [and] Frederick Law Olmsted, Jr. . . . Representatives of municipal art, social work, architectural, civil engineering, and conservationist groups also attended. The

meeting vividly reflected the many interest groups concerned with city planning at the time.

The conference papers, which summarized the state of the art of city planning in the United States and abroad, were subsequently published as a Senate document (U.S. Congress, 1910). The major theme of the conference, the evils of congestion, was articulated in the opening address by Henry Morgenthau:

> There is an evil which is gnawing at the vitals of the country, to remedy which we have come together—an evil that breeds physical disease, moral depravity, discontent, and socialism—and all these must be cured and eradicated or else our great body politic will be weakened. This community can only hold its preeminence if the masses that compose it are given a chance to be healthy, moral, and self-respecting (U.S. Congress, 1910).

Landscape architect Robert Anderson Pope perhaps best reflected the significance of the conference as a threshold in American planning history. Pope first indicted the City Beautiful movement as impractical and elitist (Daniel Burnham was not present):

> . . . We have assumed without question that the first duty of city planning is to beautify. We have made the aesthetic an objective of itself. In most of our cities the movement has characterized itself by the expenditure of huge sums for extensive park systems, with broad boulevards and bridle paths, for far-outlying reservations; inaccessible improvements designed to protect the needs of future generations, now, however, made available to but a small portion of the community—the wealthy and leisure classes, who of all society needs these advantages the least.
>
> Again, we have rushed to plan showy civic centers of gigantic cost, in design the expression of, at best, only a small group of individuals, . . . brought about by civic vanity, and bears the character of external adornment, when pressing hardby we see the almost unbelievable congestion with its hideous brood of evil, filth, disease, degeneracy, pauperism, and crime. What external adornment can make truly beautiful such a city? (U.S. Congress, 1910, p. 75).

In place of the City Beautiful, Pope argued that "city planning is of primary importance as a social and economic factor" (U.S. Congress, 1910, p. 76). Instead of "showy civic centers," Pope advocated: (1) decentralizing and more equitable distribution of land values; (2) widening of streets and establishment of radial and belt thoroughfares; and (3) the adoption of land-use zoning as practiced in Germany to regulate: ". . . building heights, depth of blocks, number of houses per acre, and land speculation with all its attendant evils" (U.S. Congress, 1910, p. 77). Pope further expressed a belief among urban reformers at this time that city planning was essential to national and ethnic survival:

> The average recruit in the German army is much taller, stronger, and heavier than the British soldier, spends less time in hospital, and has a lower death rate. . . . The

modern tendency toward congestion in cities and the increase of unhealthful living conditions have been so ably combatted in Germany that no real impairment of her manhood can be detected. . . . While it is admitted that many causes have contributed to this result, city planning is known to be a very important factor (U.S. Congress, 1910, p. 77).

The 1909 Conference was followed the next year by a second conference, with "Congestion" significantly dropped from its title, reflecting the dwindling influence of Marsh and the CCP. Subsequent national conferences on city planning provided an annual forum for the development of the planning profession. Thenceforth, fewer social workers and civic reformers attended, and the conference was more connected with ". . . data, statistics, techniques, management, standards, efficiency, and evaluation" (Kantor, 1983, p. 71).

Meanwhile, Marsh continued to lobby for anti-congestion policies in New York. In 1910, he persuaded the city to establish an official Commission on Congestion of Population and was duly appointed its secretary. The commission's 1911 report recited the litany of reformist views on congestion and city planning, including Marsh's own pet notion of a progressive tax on land value increments, which caused the commission's Chair, landowner Henry Morgenthau, to resign (Marsh, 1953, p. 29).

Social progressives such as Marsh, Howe, and Jane Addams thus joined in common cause with practitioners of city design such as Frederick Law Olmsted, Jr., John Nolen, and Charles Mulford Robinson to prepare a fertile soil for the advent of zoning. The actual planting of the seed, however, was motivated by an impulse quite distinct from either slum eradication or urban design, namely, hostility to neighborhood change. As vividly chronicled by Seymour Toll (1969), the proximate cause of zoning was the antipathy of prestigious merchants along New York's Fifth Avenue toward encroachment by garment factories and offices. (The retailers themselves had recently displaced the mansions of the city's elite.) Since the unwanted activities generally occupied taller buildings, the Fifth Avenue Association representing the merchants advocated the adoption of controls on building size by district, to protect existing commercial property values. Pursuant to the reports of two city commissions established at the behest of the Fifth Avenue Association, a zoning enabling act was adopted by the New York State Legislature, which led to the adoption of the nation's first zoning ordinance in New York City in 1916.

THE NUTS AND BOLTS OF EUCLIDEAN ZONING

While "Euclidean zoning" has undergone many refinements and embellishments since its inception in 1916, in basic outline it has remained remarkably unchanged from its early form. That form, which originated in New York City, spread rapidly across the nation and was constitutionally upheld by the U.S. Supreme Court in the landmark *Euclid* case. The following is a summary of the elements of that fundamental institution which remains in use in thousands of communities and counties today.

Authority to Adopt Zoning

Zoning is a noncompensatory exercise of the municipality's regulatory power or "police power." Although communities in many states have enjoyed considerable latitude to adopt regulations under the doctrine of home rule, the power to zone land use has been specifically delegated to local governments by state legislatures through enabling acts in every state. Once the state has adopted such an act, local land-use zoning must conform to its provisions. State zoning-enabling acts are permissive, not mandatory: Local communities are not required to adopt zoning regulations, but if they do (and most have), they must follow the requirements of the state law. These requirements are largely procedural rather than substantive—they address *how* zoning must be adopted and administered, with little attention to *what* local zoning regulations may require. Limitations on substantive content have been applied primarily by courts rather than by state legislatures.

While state zoning acts differ somewhat due to amendments over the years, most originated from a common source. The Standard Zoning Enabling Act (SZEA) was developed in 1926 by an Advisory Committee on Zoning established within the U.S. Department of Commerce by then Secretary of Commerce, Herbert Hoover. The SZEA was furnished to all state legislatures and municipalities and was widely adopted in place of earlier acts that had been haphazardly developed during the previous decade (Bassett, 1936, pp. 28–29). A parallel State Planning Enabling Act was similarly promulgated in 1927.

It is axiomatic that an exercise of the public regulatory power must serve to protect the public health, safety, and welfare. The relationship of zoning to these goals was articulated in the SZEA and most state enabling acts in the following language:

> Such regulations shall be made in accordance with a comprehensive plan and designed to lessen congestion in the streets; to secure safety from fire, panic, and other dangers; to promote health and the general welfare; to provide adequate light and air; to prevent the overcrowding of land; to avoid undue concentration of population; to facilitate the adequate provision of transportation, water, sewerage, schools, parks, and other public requirements. Such regulations shall be made with reasonable consideration, among other things, to the character of the district and its peculiar suitability for particular uses, and with a view to conserving the value of buildings and encouraging the most appropriate use of land throughout such municipality (Bassett, 1936, pp. 51–52).

This statement reflects the welter of concerns that underlay the advent of zoning, including congestion in housing and streets, fire, loss of open space, and neighborhood change. Much of the statement is rhetorical: Zoning cannot ipso facto "facilitate the adequate provision of transportation, water, sewerage, schools, and parks." It is arguable that public spending and locational decisions regarding these services influence the way land may be zoned rather than vice versa.

The Nuts and Bolts of Euclidean Zoning   179

Two institutional fixtures of American zoning practice that originated in the model acts are the planning board or commission and the zoning board of appeals. The former serves as the sounding board for assimilating expert advice and community opinion in order to formulate recommendations to the local elected body regarding the adoption and modification of zoning. The latter comprises a quasi-judicial panel that is authorized to grant exceptions from the strict rules of the zoning ordinance to individual property owners in appropriate cases.

What Is Regulated

Euclidean zoning regulates: (1) the use of land; (2) the density of population; and (3) the dimensions or "bulk" of buildings. Initially, each of these variables was addressed by a separate set of districts: The Euclid ordinance involved use, height, and area zones. This proved to be cumbersome and zoning practice soon resorted to the division of a community into a single system of zones designated by use, with bulk and lot-size requirements specified accordingly.

The principal classes of land use for zoning purposes are residential, commercial, and industrial. These are normally divided into subclasses such as single-family residence, one- or two-family residence, rural residence, neighborhood business, highway business, and so forth. Each type of zone involves specific rules on use, building size, and lot size.

Land-use regulations specify for each class of zone which activities are (1) permitted "as of right"; (2) prohibited; and in more recent ordinances, (3) permitted conditionally if a special permit is obtained. Originally, zoning was *cumulative* in its structure, resembling a pyramidal hierarchy of land-use districts (Figure 7–4). At the apex of the pyramid was the exalted single-family residence zone in which all other uses were prohibited. Next was the multiple-family zone, which allowed

Figure 7.4 Hierarchy of "Euclidean zoning" classifications.

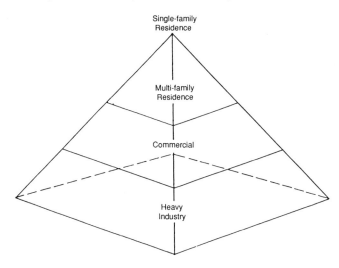

Table 7-1 Table of zoning districts. Town of Amherst, MA. Zoning bylaws, 1987.

Section 2.0			Zoning Districts
The Town is hereby divided into the following classes of zoning districts:			
	2.01	Residence Districts:	
		R-LD	Low Density Residence
		R-O	Outlying Residence
		R-N	Neighborhood Residence
		R-G	General Residence
		CR	Campus Residence
		R-F	Fraternity Residence
		PURD	Planned Unit Residential Development
	2.02	Business Districts:	
		B-G	General Business
		B-L	Limited Business
		B-VC	Village Center Business
		OP	Office Park
		COM	Commercial
	2.03	Industrial Districts:	
		LI	Light Industrial
		PRP	Professional and Research Park
	2.04	Educational Districts:	
		ED	Educational
	2.05	Conservation Districts:	
		FPC	Flood-Prone Conservancy
		WP	Watershed Protection
		ARP	Aquifer Recharge Protection
		WD	Wetlands District

either single- or multiple-family dwellings. Commercial zones allowed both businesses and residential uses. Industrial zones were essentially unrestricted, except perhaps designated nuisance-type activities. Thus, despite its instigation by Fifth Avenue merchants, zoning soon emerged as a tool to protect the residential areas from business, but not vice versa (Toll, 1969).

Cumulative zoning has been replaced in most communities by *noncumulative zoning* in which each class and subclass of zone is mutually exclusive as to use. Thus, homes cannot be built in commercial districts, just as businesses are barred from residential zones. Noncumulative zoning requires equal attention to all parts of the community, not just its prime residential areas. This imposes a much more complex burden on the planning commission, its staff and consultants, to try to envision the most appropriate use of each section of land in the community. The result has been a proliferation of classes of zones, each with its specific rules regarding use, bulk, and area. The Town of Amherst, Massachusetts for instance, with a permanent population of 27,000 has 19 classes of land-use districts (Table 7.1).

The Nuts and Bolts of Euclidean Zoning

Population density in residential areas is regulated through establishment of minimum lot sizes for dwellings or groups of dwellings. Lots for single-family homes typically range from 12,000 to 80,000 square feet (about one-quarter to two acres). Lot size roughly correlates with size and cost of the home, with less expensive dwellings normally built on land zoned for smaller sized lots. (In the 1990s, however, tract homes originally costing $20–30,000 in the 1960s may cost more than five times as much.) Developers usually prefer smaller lots which allow more homes per total project area and reduce the costs per unit of streets and other infrastructure provided by the developer.

Density in commercial or office districts is regulated by establishing *floor area ratios* (FARs). An FAR specifies the maximum enclosed floor area within a structure as a multiple of the site area. Thus an FAR of 10 allows a ten story structure to cover an entire building site, or a 20 story building on half the site. (FAR's are frequently raised in central city locations when a developer provides "amenities" such as a public plaza, off-street parking, or residential units.)

Beyond the traditional trio of use, density, and bulk regulations, zoning has encompassed other land-planning concerns. Modern zoning ordinances typically address: minimum off-street parking requirements in multi-family and nonresidential areas; rules regarding billboards and other forms of outdoor advertising signs; extraction of sand, gravel, and other minerals; secondary or accessory uses such as professional offices within homes and "mother-in-law" apartments; and special overlay districts for shorelines, floodplains, and wetlands. An overlay district imposes additional regulations beyond the basic Euclidean requirements for the applicable zone. Alternatively, special restrictions for sensitive lands such as wetlands may be expressed in a separate ordinance apart from the local zoning law.

Zoning Map and Text

Euclidean zoning differs from other forms of regulatory power such as building and sanitary codes in that its rules differ from one area of a community to another. The mapping of land-use zones is equal in legal significance to the written content of regulations. A zoning ordinance therefore consists of two legally adopted elements: the zoning map and the zoning text. One defines *where* specific regulations apply, and the latter *what* those regulations consist of.

The actual legal zoning map of a municipality or county is located in the office of the building inspector or other enforcement officer. The zoning map must be at scale large enough to be able to identify individual parcels: tax assessment maps are often used as the base for the master zoning map. Smaller-scale copies of the zoning map may be purchased from the appropriate municipal office, but where important decisions are involved, referral to the master zoning map is necessary (Figure 7–5).

The delineation of zone boundaries and assignment of areas thus defined to specific zone classifications is a legal act of the local elected body (see Figure 6–9). Boundaries are recommended by the planning board or commission in response to professional advice and public opinion. The drawing of zone boundaries in devel-

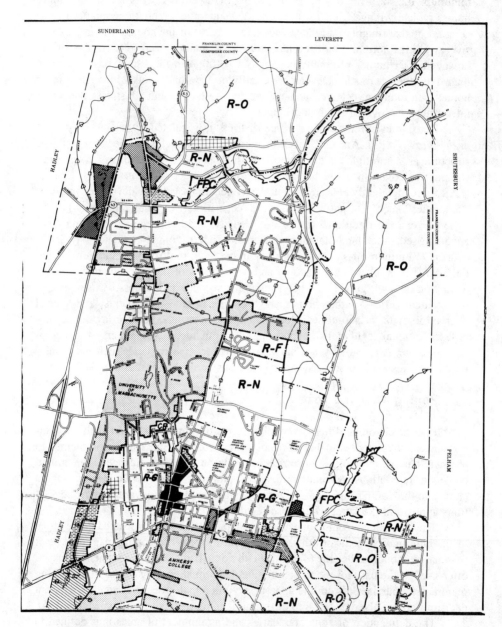

Figure 7.5 Excerpt from Town of Amherst, Massachusetts Zoning Map, 1987.

The Nuts and Bolts of Euclidean Zoning

oped communities is difficult since it is undesirable to designate existing structures as "nonconforming uses." Wherever possible, zoning boundaries are drawn to reflect pre-existing land-use patterns.

The impact of zoning is thus largely prospective. It has far more influence on the future development of vacant land than on already built-up areas, although redevelopment or conversion of older structures is subject to zoning. Also, zoning applies most forcefully to privately owned land. Land owned by the federal government is exempt from local zoning; state land may or may not be exempt depending on state law. A municipality technically must obey its own zoning in the use of local public land, but it can readily amend the zoning to meet its needs. Nevertheless, zoning maps normally cover the entire municipal jurisdiction, developed and vacant, private and public. Special classes of zones may be utilized for land that is outside the normal land market, for example, institutional, conservancy, or open space.

Much controversy surrounding zoning involves the location of zone boundaries. Where streets serve as boundaries, property on opposite sides of the street is assigned to different zones with perhaps very different economic implications (e.g., shopping versus residential). Also, boundaries may split larger parcels of land into two or more zones, sometimes to the distress of the property owner. Zoning is to a significant degree an exercise in the art of rational line drawing.

Once a given parcel is located on the master zoning map, one refers to the zoning text to determine what restrictions apply to the class of zone to which it is assigned. Each zone is designated on the map by a designator code. The designator code is located in the table of use regulations (Table 7–2) to determine the allowed, prohibited, or conditionally allowed uses applicable to the zone containing the site. The code is then located in the table of dimensional regulations to determine the lot size, FAR, or other dimensional limitations that apply to the proposed use if allowed (Table 7–3). Parking, signage, and other regulations may be specific to zones or may be community wide. These are covered in subsequent sections of the zoning text.

Flexibility

Euclidean zoning, in theory, attempts to predetermine the use of all vacant private land according to a comprehensive plan. This of course is a quixotic venture that continually encounters the windmills of reality. One of the claims made on behalf of zoning from its inception was that it would provide stability and predictability for the benefit of land owners and investors. While it has done this to some extent, there is a contradictory trend in the evolution of zoning—toward greater flexibility to cope with the unexpected and the unfair. From its earliest days, zoning involved two means of addressing those concerns, namely, amendments and variances. Later these were joined by special or conditional use permits, floating zones, cluster zoning, planned unit developments (PUDs), and transfer of development rights (TDRs), among other innovations.

Table 7–2 Excerpt from Town of Amherst Zoning By-law: Table of use regulations. (Y = allowed; N = not allowed; SP = allowed by special permit)

Bylaw Number	Land-Use Classification	Standards & Conditions	Zoning District										
			R-O R-LD	R-N	R-G	CR	R-F	B-G	B-L B-VC	COM	OP	LI	PRP	FPC
SECTION 3.31	EXTENSIVE USES													
3.310	Forestry and the harvesting of forest products.		Y	Y	Y	Y	Y	Y	Y	Y	Y	Y	Y	Y
3.311	Orchard, market garden, nursery, or other use of land for commercial agricultural production.		Y	Y	Y	Y	Y	Y	Y	Y	Y	Y	Y	Y

3.312	Commercial greenhouse; salesroom or stand for the sale of nursery, garden, or other agriculture produce (including articles of home manufacture from such produce).	The majority of the produce and products sold shall be produced by the owner of the land or made from the products so produced. The farm shall be a minimum of five acres in size.	Y	Y	N	N	N	N	N	Y	N	Y	SP	SP
3.313	Commercial poultry or livestock farm, or the raising of pets for gainful purposes.		Y	N	N	N	N	N	N	N	N	SP	SP	SP
3.314	Reservation, wildlife preserve, or other conservation use.		Y	Y	Y	Y	Y	Y	Y	Y	Y	Y	Y
3.315	Country club, organized camp, sportsing grounds, or other predominantly outdoor recreational use.	Not to be conducted as a gainful business.	SP	SP	SP	SP	SP	SP	SP	SP	N	SP	SP

Table 7-3 Amherst table of dimensional regulations.

Zoning District	Basic Minimum Lot Area (sq ft)	Cluster Minimum Lot Area (sq ft)	Additional Lot Area per Family (sq ft)	Basic Minimum Lot Frontage (ft)	Cluster Lot Frontage (ft)	Minimum Front Setback (feet)	Min. Side & Rear Yards (ft)[g]	Cluster Minimum Side & Rear Yards (ft)	Maximum Building Coverage (%)	Maximum Lot Coverage (%)	Maximum Floors[a]	Maximum Height[a] (feet)
R-LD	80,000	25,000	10,000	200	100[a]	40	20	15[a]	10	15	2½	35
R-O[i]	30,000	15,000	10,000	150	100[a]	40	25	15[a]	15	25	2½	35
R-N[i]	20,000	10,000	6,000	120	80[a]	30	15[d]	15[a]	20	30	3	35
R-G	12,000	6,000	2,500[a]	100	50[a]	25	10[d]	10[a]	25	40	3	40
CR	12,000		2,500[a]	100		25	10		25	70	3	40
R-F	40,000			150		40	20		20	45	3	40
B-G	12,000[b]		2,500[a]	100[b]		20[c]	e		70[a]	95[a]	3	50
B-L COM	20,000[b]		4,000	125[b]		20	25[a]		35	70/85[j]	3	35
B-VC	20,000[b]		4,000	125[b]		20	25[a]		20	50	3	35

OP	40,000[a]		100[a]	40	20	70	2½	35
LI				[f]	20[e]	25[f]	65	3	50
PRP	30,000[a]		100[a]	30	25	70	3	35
FPC	80,000		200	40	10	15	1	20
ED	SEE SECTION 3.213							

[a] Requirement may be modified under a Special Permit, issued by the Board of Appeals.

[b] Applies to Residence Uses only (Section 3.32).

[c] Applies to any part of a building which is within 200 feet of the side boundary of a Residence District abutting on the same street within the same block; otherwise, a 5 foot setback is required.

[d] A side yard need not be provided on one side of a single family dwelling if it shares a party wall or double wall with a single-family dwelling on the next lot built at the same time.

[e] Rear and side yards shall be at least 20 feet when adjoining a Residence District. Otherwise, rear and side yards are not required, but if provided, shall be at least 10 feet.

[f] Rear and side yards shall be at least 50 feet when adjoining a Residence District. Otherwise, rear and side yards shall be at least 10 feet.

[g] See Section 6.15 for Interpretation.

[h] A buildable lot shall contain either 90% of its total lot area, or 20, 000 square feet, in contiguous upland acreage.

[i] Substitute with dimensional requirements in Section 4.322 for 10% affordable projects.

[j] 85% in B-L zone adjacent to B-G and along University Drive; 70% in more outlying B-L and in COM.

Amendments are formal changes to the zoning ordinance that are adopted, like the original ordinance, by the local elected body in response to advice of the planning board and public testimony. Amendments may alter either the text of the zoning ordinance, the zoning map, or both. A map amendment involves a change of boundaries and possibly the redesignation of certain areas to a different class of zone. If a new zone class is created through a text amendment, it is usual (but not absolutely necessary) to rezone some land on the map to that class. If no land is immediately rezoned for that purpose, the new class is said to be a "floating zone," which may come down to earth in a later map amendment. Floating zones may be used to implement planned unit developments (Dawson, 1982, pp. 41–42).

Amendments of either text or map are intended to remedy area-wide needs or problems and should normally involve more than one parcel of land. Amendments that benefit only one property owner are subject to challenge (usually by neighbors) as unconstitutional *spot zoning*. However, rezoning of a single parcel is sometimes upheld. The Connecticut Supreme Court in *Bartram* v. *Zoning Commission of Bridgeport* 68 A.2d 308, 1949, sustained a zoning change of one parcel from residential to business, making the following distinction:

> Action by a zoning authority which gives to a single lot or small area privileges which are not extended to other land in the vicinity is in general against sound public policy and obnoxious to the law. It can be justified only when it is done in furtherance of a general plan properly adopted for and designed to serve the best interests of the community as a whole (68 A.2d, at 310).

The Court, in *Bartram,* held that the zoning commission had an adequate planning basis to justify the zone change in terms of providing convenience shopping to a residential neighborhood. Many cases hold the opposite, however, in challenges to amendments that favor only one parcel of land.

By contrast, *variances* almost always involve a single parcel. A variance is an administrative exception granted by the zoning board of appeals to relieve a "hardship" to a property owner who cannot make reasonable use of his or her land if the applicable zoning regulations are strictly enforced. The burden of proof is on the property owner to demonstrate such hardship. The variance is essentially a constitutional safety valve to remedy cases in which zoning unreasonably restricts the use of a particular parcel.

Since variances invite favoritism, courts usually view them sternly as compared with amendments. No presumption of validity operates in favor of a variance. According to the Pennsylvania Supreme Court in *Devereaux Foundation Inc. Zoning Case,* 41 A.2d 744, 1945:

> The strict letter of the ordinance may be departed from only where there are practical difficulties or unnecessary hardships in the way of carrying it out; and in such manner that the spirit of the ordinance may be observed, the public health, safety and general welfare secured and substantial justice done. No other considerations should enter into the decision (41 A.2d, at 745).

The Massachusetts Zoning Act as revised in 1975 limits the availability of variances to cases involving:

> . . . the soil conditions, shape, or topography of such land or structures and especially affecting such land or structures but not affecting generally the zoning district in which it is located . . . and desirable relief may be granted without substantial detriment to the public good and without nullifying . . . the intent or purpose of such ordinance . . . (Mass. G.L.A. Ch. 40A, sec. 10).

This act allows local towns and cities to abolish *use variances* within their jurisdictions. Variances to permit property to be used for a different purpose than specified in the zoning rules are considered to be more pernicious than *dimensional variances*. The latter involve merely an adjustment of minimum frontage, setback, lot size, or height requirements to accommodate odd-shaped lots or buildings.

Either type of variance may be denied by a board of appeals or invalidated by a court if the alleged hardship is viewed as self-created. Property owners who recently bought a parcel and seek a variance from the zoning in force at the time of purchase is viewed as "creating" their own hardship. Similarly, if they sell off part of a parcel, they cannot later claim that the remainder deserves a minimum-area variance.

Special permits entered zoning practice after World War II to provide municipalities with greater discretion in dealing with development proposals than amendments or variances could afford (Babcock, 1966, p. 7). The special permit (sometimes called a conditional use permit or special exception) is essentially a maybe. Rather than list a particular use as either allowed or prohibited in specific zones (e.g., a branch bank or convenience store in a residential zone), it is listed "SP" in the Table of Use Regulations. This means that the use may be permitted in that zone if approved by the board of appeals. As with a variance, a public hearing is held with notice to neighboring property owners. In the absence of strenuous objection or countervailing planning reasons, the special permit is likely to be granted. Unlike a variance, the applicant need not allege a hardship.

Some communities use special permits for virtually all development, allowing each proposal to be scrutinized in light of existing circumstances. This process resembles the system in Great Britain and many other countries where development approval must be obtained for any new construction. *Special permits,* however, should not be confused with *building permits,* which are required in most American communities for new construction or alterations of existing structures. If a proposed improvement conforms with applicable zoning and building codes, a building permit is routinely issued. If a zoning problem exists, an amendment, variance, or special permit must be obtained to remedy the problem before a building permit can be issued.

Cluster zoning and planned unit developments (PUDs) are products of the public concern about urban sprawl and loss of open space which emerged in the 1960s. Both techniques involve a relaxation of minimum lot size requirements in

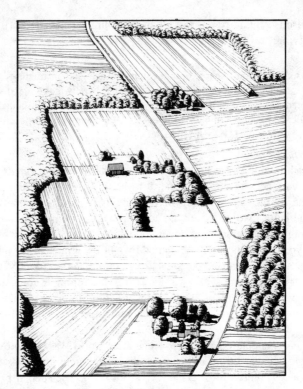

Figure 7.6A A hypothetical rural New England landscape before development. Reproduced by permission of Center for Rural Massachusetts.

exchange for the preservation of a portion of the project site as natural or recreational open space. A PUD also involves the possibility of mixed land uses and house types. In effect, it throws out the existing zoning rules for a major development and substitutes a set of special rules negotiated between the municipality and the developer. The goal is to achieve a higher quality of development with diversity of uses and retention of open land. While a number of excellent PUDs have been constructed, the technique is applicable primarily to large developments. Small-scale developers cannot afford the front-end legal and design costs of a PUD and opt instead to follow the path of least resistance under the prevailing zoning rules (Figures 7–6A, B, and C).

Many states and cities incorporate various types of density bonus incentives into their zoning laws. Massachusetts, for instance, since 1975 has authorized municipalities to issue special permits allowing "increases in the permissible density of population or intensity of a particular use in a proposed development" in cases where the developer undertakes to provide:

The Nuts and Bolts of Euclidean Zoning   191

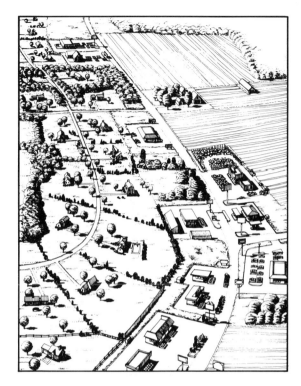

Figure 7.6B The same landscape as developed according to conventional zoning rules. Reproduced by permission of Center for Rural Massachusetts.

. . . open space, housing for persons of low or moderate income, traffic or pedestrian improvements, installation of solar energy systems, protection for solar access, or other amenities (Mass. G.L.A., Ch. 40A, sec. 9).

Transfer of development rights (TDR), another non-Euclidean technique to provide flexibility and greater community control over open space and landmark sites, is discussed in Chapter 9.

Nonconforming Uses

Uses already in existence upon the adoption of zoning or rezoning may continue as *legal nonconforming uses*. (In zoning jargon, they are "grandfathered.") According to Bassett (1936, p. 105): ". . . buildings erected according to law, even if out of place, should be allowed to stand indefinitely, . . . Zoning seeks to stabilize and protect and not to destroy." But such uses, while tolerated, are second-class citizens as compared with activities that conform to zoning. Segrega-

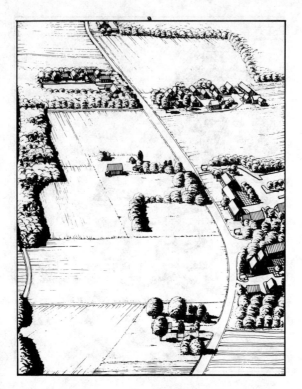

Figure 7.6C The same landscape as developed under cluster zoning to preserve landscape. Reproduced by permission of Center for Rural Massachusetts.

tion of uses was an article of faith in traditional Euclidean zoning, and nonconforming uses have therefore long been regarded as a necessary evil to be eliminated if and when possible.

State laws modeled on the Standard Zoning Enabling Act (SZEA) have placed limits on nonconforming uses. They may not be enlarged physically or converted to a new use as a matter of right. Typically, a variance must be sought to expand or change a nonconforming use, with the burden of proving hardship on the owner as with any variance. It is therefore difficult to change the use of nonconforming commercial space—say, a drug store to a dry cleaner—even though there may be a natural market for the services of the latter in surrounding residential areas. This frustrates entrepreneurship and may cause a marginal use to continue in a building that is increasingly dilapidated for lack of maintenance.

Another limitation is that nonconforming uses that are discontinued may be deemed to be abandoned after a given period of time, usually two years, and may not thereafter be revived. Nonconforming structures that are substantially destroyed by fire, flood, or other calamity may have to be replaced in conformity with zoning use and dimensional regulations.

Efforts to phase out nonconforming uses through *amortization* (loss of legal protection after a specified period of years) have been attempted in several cities (Delafons, 1969, pp. 63–64), but without notable success. The only exception has been the amortization of billboards in certain states and communities.

Clearly, the rezoning of a community should endeavor to avoid placing existing structures into nonconforming status. This is difficult in an older community with great diversity of land use and structural type. The rigidity of Euclidean zoning regarding segregation of uses is incompatible with such heterogeneous but often thriving neighborhoods. A revival of the cumulative approach to zoning would resolve some of these difficulties where some areas could be essentially unrestricted. Alternatively, certain zones may be deliberately zoned for multiple uses, as in cases of older cities that zone their downtown business districts to encourage residential dwellings above streetfront commercial space.

REFERENCES

BABCOCK, R. F. 1966. *The Zoning Game*. Madison: University of Wisconsin Press.

BASSETT, E. M. 1936. *Zoning*. New York: Russell Sage Foundation.

BURNHAM, D. H. AND E. H. BENNETT. 1909. *Plan of Chicago*. Chicago: The Commercial Club.

CIUCCI, G. ET AL. 1979. *The American City: From the Civil War to the New Deal*. Cambridge: The M.I.T. Press.

DAWSON, A. D. 1982. Land-Use Planning and the Law. Hadley, MA: The Author.

DELAFONS, J. 1969. *Land-Use Controls in the United States* (2nd ed.). Cambridge: The M.I.T. Press.

FREUND, E. 1904. *The Police Power: Public Policy and Constitutional Law*. Chicago: Callaghan.

GOODMAN, W. I. 1968. *Principles and Practice of Urban Planning*. Chicago: International City Managers Association.

GUTHEIM, F. 1976. *The Federal City: Plans and Realities*. Washington, D.C.: Smithsonian Institution Press.

HINES, T. S. 1979. *Burnham of Chicago: Architect and Planner*. Chicago: University of Chicago Press.

HOFSTADTER, R. 1955. *The Age of Reform*. New York: Knopf.

HOWE, F. C. 1905. *The City: The Hope of Democracy*. New York: Charles Scribner's.

KANTOR, H. A. 1983. "Benjamin C. Marsh and the Fight Over Population," in *The American Planner: Biographies and Recollections* (D. A. Krueckeberg, ed.). New York: Methuen.

MARSH, B. C. 1909. *An Introduction to City Planning*. New York: Benjamin C. Marsh.

———. 1953. *Lobbyist for the People*. Washington, D.C.: Public Affairs Press.

MUMFORD, L. 1955. *Sticks and Stones: A Study of American Architecture and Civilization*. New York: Dover.

RIIS, J. A. 1890/1972. *How the Other Half Lives*. Williamstown, Mass.: Corner House Publishers.

SCHAFFER, D. 1982. *Garden Cities for America: The Radburn Experience*. Philadelphia: Temple University Press.

SINCLAIR, U. 1906/1985. *The Jungle*. New York: Penguin Books.

STEFFENS, L. 1904/1957. *The Shame of the Cities*. New York: Hill and Wang.

STRONG, J. 1898. *The Twentieth Century City*. New York: Baker and Taylor.

TOLL, S. 1969. *Zoned American*. New York: Grossman Publishers.

U.S. CONGRESS. 1902. *The Improvement of The Park System of the District of Columbia*. Senate Report no. 166. (57th Congress, 1st Session). Washington, D.C.: U.S. Government Printing Office.

_____. 1910. *City Planning*. Senate Doc. no. 422 (61st Congress, 2nd Session). Washington, D.C.: U.S. Government Printing Office.

WILLIAMS, N. 1975. *American Planning Law: Land Use and the Police Power*. Chicago: Callaghan.

WILSON, W. H. 1990. *The City Beautiful Movement*. Baltimore: The Johns Hopkins University Press.

WRIGLEY, R. L., JR. 1983. "The Plan of Chicago," in *Introduction to Planning History in the United States* (D. A. Krueckeberg, ed.). New Brunswick: Rutgers Center for Urban Policy Research.

8

POLICY ISSUES OF ZONING

INTRODUCTION

The nuts and bolts of zoning described in the preceding chapter scarcely suggest the storm clouds of policy debate and controversy that have swirled around zoning since its advent. Zoning touches the property and livelihood of tens of millions of Americans in thousands of communities. It is administered at the grassroots level by tens of thousands of volunteer board members, many of whom have little professional knowledge of zoning but deep roots in their communities. Zoning actions therefore often reflect personal instincts or personal loyalties, rather than abstract planning theory. Also, zoning is defiantly localistic, and despite efforts to broaden its geographic area of concern, it continues to obstruct the provision of regional services and needs.

Zoning has been called on to serve diverse and sometimes conflicting objectives, such as environmental protection and economic development, often failing to serve any of them well. However, zoning is but one tool, and a weak one at that, amid an increasingly complex array of local, state, and federal programs and policies affecting land use, for better or worse.

This chapter examines certain policy issues arising from the practice of zoning over the past seventy years. These include: (1) the "taking" issue; (2) reasonableness; (3) the role of planning; (4) exclusionary zoning; and (5) aesthetics.

THE TAKING ISSUE

The earliest and most enduring issue confronting the exercise of zoning is *the "taking" issue*—how far can zoning reduce the value of private property without compensation to the owner? The issue, it should be noted, is not confined to municipal zoning but applies to any use of the police power by any level of government.

The taking issue arises from the final clause of the Fifth Amendment to the U.S. Constitution (and its counterparts in most state constitutions), which reads: ". . . nor shall private property be taken for public use without just compensation." Where private property is deliberately taken for a public use or purpose, as for streets, parks, or schools, the public authority clearly must pay the private owner "just compensation." This means fair market value as established by a jury. Such an action involves the public power of eminent domain, also referred to as "condemnation." Eminent domain is legal as long as the correct procedures are followed, the public purpose is legitimate, and compensation to the private owner is "just."

But where zoning is concerned, the public neither seeks legal ownership of the property nor pays any compensation to the owner. Instead, the use of the property is restricted in the interest of the public health, safety, and welfare. In the process, zoning encounters a paradox. On the one hand, one of its long-standing purposes, dating back to the Fifth Avenue merchants, is to protect and enhance property values, a somewhat strained but long-accepted interpretation of the "general health, safety, and welfare." On the other hand, zoning necessarily reduces some property values by limiting the range of choice and manner in which the property may be developed and utilized. Should such reduction in the value of some property for the benefit of others be compensable? In other words, is such a reduction in value a taking for public benefit within the scope of the Fifth Amendment?

At the time of zoning's birth, the police power as applied to the abatement of nuisances was enjoying a high degree of judicial favor, as recounted in the preceding chapter. *Welch* v. *Swasey* (1909) and *Hadachek* v. *Sebastian* (1915) lent confidence to those such as planning lawyer Edward M. Bassett who defended the constitutionality of zoning as a noncompensatory exercise of the police power, even in the face of substantial property value reduction.

But this optimism was to be dampened by the 1922 Supreme Court decision in *Pennsylvania Coal Co.* v. *Mahon* (260 U.S. 393). This case did not involve zoning per se, which had not yet come before the high Court. It did involve the police power in the form of a Pennsylvania statute that limited the right of coal producers to undermine developed areas if surface subsidence would result. The plaintiff, which had purchased the mineral rights to all the coal under the defendant's land, claimed that the statute "took" their property right in the coal without compensation. The Supreme Court, in a landmark opinion by Justice Oliver Wendell Holmes, agreed. Holmes's test has been cited by thousands of irate property owners:

> The general rule at least is, that while property may be regulated to a certain extent, if regulation goes too far, it will be recognized as a taking (260 U.S. at 415).

In the view of Bosselman, Callies, and Banta (1973, p. 134), Holmes in this statement "rewrote the Constitution" and interpreted the Fifth Amendment to imply that the difference between regulation and taking is a "difference of *degree not kind*" (emphasis original). Justice Louis Brandies in dissent urged that the police power may properly protect against threats to life and property without payment of compensation, but Holmes' majority opinion was the "law of the land" when zoning arrived in the Supreme Court four years later.

The circumstances involved in *Euclid* v. *Ambler Realty Co.* have prompted one foreign commentator to write: "a severer test for zoning could hardly have been devised. The merits of the case were certainly dubious and the damage to private property values were impressive" (Delafons, 1969, p. 26). The site in question consisted of 68 acres of vacant land bordered by rail lines on the north and a major avenue on the south. It was divided by Euclid's zoning ordinance into two principal zones and a buffer strip between them. The northern portion was zoned for industry and (cumulatively) anything else. The problem for the owner arose from the residential classification of the southern portion of the tract. It was agreed that the value of the latter would be $10,000 per acre for industry but only $3,500 for residential use.

After a decision of a federal district court holding the ordinance to be unconstitutional, the case was appealed directly to the Supreme Court. It was thereby selected from many possible candidates as the test case to determine the constitutionality of land-use zoning in the United States. To bolster the municipal position, Alfred Bettman, a Cincinnati planning lawyer, was retained by the National Conference on City Planning and other pro-zoning organizations to submit a brief on their behalf as *amici curiae* ("friends of the court"). Bettman's brief, one of the seminal documents in the history of zoning, sidesteps the facts of the *Euclid* case in favor of broadly addressing the constitutionality of zoning, including the issue of compensation. (The brief is reprinted in its entirety in Comey, 1946, pp. 157–193).

Bettman argued that reduction of value per se cannot be the test of constitutionality since that "begs the question." Any police power measure, including many already approved by the Supreme Court as in *Hadacheck* v. *Sebastian,* had involved measurable and perhaps severe economic impact to affected property owners. If the purpose is appropriate and necessary, Bettman maintained, loss of value is constitutionally tolerable.

Regarding the necessity of zoning, Bettman compared (but did not equate) it with the familiar governmental function of abating public nuisances. This was an easier task in relation to the height, bulk, and density provisions of zoning, since decisions such as *Welch* v. *Swasey* had already approved building regulations on a districting basis (Bassett, 1936, p. 46). Regulation of use was a more difficult proposition since the separation of homes and commerce could not be viewed as abating a nuisance in the common law sense.

Bettman argued that modern urban development was producing unprecedented problems of congestion and inefficiency. He argued that orderliness of land usage pursuant to a master plan is a proper use of the police power. In effect, he urged in Holmes's terms that zoning did not go "too far" and did not amount to a compensable taking.

Bettmann prevailed. In a 6–3 decision (Holmes voted with the majority) the opinion declared:

Building zone laws are of modern origin. They began in this country about twenty-five years ago. Until recent years, urban life was comparatively simple; but with the great increase and concentration of population, problems have developed, . . . which require . . . additional restrictions in respect of the use and occupation of private lands in urban communities. Regulations, the wisdom, necessity and validity of which as applied to existing conditions, are so apparent that they are now uniformly sustained, under the complex conditions of our day. . . .

And the law of nuisances, . . . may be consulted, not for the purpose of controlling, but for the helpful aid of its analogies in the process of ascertaining the scope of the [police] power. Thus the question whether the power exists to forbid the erection of a building of a particular kind or for a particular use, . . . is to be determined, not by an abstract consideration of the building or other thing considered apart, but by considering it in connection with the circumstances and the locality.

A nuisance may be merely the right thing in the wrong place,—like a pig in the parlor instead of the barnyard (272 U.S., at 370).

The theory of land-use zoning was thus held to be constitutional. But the Supreme Court left open the possibility that individual applications of zoning might be rejected if a property owner proved that a restriction was arbitrary or unreasonable as applied to his or her property. While most cases of individual hardship have been remedied through the variance procedure, landowner challenges to local regulations based on the taking issue have persisted to the present time. The Supreme Court itself has left the resolution of these issues largely to the state courts, providing only the sketchy guidance of Holmes's *Mahon* test as a basis for decision.

Based on a review of decades of case law, Bosselman, Callies, and Banta (1973, p. 141) found that most taking-issue challenges address nontraditional police power restrictions. These included mining restrictions (especially sand and gravel); open space preservation; floodplain, wetland, and coastal areas; and urban growth limitations. In resolving these disputes, courts were found to apply a balancing test incorporating Holmes's rule: The public benefits of a regulation were weighed against the private harm. The more vital the public need for a restriction, the greater the impact on private property values that would be upheld judicially. Thus as public recognition of the importance of a particular type of regulation grew more widespread, so too did the courts' tolerance of more severe impacts on property values as the unavoidable price of achieving such regulation. (See discussion of flood plain and wetland cases in Chapter 9).

In 1987, the U.S. Supreme Court revisited the taking issue in two important decisions. In *Keystone Bituminous Coal Assn. v. DeBenedictis,* 107 S.Ct 1232, it upheld a Pennsylvania state law resembling the one invalidated in the 1922 *Mahon* decision. The *Keystone* opinion distinguished the earlier decision (rather than directly overruling it) in holding the new Pennsylvania Subsidence Act to serve public purposes not mentioned in the earlier version. But the impact of a change in judicial perception and cultural values is admitted:

The Subsidence Act is a prime example that "circumstances may so change in time . . . as to clothe with such a [public] interest what at other times . . . would be a matter of purely private concern" (107 S.Ct. at 1243).

Thus, the substance of the *Mahon* decision was overturned after 65 years, but not its rationale: The Holmes balancing test is alive and well in the 1990s.

Another 1987 Supreme Court opinion cast a different perspective on the taking issue. In a 5–4 split decision in *First English Evangelical Lutheran Church v. County of Los Angeles,* 107 S.Ct. 2378, the conservative majority sustained a theory of *inverse condemnation* whereby a property owner whose land is restricted by zoning later held to be invalid is entitled to monetary damages for the loss of value during the time the restriction was in effect. This ruling poses a threat to local governments whose zoning laws may later be held invalid as to particular land. Of course, the issuance of a variance remains the customary way to address individual hardship and a right to compensation presumably would not commence until after a variance was denied. Furthermore, the Supreme Court did not redefine the balancing test and did not determine whether a taking had actually occurred in *First English*. Thus, the prevailing standards for a taking remain unchanged. In fact, the possibility that a local government may be held liable may cause lower courts to be even more reluctant to determine that a taking has occurred.

REASONABLENESS

The Fourteenth Amendment to the U.S. Constitution adds further cautions to the "taking" provision of the Fifth Amendment:

> . . . nor shall any State deprive any person of life, liberty, or property, without *due process of law,* nor deny to any person within its jurisdiction the *equal protection of the laws* (emphasis added).

These requirements of due process and equal protection each represent major branches of constitutional law beyond the scope of this discussion. As applied to zoning, the due process clause has two connotations: procedural and substantive. The first means that the formal requirements of state and local law concerning procedures (for example, public notice, hearings, voting) must be complied with. Violation of these requirements is a ground for invalidation of the zoning action.

Substantive due process is closely related to the taking issue. Courts generally apply a two-fold test. First, do the regulatory objectives serve to protect the "public health, safety, and welfare?" Second, does the regulation reasonably relate to the achievement of those objectives? For instance, a regulation that limits building height to two stories as a means of alleviating traffic congestion might be held invalid under the second test. A regulation that bans all pets from the city of *X* would probably flunk the first test. Courts, however, are reluctant to invalidate police power measures under substantive due process unless "fundamental constitutional rights" are threatened. They normally apply a presumption that the action of a state or local legislative body is valid unless the plaintiff proves that the action is "arbitrary, capricious, and unreasonably discriminatory" or words to that effect (Hamann, 1986, pp. 9–10).

The equal protection clause poses the issue of equity or fairness in zoning treatment. The Constitution does not forbid all discrimination among property owners: that would nullify zoning altogether. It does require, however, that rules be uniform within zoning districts and that the boundaries between districts be drawn objectively. Equal protection requires that "similarly situated property owners must be treated similarly." The determination of similarity or dissimilarity for zoning purposes must be based on objective criteria so as not to be deemed arbitrary or capricious.

In actual zoning litigation, the various constitutional grounds mentioned in the foregoing tend to become blurred. The plaintiff (party challenging the zoning action) typically alleges that all possible constitutional guarantees have been violated as in the following excerpt from *Vernon Park Realty Inc.* v. *City of Mount Vernon,* 121 N.E.2d 517, 1954:

> The amended complaint alleges that the ordinance . . . [and a later amendment] work an undue hardship as to use, destroy the greater part of its value, are discriminatory as a denial of the equal protection of the law, and amount to a taking of private property without just compensation contrary to due process and, as such, are invalid and void (121 N.E.2d, at 519).

Such a shotgun approach invites a nonanalytical, all-purpose response by the courts. Rather than examining each allegation in detail, the court applies a single litmus test: reasonableness. The court in *Mount Vernon* simplified the plaintiff's allegations as follows:

> While the common council has the unquestioned right to enact zoning laws respecting the use of property . . . it may not be exerted arbitrarily or unreasonably.

"Reasonableness" is thus used as a surrogate for "constitutionality." Although it may beg the question, reasonableness has provided a convenient rubric for courts to resolve zoning challenges. The inquiry takes the following form: Is the ordinance *reasonably* related to a valid purpose of the police power and does it reserve for the owner some *reasonable* form of use (although not necessarily the most profitable one)? In *Mount Vernon,* the city had zoned the plaintiff's property, which adjoined the railroad station, for residence or for a parking lot. The court held that parking lots should be provided by public acquisition, not zoning, and that the owner ". . . has met the burden of proof . . . that the property . . . has no possibilities for [reasonable use as zoned]." The court accordingly held the zoning as applied to the plaintiff's property to be ". . . so unreasonable and arbitrary as to constitute an invasion of property rights, contrary to constitutional due process and, as such, [is] invalid, illegal, and void . . . " (121 N.E.2d, at 521).

Most zoning challenges appealed to higher courts are not as clearly favorable in their facts to the property owner as *Mount Vernon*. Nine times out of ten, courts will hold a challenged zoning ordinance to be reasonable—not necessarily demonstrating "wisdom or sound policy in all respects" as the *Euclid* court noted—but at least arguable: "If the validity of the legislative classification for zoning purposes be fairly debatable, the legislative judgment must be allowed to control" (272 U.S., at 375).

But even though the plaintiff bears the burden of proving that a zoning measure is unreasonable and arbitrary, this proof may be easy to provide unless the municipality can present some valid reasons for its action. The lack of such reasons is strong evidence of arbitrary and capricious use of the zoning power. In short, there must be a *planning justification* for the zoning. The reasonableness (read constitutionality) of a zoning measure is a function of its basis in planning.

THE ROLE OF PLANNING

It is an axiom of planning that zoning is merely one technique among several others to effectuate a community plan, and that zoning is subordinate to planning (Babcock, 1966, p. 120). This doctrine dates back to the pioneers of zoning: Marsh, Nolen, Robinson, and the 1909 City Planning Conference. The argument that zoning is legitimized by planning was articulated by Bettman in his *Euclid* brief:

> Zoning is based upon a thorough and comprehensive study of developments of modern American cities, with full consideration of economic factors of municipal growth, as well as the social factors . . .
>
> . . . the zone plan is one consistent whole, with parts adjusted to each other, carefully worked out on the basis of actual facts and tendencies, including actual economic factors, so as to secure development of all the territory within the city in such a way as to promote the public health, safety, convenience, order, and general welfare (quoted in Comey, 1946, p. 174).

The role of planning was codified in state laws which followed the standard zoning and planning enabling acts (which Bettmann helped to draft) but with a curiously bifurcated result. State *zoning* enabling acts generally include the phrase from the SZEA that zoning should be ". . . in accordance with a comprehensive plan." The question as to what constitutes such a plan and how it should be adopted as a basis for zoning is not usually further mentioned in the zoning acts.

Planning enabling acts were also adopted by many, but not all, states that adopted zoning acts. The planning acts specify the powers and duties of the local planning commission, for example:

> It shall be the function of the [planning] commission to make and adopt a master plan for the physical development of the municipality . . . Such a plan, with the accompanying maps, plates, charts, and descriptive matter, shall show the commission's recommendations for the development of said territory. . . .

In the preparation of such plan the commission shall make careful and comprehensive surveys and studies of present conditions and future growth of the municipality and with due regard to its relations to neighboring territory (Mass. G.L.A. Ch. 41, Sec. 81).

This would appear to supply the mandate to prepare a "master plan" with which local zoning would be in accord that Haar (1955a) argues is constitutionally necessary. But the two acts are not tied together. The "comprehensive plan" mentioned in the zoning act does not necessarily refer to the "master plan" of the planning act. Furthermore, each act is permissive, not mandatory. A community may engage in zoning without a master plan, or may theoretically prepare a master plan without zoning. The two functions were certainly connected in the minds of Bettman and Bassett, but practice has widely diverged from theory.

The relationship of planning and zoning from 1905 to 1935 is depicted graphically in Figure 8–1. A modest number of "comprehensive plans" such as those prepared for Chicago and San Francisco by Daniel Burnham appeared earliest. Ongoing city planning commissions first emerged about 1912 and increased rapidly until the Depression. Zoning ordinances followed a parallel rapid rate of growth after 1920. But comprehensive plans, with which zoning is supposedly "in accord," lagged far behind during the 1920s. Whatever the planning commissions were doing, it apparently was not the preparation of comprehensive plans.

After nearly four decades of national experience with zoning, Haar (1955a, p. 1157) reported:

> For the most part, however, zoning has preceded planning in the communities which now provide for the latter activity, and indeed, nearly one-half the cities with comprehensive zoning ordinances have not adopted master plans at all.

Clearly, judicial acceptance of zoning laws was not dependent on the existence of a master plan. Haar found that the statutory (and constitutional) requirement that zoning be "in accordance with a comprehensive plan" often is construed to mean that zoning be ". . . logically related to something broader than and beyond itself" (Haar, 1955a, p. 1167). The meaning of "comprehensive plan" therefore differs from one court and jurisdiction to another:

> It may be the basic zoning ordinance itself, or the generalized "policy" of the local legislative or planning authorities in respect to their city's development—or it may be nothing more than a general feeling of fairness and rationality. Its identity is not fixed with any precision, and no one can point with confidence to any particular set of factors, or any document, and say that there is the general plan to which the zoning enabling act demands fidelity.

> The analysis of the "comprehensive plan" requirement in terms of itself is a common judicial phenomenon. The reasoning seems to be that a comprehensive ordinance, one which blankets the entire area and is internally consistent, is automatically "in accordance with a comprehensive plan." The plan is the ordinance and the ordinance the plan, . . . (Haar, 1955a, p. 1167).

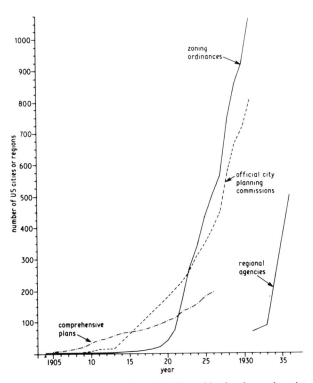

Figure 8.1 Cumulative number of municipalities with planning and zoning, and of regional planning agencies. *Source:* Kreuckeberg, 1983, Fig. 1.6. Reproduced by permission of Methuen.

In a parallel article, Haar (1955b) examined the purposes and need for refinement of the master plan. He first identified a dichotomy in the purposes of a master plan as used by planners and lawyers respectively: "one part is largely didactic and deals with the virtues of planning; another—and quite distinct—portion moves away from speculation and is concerned with directing the application of human energies in land development" (p. 355). Curiously, Haar, a law professor, urged a less detailed and legalistic concept of the master plan in favor of a more rhetorical and policy-oriented product. Critical to developing such a plan is involvement of the general public:

> Planning is a leading to understanding and the possibility of community acceptance of the master plan, not, as it has often tended to be, a holier-than-thou attitude with respect to *the* plan. The idea of experts who prepare *the* plan is a static one. Those who are affected by the plan must participate in its making and in carrying it out. And as it changes, as it must, to cope with new conditions and to introduce new concepts, the different interest groups must be won over, or reconciled (p. 373).

Haar's analysis marked a threshold in U.S. planning history. Until the late 1950s, the master plan was a particularistic document, heavily oriented toward physical land-use patterns and public infrastructure. Although it did not necessarily serve as the "comprehensive plan" for zoning purposes, it did guide other local land-related functions such as the provision of streets, parks, schools, water and sewer services, and other infrastructure. Beginning in the early 1960s, the master plan—and indeed zoning itself—was viewed with increasing skepticism and indignation by reputable critics, such as William H. Whyte, Richard Babcock, John Reps, and Norman Williams, Jr. Much of the impact of this criticism affected the form and process of planning which became more open and issue-oriented as Haar had urged. It is, however, difficult to discern any immediate improvement in the use of zoning in response to these changes in planning. Except where jolted in new directions by court orders or new state and federal legislation (see Chapters 11–13), the piecemeal practice of zoning in some 10,000 communities has chugged along as before. As Delafons (1969, p. 24) has written:

> The lack of any substantial relationship between the legal machinery and a clear concept of city planning is the firmest impression left by the origin and later course of land-use control in America.

METROPOLITAN NEEDS AND EXCLUSIONARY ZONING

The administration of zoning by local governments in the United States for seven decades has been overwhelmingly parochial. This has taken at least three forms, each of which has been widely criticized in professional literature:

- *Exclusionary zoning*—the use of zoning to deter construction of homes or siting of mobile homes for moderate-income families or members of racial or other minorities.
- *Fiscal zoning*—the use of zoning to minimize local property taxes by encouraging revenue generating activities such as shopping centers and industrial parks while discouraging revenue demanding uses such as lower-cost homes for families with children (regardless of race).
- *NIMBY*—the use of zoning and other legal means to resist the location of unwanted uses, facilities, or activities within the municipality (e.g., regional incinerators or toxic waste disposal sites, prisons, mental health facilities, oil refineries, public housing, etc.). The acronym stands for "not in my backyard."

These forms of parochialism in local zoning arise from somewhat different motivations such as racial or ethnic snobbery, economic self-interest, environmental concerns, neighborhood aesthetics, or a sense of security and peace of mind. These motives are not mutually exclusive and they frequently overlap. More socially acceptable concerns such as environment and aesthetics sometimes serve as the "sheep's clothing" for the "wolf" of racial or economic motives (Bosselman, 1973).

Localism in zoning is not necessarily a bad thing. The local government is most familiar with the circumstances of particular land-use decisions and has a valid interest in protecting the public health, safety, and welfare of its inhabitants. It is arguably the best-qualified level of government to plan and regulate the use of private land within its borders. But in the hands of the multifarious municipalities, many of them minuscule in territory and population, localism in zoning can and does obstruct broader regional or national objectives. The essence of the issue is *which* public is implied in constitutional protection of the "general welfare"—the strictly local public, or the larger publics of the region, state, nation, or globe? (see Clark, 1984; 1985).

Richard F. Babcock (1966) was one of the first critics to challenge the proposition, argued so forcefully by Haar (1955a and 1955b), that zoning is necessarily constitutional if it is in accordance with a local comprehensive or master plan. Babcock asserted that such accordance is a necessary but not sufficient condition of constitutionality:

> The local plan . . . is imperative as a device to bring some consistency and impartiality in local administrative decisions among residents of the same municipality.
>
> It is an error, however, to dignify the municipal plan with more authority than this limited function. But to measure the validity of zoning by the degree to which it is consistent with a municipal plan does just that. The municipal plan may be just as arbitrary and irresponsible as the municipal zoning ordinance if that plan reflects no more than the municipality's arbitrary desires. If the plan ignores the responsibility of the municipality to its municipal neighbors and to landowners and taxpayers who happen to reside outside the municipal boundaries, and if that irresponsibility results in added burdens to other public agencies and to outsiders . . . then a zoning ordinance bottomed on such a plan should be as vulnerable to attack as a zoning ordinance based on no municipal plan (Babcock, 1966, p. 123).

The "Metropolitan Factor"

The possibility of conflict between municipal and larger regional objectives in zoning policy was recognized as early as the *Euclid* decision. By the 1920s, suburbanization and metropolitan expansion had been in progress for decades since the advent of horse-drawn streetcars and electric railways. Euclid, Ohio itself was a suburb of Cleveland. Bettman's brief somewhat disingenuously extolled zoning as a tool for bringing order to the "modern American city" while ignoring that a "city" such as Euclid comprised only a fragment of its metropolitan region. The Court's opinion however included a portentous dictum on this issue:

> It is not meant . . . to exclude the possibility of cases where the general public interest would so far outweigh the interest of the municipality that the municipality would not be allowed to stand in the way (272 U.S., at 390).

Bettman himself addressed the "metropolitan issue" the year after *Euclid:*

> The last word has not as yet been uttered on this question of the relationship of the metropolitan factor to the validity of specific zone plans. . . . Insofar as the fact of the location of a municipality within a metropolitan urban area has a bearing upon these factors of development trends, land values, and appropriateness of use, such fact has a relation to the social validity and, consequently, in the last analysis, to the constitutional validity of the zone plan (quoted in Comey, 1946, p. 55).

The 1928 Department of Commerce Standard City Planning Enabling Act which Bettman helped to write included in the coverage of a municipal master plan: ". . . any areas outside of its boundaries which, in the planning commission's judgment, bear relation to the planning of such municipality." Also, the purpose of such plan was stated to be "the harmonious development of the municipality *and its environs* (emphasis added) (quoted in Beuscher et al., 1976, pp. 272–3).

Many state planning acts today permit municipalities to exercise limited planning and regulatory control over unincorporated land within a specified distance (e.g., two miles outside their corporate boundaries). But this merely reflects the expectation that such land will eventually be annexed to that municipality. It is scarcely a mandate to consider external implications of local zoning policies, particularly to other municipalities and their inhabitants.

The *Cresskill* Decisions

Surprisingly few cases addressed the regional context of zoning before the 1960s. Two early landmark decisions involved the obscure New Jersey Borough of Cresskill. In *Duffcon Concrete Products, Inc.* v. *Borough of Cresskill* 64 A.2d 347, 1949, which upheld Cresskill's ban on industry, the court stated:

> What may be the most appropriate use of any particular property depends not only on all the conditions, physical, economic and social, prevailing within the municipality and its needs, present and reasonably prospective, but also on the nature of the entire region in which the municipality is located and the use to which the land in that region has been or may be put most advantageously (64 A.2d, at 349–50).

The court noted the availability of industrial land nearby but outside Cresskill's jurisdiction.

The principle was restated in *Borough of Cresskill* v. *Borough of Dumont* 104 A.2d 441, 1954 in which three neighboring boroughs challenged Dumont's rezoning for a small shopping center a single parcel that coincidentally abutted the boundaries of all three plaintiff jurisdictions. In response to Dumont's assertion that "the responsibility of a municipality for zoning halts at the municipal boundary lines," the court stated:

Such a view might prevail where there are large undeveloped areas at the borders of two contiguous towns, but it cannot be tolerated where, as here, the area is built up and one cannot tell when one is passing from one borough to another. Knickerbocker Road and Massachusetts Avenue are not Chinese Walls separating Dumont from the adjoining boroughs. At the very least Dumont owes a duty to hear any residents and taxpayers of adjoining municipalities who may be adversely affected by proposed zoning changes . . . To do less would be to make a fetish out of invisible municipal boundary lines and a mockery of the principles of zoning (104 A.2d, at 445–6).

These two New Jersey decisions have long been praised for their recognition of the regional context of zoning. Haar (1963, p. 204), for instance, referred to the quoted statement from *Duffcon* as "probably the best judicial statement on regional planning." But both cases are exclusionary, rather than inclusionary. One holds that a municipality may, and the other that it must, *exclude* land uses that conflict with the interests of adjoining areas. That was step one toward recognition of a metropolitan dimension to zoning. Step two, beginning in New Jersey and Pennsylvania in the 1960s, would assert that a municipality must also zone to *include* those land uses necessary to meet regional deficiencies.

Opening the Suburbs

The exclusion of apartments from single-family neighborhoods was approved by the U.S. Supreme Court in a widely cited dictum in *Euclid:*

> . . . the development of detached house sections is greatly retarded by the coming of apartment houses . . . very often the apartment house is a mere parasite, constructed in order to take advantage of the open spaces and attractive surroundings. Moreover, the coming of one apartment house is followed by others, . . . until, finally, the residential character of the neighborhood and its desirability as a place of detached residences are utterly destroyed. Under these circumstances, apartment houses, which in a different environment would be . . . highly desirable, come very near to being nuisances (272 U.S., at 394–5).

If "municipality" is substituted for "neighborhood," this would appear to legitimize zoning that excludes apartments and other multi-family housing from entire communities. This issue probably did not occur to the Court, nor did the plaintiff in *Euclid* seek to build apartments. The Court was merely airing its views on the bulky, nonsetback apartment buildings that were invading single-family neighborhoods in the 1920s (Williams, 1975, Vol. 6, Plates 6 and 7). Since the Court's statement was "dictum" and not central to its holding in the case, the issue was left open for other courts to consider.

Municipal exclusion of apartments was to remain virtually unchallenged until the Pennsylvania Supreme Court addressed the issue in 1970. In *Appeal of Girsh* 263 A.2d 395, the plaintiff, who sought to construct "two nine-story luxury apartments, each containing 280 units," challenged the local zoning ordinance, which designated no land within the township for apartments. In upholding the plaintiff, the court quoted extensively from its own prior decision which invalidated a four-acre minimum lot requirement in *National Land and Investment Co. v. Easttown Twp. Board of Adjustment* 215 A.2d 597, 1965:

> The question posed is whether or not the township can stand in the way of the natural forces which send our growing population into hitherto undeveloped areas in search of a comfortable place to live. We have concluded not. A zoning ordinance whose primary purpose is to prevent the entrance of newcomers in order to avoid future burdens, economic and otherwise, upon the administration of public services and facilities cannot be held valid. . . .
>
> Zoning is a tool in the hands of governmental bodies which enables them to more effectively meet the demands of evolving and growing communities. It must not and cannot be used by those officials as an instrument by which they may shirk their responsibilities. Zoning is a means by which a governmental body can plan for the future. . . . Zoning provisions may not be used . . . to avoid the increased responsibilities and economic burdens which time and natural growth invariably bring (263 A.2d, at 397).

Girsh, National Land, and another 1970 Pennsylvania decision, *Appeal of Kit-Mar Builders* 269 A.2d 765 (invalidating a 2–3 acre minimum lot size) reflected a recognition by the Pennsylvania court that the geographical functions and morphology of suburbs were changing. No longer simply bedrooms for central city executives, suburbs were increasingly attracting new jobs, thereby creating a demand for a wider range of housing opportunities. According to these cases, suburbs that welcome new commercial investment may not use zoning to avoid the burden of accommodating new residents and building types. The concept of a "fair share" of regional housing needs was thereby introduced into Pennsylvania zoning law. A portentous footnote to the *Girsh* opinion stated:

> . . . as long as we allow zoning to be done community by community, it is intolerable to allow one municipality (or many municipalities) to close its doors at the expense of surrounding communities and the central city (263 A.2d, at 399).

Girsh fired a warning shot across the bow of exclusionary-minded communities in Pennsylvania and, while not directly applicable to other states, it bolstered similar challenges elsewhere through its constitutional rationale. But *Girsh* was vague as to what a municipality must do to avoid exclusionary challenges and it did not involve lower-cost housing. Indeed the luxury apartments proposed by Girsh were perhaps more akin to the "parasitic" apartments invading single-family districts of the *Euclid* era than to subsidized housing of the late 1960s. Also, the Pennsylvania cases including *Girsh* did not specify which geographic types of communities were "denying the future"—developing suburbs, central cities with remaining vacant land, rural townships? How were each of these to be judged as to the adequacy of their zoning?

The onerous task of applying a *Girsh*-type rationale to the specific circumstances of regional housing markets was assumed by the New Jersey Supreme Court in its landmark 1975 unanimous decision in *Southern Burlington County NAACP* v. *Township of Mount Laurel* 336 A.2d 713 (*"Mount Laurel I"*). As early as 1962, Justice Hall, who wrote this opinion, had signaled the judicial revolution to come in a famous dissent in *Vickers* v. *Township Committee of Gloucester Township* 181 A.2d 129. In objection to the majority's upholding of a total ban on mobile homes by a large rural township, Hall had declared:

> The import of the holding gives almost boundless freedom to developing municipalities to erect exclusionary walls on their boundaries, according to local whim or selfish desire, and to use the zoning power for aims beyond its legitimate purposes (181 A.2d, at 140).

The conversion of his viewpoint from dissent in Vickers to majority opinion in *Mount Laurel I* was facilitated by research findings on exclusionary zoning in New Jersey and elsewhere during the 1960s and early 1970s. A fundamental contribution was a Rutgers study (Williams and Norman, 1974) that documented the practice of exclusionary zoning in four northeastern New Jersey counties. Of 474,000 acres of vacant buildable land in those counties, 99.5 percent was zoned for single-family use and no land was available for mobile homes. Minimum lots of more than 40,000 square feet (about one acre) were required for 77 percent of suitable land. Only 5.1 percent of such land was zoned for less than 20,000 square feet (about a half acre). Only 2,262 acres (0.5 percent) were zoned for multi-family dwellings.

Mount Laurel, New Jersey is a flat, sprawling, 22-square-mile township of mixed developed and agricultural land uses within commuting distance of Camden and Philadelphia. Between 1960 and 1970, its population more than doubled to 11,221. Most of the vacant land remaining at the time of the lawsuit was zoned for industry. In the court's opinion:

> The record thoroughly substantiates the findings of the trial court that over the years Mount Laurel "has acted affirmatively to control development and to attract a selective type of growth" . . . and that "through its zoning ordinances has exhibited economic discrimination in that the poor have been deprived of adequate housing, and has used federal, state, county, and local finances and resources solely for the betterment of middle- and upper-income persons."

> There cannot be the slightest doubt that the reason for this course of conduct has been to keep down local taxes on *property* . . . and that the policy was carried out without regard for non-fiscal considerations with respect to *people,* either within or without its boundaries (336 A.2d, at 723; emphasis is Justice Hall's).

The opinion explicitly raises for perhaps the first time the constitutional issue as to "whose general welfare must be served or not violated in the field of land-use regulation" (336 A.2d, at 726). The court answered its own question by declaring that the constitutionality of zoning requires that in the case of "developing municipalities" (that is, growing suburban jurisdictions with available vacant land):

> . . . every such municipality must, by its land-use regulations, presumptively make realistically possible an appropriate variety and choice of housing . . . . at least to the extent of the municipality's fair share of the *present and prospective regional need* therefor (336 A.2d, at 724; emphasis added).

Thus it was no longer constitutional in New Jersey for communities in the "developing" category to use zoning to serve only their own parochial objectives. They were now required to accommodate a "fair share" of the regional demand for lower-cost housing. The court ordered Mount Laurel and other developing municipalities to revise their zoning accordingly.

The constitutional reverberations of *Mount Laurel* thundered across the land. Although a state court decision, it was widely regarded as a national precedent. An appeal to the U.S. Supreme Court was dismissed for "want of jurisdiction" (98 Sup. Ct. 18) since the Hall opinion was deliberately based on state law, not federal constitutional grounds (although both involve the same principles): *Mount Laurel* thus avoided possible reversal by the U.S. Supreme Court.

The reaction in New Jersey was a deluge of lawsuits against other municipalities based on *Mount Laurel* grounds as well as political hostility to the court's intervention in local affairs (Babcock and Siemon, 1985, ch. 11). The Hall opinion had not provided detailed guidance as to the meaning of key concepts, for example, "region," "developing municipality," and "fair share." Nor was it clear whether a municipality must do more than merely rezone land for lower-cost housing and wait to see if a developer comes along. In a lengthy 1977 opinion in *Oakwood at Madison v. Twp. of Madison* 371 A.2d 1192, the New Jersey court advised trial courts to examine the substance of challenged zoning ordinances and to look for bona fide efforts to meet *Mount Laurel* obligations. It declined to specify a particular numerical approach:

> We do not regard it as mandatory for developing municipalities whose ordinances are challenged as exclusionary to devise specific formulae for estimating their precise fair share of the lower income housing needs of a specifically demarcated region . . . . firstly, numerical housing goals are not realistically translatable into specific substantive changes in a zoning ordinance . . . . Secondly, the breadth of approach by the experts to the factors of the appropriate region and to the criteria for allocation of regional housing goals to municipal 'sub-regions' is so great and the pertinent economic and sociological [and geographical, ed.] considerations so diverse as to preclude judicial dictation or acceptance of any one solution as authoritative (371 A.2d, at 1200).

The court, however, could not long avoid the task of bringing order to the legal and planning chaos resulting from *Mount Laurel I*. In 1983, it responded in a

270-page unanimous decision in *Southern Burlington County NAACP* v. *Township of Mount Laurel* 456 A.2d 390 (*"Mount Laurel II"*). It began by stating that the court was ". . . more firmly committed to the original *Mount Laurel* doctrine than ever" but recognized a "need to put some steel into that doctrine." In a level of detail more characteristic of a legislature or the federal courts, the New Jersey court articulated a series of policies and standards for the resolution of the myriad Mount Laurel cases then clogging the state's lower courts. Among these were the following:

1. Every municipality must provide lower cost housing opportunities for its resident poor;
2. The concept of "developing municipality" was replaced by "growth areas" designated in the State Development Guide Plan (SDGP);
3. Municipalities must demonstrate that they are providing specific numbers of lower-cost housing units to meet their fair share of immediate and prospective regional needs. "Numberless" determinations based on provision for *some* lower-cost units will be insufficient;
4. A special panel of judges was to be designated to hear *Mount Laurel* cases;
5. Municipalities must do more than merely rezone land for lower-cost housing. Affirmative action may be required such as subsidies, tax incentives, density bonuses, and mandatory set-asides of lower cost units in new developments.

Two years later, the New Jersey Legislature enacted a state Fair Housing Act (*N.J. Laws* 1985, ch. 222) which codified the *Mount Laurel II* approach with some modification. The Act established a Council on Affordable Housing which is empowered to determine housing regions and calculate regional housing needs and municipal fair-share allocations. The Act also provides a mediation and review process to resolve *Mount Laurel* litigation, a procedure for "substantive certification" of municipal zoning ordinances, authority for "regional contribution agreements" among municipalities, amendment of the state zoning law to require a housing element, and a program of financial assistance to help municipalities meet their fair-share allocations (Rose, 1987, pp. 448–9).

In another mammoth opinion, *Hills Development Co.* v. *Somerset County* 510 A.2d 621, 1986 (*"Mount Laurel III"*), the New Jersey Supreme Court held the Fair Housing Act to be constitutional despite objections that it diluted the impact of the court's earlier decision. Rose (1987) welcomed the assumption of responsibility for housing policy by the state legislature in place of the courts, which he considers ill-suited and inappropriate forums to resolve technical housing issues. On the other hand, he believes that the legislature was too "servile" to the "fair-share" concept established by the court which

> . . . has converted a generally acceptable principle of justice and equity to a commandment that requires the creation of a government megastructure and an army of

government officials to do battle with the partisans of real estate development in the statistical warfare provoked by that concept (Rose, 1987, p. 473).

Overall, Rose appraises the Act as follows:

> Although it is significantly less than a perfect program to provide housing for all the low- and moderate-income persons in the state, the law nevertheless establishes several principles and creates a number of programs and procedures that put the state of New Jersey in the forefront of creative housing policy in the United States (Rose, 1987, p. 458).

Babcock and Siemon (1985, p. 233), in contrast to Rose, praise the New Jersey court for its activism in *Mount Laurel:*

> We believe New Jersey should be applauded for this daring step. Without such experiments our federal system would lose much of its resistance and its capacity to survive. It is hard to imagine Congress or the U.S. Supreme Court daring to venture into such socially and politically treacherous waters.

By late 1988, the New Jersey experiment was making slow progress according to *The New York Times* (Nov. 13, 1988, Sec. 4, p. 5):

> . . . Progress toward the goals of the Mount Laurel decisions which required the states' 567 communities to provide zoning for low and moderate income housing, has been slow and hampered by lawsuits, experts say. Only about 30 percent of the cities and towns have submitted affordable housing plans . . .

The last word on the New Jersey approach to the problem of exclusionary zoning has certainly not been uttered. It will require many years and certainly more litigation before the results of this judicial/legislative revolution may be fairly evaluated. It is clear, however, that the *Mount Laurel* experience is consistent with the model set forth in Chapter 2, that legal innovation in land-use control arises from the changing perception of those in authority that the existing legal rules are yielding undesirable consequences in the real world. *Mount Laurel* demonstrates the power of empirical research of a strongly geographical flavor to influence public decision making. The case also exemplifies the role of states as laboratories for legal innovation that may subsequently diffuse to other jurisdictions.

AESTHETICS AND HISTORIC PRESERVATION

Whether zoning may validly seek to protect or enhance aesthetic considerations in land use has long posed a dilemma in the development of zoning law. The concept of aesthetics as "beauty" is virtually impossible to define in terms of legislative criteria sufficient to provide objective guidance to zoning officials. In the words of one court:

[A] primary objective to aesthetic zoning is founded upon its subjective nature, for what may be attractive to one many may be an abomination to another. . . . Therefore many courts have long been unwilling to act as super art critics by ruling on the reasonableness of ordinances which are essentially based on subjective aesthetic considerations, and they have held all such ordinances invalid (*Naegele Outdoor Advertising Co. v. Village of Minnetonka*, 162 N.W.2d 206, at 212, 1962).

The early period of zoning was marked by a paradox concerning aesthetics. On the one hand, the basic terms of Euclidean zoning—segregation of uses, minimum lot size, setbacks and so forth—are inherently aesthetic in nature. They embody the aesthetic of the turn-of-the-century garden suburb: single-family detached homes with spacious, landscaped front yards set back evenly from the street with nonresidential activities banished from the area.

Yet the proponents of Euclidean zoning steadfastly denied that zoning was intended to serve aesthetic purposes. In his *Euclid* brief, Bettman refuted the plaintiff's contention that zoning was motivated by "aesthetic considerations and the promotion of beauty." Bettman urged that the Court view zoning as promoting the "good order of cities" which is ". . . quite different from the artistic or the beautiful":

> The essential object of promoting what might be called orderliness in the layout of cities is not the satisfaction of taste or aesthetic desires, but rather the promotion of those beneficial effects upon health and morals which come from living in orderly and decent surroundings (quoted in Comey, 1946, p. 172).

The 1926 Standard Zoning Enabling Act made no reference to aesthetics or beauty in its section on purposes. The justification for zoning has subsequently been expressed in terms of "public health, safety and welfare" as reflected in reduction of congestion, orderly location of land uses and avoidance of fire, flood and other public perils.

Early decisions involving regulations of billboards and other signs struggled to find a nonaesthetic basis for such laws. In 1919, the Illinois Supreme Court upheld such an ordinance, citing evidence that:

> . . . nuisances were permitted to exist in the rear of surface billboards, and physicians testified that deposits found behind billboards breed disease germs which may be carried and scattered in the dust by the wind and by flies and other insects. It was shown that dissolute and immoral practices were carried on under the cover and shield furnished by these billboards (*Thomas Cusack Co. v. City of Chicago* 108 N.E. 340).

This decision may be contrasted in rationale with a 1975 opinion of the Massachusetts Supreme Judicial Court which upheld a total ban on off-premises billboards in the Town of Brookline, stating:

> We live in a changing world where the law must respond to the demands of a modern society. [*Euclid* quote omitted]. What was deemed unreasonable in the past may now

be reasonable due to changing community values. Among these changes is the growing notion that towns and cities can and should be aesthetically pleasing; that a visually satisfying environment tends to contribute to the well-being of its inhabitants (*John Donnelly and Sons, Inc.* v. *Outdoor Advertising Board* 339 N.E. 2d 709, at 717).

Although *Donnelly* applies only to Massachusetts, the acceptance of aesthetics as a purpose of zoning is widely shared in other jurisdictions, at least with respect to billboards. According to Williams (1975, Sec. 11.02): "In no other area of planning law [than aesthetics] has the change in judicial attitudes been so complete." The evolution of this change is frequently divided into three stages: (1) early period—aesthetics impermissible as a basis for zoning; (2) middle period—aesthetics allowable if other valid grounds are present; and (3) modern period—aesthetics acceptable per se without the need for additional grounds.

A major turning point in the recognition of aesthetics in planning law was the 1954 U.S. Supreme Court decision in *Berman* v. *Parker* (75 S.Ct. 98). The opinion by Justice William O. Douglas, an ardent conservationist, declared:

> The concept of the public welfare is broad and inclusive . . . The values it represents are spiritual as well as physical, aesthetic as well as monetary. It is within the power of the legislature to determine that the community should be beautiful as well as healthy, spacious as well as clean, well-balanced as well as carefully patrolled . . . If those who govern the District of Columbia decide that the Nation's Capital should be beautiful as well as sanitary, there is nothing in the Fifth Amendment that stands in the way (75 S.Ct., at 102–3).

Strictly speaking, Douglas's sweeping statement did not apply to zoning. *Berman* v. *Parker* involved a challenge to the constitutionality of the federal urban renewal program in the District of Columbia. It involved the use of the eminent domain power, not the police power or zoning. The plaintiff's store was being condemned and razed by the District as part of a widespread program to eliminate "blighted areas" to make way for modern and expensive new homes and businesses. (The project was located in the Southwest quadrant of Washington where the Department of Housing and Urban Development now sits on land cleared by urban renewal). Where government pays for land, much broader latitude exists as to public purpose than when the land is restricted under the police power. Thus, Douglas's statement really applies to the eminent domain function of government, not zoning.

Nevertheless, Douglas's statement has been widely quoted in zoning cases involving aesthetics, including the 1975 *Donnelly* case cited above. Douglas in fact quoted the statement himself in a zoning case which upheld community limitations on the occupancy of dwelling units by more than two unrelated adults (*Village of Belle Terre* v. *Borass*, 94 S.Ct. 1536, 1974).

But the importance of the Douglas dictum should not be overstated. A zoning measure that forbids building on scenic land in the interest of maintaining a "beautiful community" would almost certainly be held invalid. The taking clause of the Fifth Amendment certainly "stands in the way" not withstanding Douglas's sweeping disclaimer.

In practice, aesthetic zoning has not generally attempted to legislate "beauty" or attractiveness in the general urbanizing landscape. As stated earlier, such values may be implicit to Euclidean zoning but seldom are stated explicitly due to the difficulty of drafting suitable legal criteria. Instead, aesthetic zoning has largely addressed two special cases of land-use regulation, namely advertising signs and historic landmark preservation. Each of these involve marginal restraints on property that is otherwise economically productive. Even in these special areas of regulation, aesthetic zoning measures may be held invalid on other grounds.

Billboard regulations are authorized in all 50 states and are widely in effect. Restraints and even total bans of billboards have been upheld in many states. There appears to be widespread consensus that billboards and other types of outdoor advertising are per se unattractive, distracting, and exploitive of the captive audience represented by motorists on public thoroughfares. As a class of aesthetic nuisance, billboards have been categorically disfavored though by no means eliminated in many states.

But several issues arise when local governments begin to regulate billboards and other signs. Considerations include the size and characteristics of the sign, whether it is on-site or off-site (i.e., advertises the business on whose premises the sign is located), lighted or unlighted, freestanding or supported by a building, commercial or noncommercial. Signs may occasionally assume the status of landmarks, such as the "Citgo" sign that looms over the left field wall at Boston's Fenway Park. Other signs may be obstructions to traffic, pornographic, excessively distracting, or otherwise nuisance-like. Thus, communities do not usually prohibit all signs. But discrimination among different types of signs may generate constitutional difficulty.

In 1981, the U.S. Supreme Court addressed the problem of potential conflict between sign regulation and the First Amendment protection of freedom of speech. The court endorsed the general principle that signs may be regulated in the public interest, but it invalidated an ordinance that banned noncommercial advertising throughout the city while allowing on-site commercial signs, viewing this as impermissible discrimination against noncommercial (for example, political) signs (*Metromedia Inc.* v. *City of San Diego* 101 S.Ct. 2882).

Historical preservation districts are another instance of aesthetic zoning that has received widespread but often qualified judicial support. The concept of the historical district originated with the designation of the Vieux Carre District in New Orleans through an amendment to the Louisiana Constitution in 1921. Other early districts were established in Charleston, South Carolina, San Antonio, Texas, and Nantucket, Massachusetts. The distinctive architecture and renowned atmosphere of these prototype districts facilitated judicial approval of regulations to control construction, demolition, and renovation. For instance, in 1941, the Louisiana Supreme Court upheld the Vieux Carre restrictions in *City of New Orleans* v. *Impastato* 3 S.2d 559 and *City of New Orleans* v. *Pergament* 5 S.2d 129. The latter opinion stated:

> There is nothing arbitrary or discriminating in forbidding the proprietor of a modern building, as well as the proprietor of one of the ancient landmarks, in the Vieux Carre

to display an unusually large sign upon his premises. The purpose of the ordinance is not only to preserve the old buildings themselves, but to preserve the antiquity of the whole French and Spanish quarter, tout ensemble, so to speak, by defending this relic against iconoclasm or vandalism. Preventing or prohibiting eyesores in such a locality is within the police power . . .

Historic district regulations typically go beyond traditional zoning controls in requiring approval from a local review board for any alteration or demolition. Even an affirmative duty to maintain the exterior appearance of the premises may be imposed (*Maher* v. *City of New Orleans* 37 F. Supp. 653).

But given the stringency of these types of regulations, the applicability to particular property owners may be harsh. Many historical districts are less homogeneous or distinctive than the Vieux Carre. Issues arise concerning the status of nonhistoric structures within a designated district, accessory structures (e.g., solar panels, satellite dishes), mixture of architectural styles, approval of new construction within a district, the need for owner consent, 11th-hour designations to obstruct a redevelopment plan, and the status of religious property in a district (Duerksen, 1985, sec. 1503). As with other zoning, objective standards must be provided by the local or state legislative body to guide the local review body in its permitting process.

The validity of an historic landmark designation was upheld by the U.S. Supreme Court in *Penn Central Transportation Co.* v. *City of New York* 98 S.Ct. 2646, 1978). The Court broadly endorsed the principle of historic preservation:

> States and cities may enact land-use restrictions or controls to enhance the quality of life by preserving the character and desirable aesthetic features of a city (98 S.Ct., at 266).

The opinion then upheld the designation of Grand Central Station as an historical landmark and the prohibition by the city of either demolition or the construction of a modern office tower in the air rights column above the station. The Court noted that the station provided some economic return to the owner, albeit not optimal, and that the ordinance permitted the owner to transfer its unusable development rights to other parcels of real estate in the vicinity. (See Chapter 9). This TDR feature and the special nature of Grand Central Station as a quasi-public facility perhaps limit the applicability of the *Penn Central* decision to more mundane historic structures. Public acquisition may still be required in the case of structures that have little economic use and lack a market for transferable development rights.

Nationally, more than 1,000 local historic preservation laws were in effect as of the mid-1980s (Duerksen, 1985). This branch of aesthetics zoning, like billboard control, is thriving.

Architectural controls in a nonhistorical context pose a more difficult problem. How can a community regulate the appearance of new structures in relation to an existing potpourri of styles? Many communities have established architectural review boards to certify that a new structure will be compatible with the surrounding area. The problem is to define "compatibility" or "suitability" of a design in the absence of a clear-cut historical or otherwise homogeneous style. Response to this problem is approached in two ways, namely, by attempting to write objective standards to guide the architectural board and by designating certain types of experts to be represented on the board. The Ohio Court of Appeals relied on the presence of both these safeguards in *Reid v. Architectural Review Board of City of Cleveland Heights* (192 N.E.2d 74, 1963). It upheld the board's disapproval of a modernistic "flat-roofed complex of twenty modules" in a "well-regulated and carefully groomed community." Some communities have even sought to regulate against "excessive similarity" of building styles.

The regulation of the appearance of new structures is a somewhat elitist exercise practiced chiefly in affluent communities where new homes may be constructed independently. Most residential construction today occurs in subdivisions where either new homes are built by the developer or are constructed by lot buyers in accordance with deed restrictions (covenants). Such private land-use regulations to which the buyer voluntarily agrees are more potent vehicles for aesthetic control within a subdivision than public zoning and architectural review.

Permissiveness in aesthetic regulation is not necessarily an unmixed blessing. According to Costonis (1982, p. 367):

> Aesthetics has been transformed from an idea into an ideology that is being employed by preservationists, environmentalists, and developers alike to rationalize pursuits that are at best tenuously related to visual beauty.

The notion of aesthetics, Costonis notes, is inherently vague and expansive. Judicial approval of the objectives of aesthetic regulations in the context of signs and historic preservation should not necessarily be extended to any public action nominally taken in the spirit of aesthetic protection.

REFERENCES

BABCOCK, R. F. 1966. *The Zoning Game*. Madison: University of Wisconsin Press.

BABCOCK, R. F. AND C. L. SIEMON. 1985. *The Zoning Game Revisited*. Boston: Oelgeschlager, Gunn and Hain.

BASSETT, E. M. 1936. *Zoning*. New York: Russell Sage Foundation.

BEUSCHER, J. H., R. R. WRIGHT, AND M. GITELMAN. 1976. *Land Use: Cases and Materials*. St. Paul: West Publishing Co.

BOSSELMAN, F. P., 1973. "Can the Town of Ramapo Pass a Law to Bind the Rights of the Whole World?" *Florida State University Law Review*. Vol. *I* (Spring): 234–259.

———., D. CALLIES, AND J. BANTA, 1973. *The Taking Issue*. Washington, D.C.: U.S. Government Printing Office.

CLARK, G. 1984. "A Theory of Local Autonomy." *Annals of the Association of American Geographers* 74(2): 195–208.

———. 1985. *Judges and the Cities*. Chicago: University of Chicago Press.

COMEY, A. C. (ed.). 1946. *City and Regional Planning Papers of Alfred Bettman*. Cambridge: Harvard University Press.

COSTONIS, J. J. 1982. "Law and Aesthetics: A Critique and a Reformulation of the Dilemmas." *Michigan Law Review* 80 361–461.

DELAFONS, J. 1969. *Land-Use Controls in the United States*. (2nd ed.). Cambridge: M.I.T. Press.

DUERKSEN, C. J. 1985. "Administering Historic Drafting and Ordinances" in *1985 Zoning and Planning Handbook* (J. B. Gailen, ed.). New York: Clark Boardman.

HAAR, C. M. 1955a. "The Master Plan: An Impermanent Constitution," *Harvard Law Review* 68: 353–377.

———. 1955b. "In Accordance with a Comprehensive Plan," *Harvard Law Review* 68: 1154–1175.

———. 1963. "The Social Control of Urban Space" in *Cities and Space: The Future Use of Urban Land* (L. Wingo, Jr., ed.). Baltimore: Johns Hopkins Press.

HAMANN, R. G., 1986. *Constitutional Issues in Local Coastal Resource Protection*. Report no. 85. Gainesville: Florida Sea Grant College.

KRUECKEBERG, D. A., (ed.). 1983. *The American Planner: Biographies and Recollections*. New York: Methuen.

ROSE, J. 1987. "New Jersey Enacts a Fair Housing Law" in *1987 Zoning and Planning Law* (N. J. Gordon, ed.). New York: Clark Boardman.

WILLIAMS, N., JR. 1975. *American Planning Use: Land Use and the Police Power*. Chicago: Callaghan.

——— and T. NORMAN. 1974. "Exclusionary Land Use Controls: The Case of North-Eastern New Jersey" in *Land Use Controls: Present Problems and Future Reform* (D. Kistokin, ed.). New Brunswick: Rutgers University Center for Urban Policy Research.

9

BEYOND ZONING: INNOVATIONS IN LOCAL LAND USE CONTROL

INTRODUCTION

Euclidean land-use zoning remains pervasive in the United States in the 1980s. Virtually all incorporated cities, towns, and villages that are authorized under state laws to zone do so. The most notable exception is Houston, Texas which has steadfastly declined to adopt a zoning ordinance. Some municipalities were created explicitly to gain the right to zone land. (See case study of Sanibel, Florida in Babcock and Siemon, 1985, Ch. 6.) Most of the nation's metropolitan counties and many rural counties also zone land in unincorporated areas within their borders. Yet this nearly ubiquitous and venerable institution continues to provoke disappointment, frustration, and sometimes outrage among practitioners and scholars of land-use planning.

Since at least the early 1960s, zoning has been faulted for tolerating, and even fostering, wasteful and unsightly patterns of land use. In 1964, John Reps called for a "requiem for zoning":

> Zoning is seriously ill and its physicians—the planners—are mainly to blame. We have unnecessarily prolonged the existence of a land-use control device conceived in another era when the true and frightening complexity of urban life was barely appreciated. We have, through heroic efforts and with massive doses of legislative remedies, managed to preserve what was once a lusty infant not only past the retirement age but well into senility. What is called for is legal euthanasia, a respectful requiem, and a search for a new legislative substitute sturdy enough to survive in the modern urban world (excerpted in Listokin, 1974, p. 29).

Zoning has particularly been criticized for procedural inadequacies: lax enforcement, favoritism, nonconsistency with planning, and excessive rigidity in some cases and undue flexibility in others. According to Babcock (1966, p. 154):

> The running, ugly sore of zoning is the total failure of this system of law to develop a code of administrative ethics. Stripped of all planning jargon, zoning administration is exposed as a process under which multitudes of isolated social and political units engage in highly emotional altercations over the use of land, most of which are settled by crude tribal adaptations of medieval trial by fire, and a few of which are concluded by confessed ad hoc injunctions of bewildered courts.

Some of the administrative chaos described by Babcock has been remedied through open meeting laws, conflict of interest penalties, and more aggressive coverage of zoning proceedings by the media. But the fragmented geographical scale, reliance on lay members of the public with diverse vested interests, and the uncertain planning basis for many zoning decisions continue to undermine the credibility and efficacy of zoning as a mainstay of municipal land-use control. The painful experience of the New Jersey Supreme Court in attempting to remedy a statewide shortage of affordable housing through zoning reform by 567 communities is illustrative. During the 1980s, proposals to abolish local zoning prerogatives have continued to be expressed, albeit quixotically (e.g., Krasnowiecki, 1980; Delogu, 1986).

To a certain degree, the widespread dissatisfaction with the performance of zoning is misplaced. As Reps suggested, zoning was designed for the development process and political geography of an earlier, simpler era. Its geographic focus on the local municipality is both too small and too large to correspond with the actual scales of contemporary urbanization. The latter occurs, on the one hand, at a metropolitan or regional scale within which individual municipalities only "control" a small piece of the total affected land area. On the other hand, development is the result of countless piecemeal actions affecting only tiny fragments of a given municipal territory. Thus, the scale of the zoning process corresponds poorly with both the macro- and the micro-dimensions of the land-development process. Furthermore, the Euclidean zoning practice of designating the use and density of all buildable land within the community, sometimes years before many sites are ripe for development, gives rise to premature constitutional challenges, demands for individual relief through ad hoc adjustments, and disputes over nonconforming uses.

The many shortcomings of municipal-level Euclidean zoning—procedural, constitutional, and practical—have fostered various expedients to better adapt public control of land use to the macro- and micro-scales of the development process. At the submunicipal scale, measures to supplement zoning include subdivision regulations, growth management, density transfer, and regulation of sensitive lands such as floodplains, wetlands, and coastlines.

At the supra-municipal level, state, regional, or metropolitan plans guide municipalities in land-use zoning and the development of physical infrastructure such as highways, parks, sewer and water services, and transportation. Since local governments have not been required to surrender their zoning prerogatives to state or regional planning agencies, such bodies usually operate in an advisory capacity, with variable results.

SUBDIVISION REGULATION AND EXACTIONS

In 1983, 67.1 percent of U.S. housing units were single-family homes. This figure had declined from 77.3 percent in 1960, reflecting the expansion of multi-family rental and ownership units during the intervening period. Nevertheless, 62.3 percent of new residential units in 1983 were single-family homes on separate lots—an astonishing contrast to the rest of the world where new housing is predominantly in high-rise apartments. (U.S. Bureau of the Census, 1986, Tables 1308 and 1296.)

Single-family homes by definition are freestanding structures situated on a separate parcel or lot that satisfies the minimum lot size required by the local zoning law. Most such homes are built on lots resulting from the subdivision of a larger parcel of land into a number of legal building lots, each intended to be owned separately as a homesite. Houses may be constructed on each lot by the subdivider (also known as the "developer") who then sells the home and lot as a package. Or the lots may be sold to speculative builders or housing consumers who erect homes on them individually. In either case, since the 1920s, it has been customary to require the subdivider to install certain on-site facilities to serve the needs of the future home occupants, for example, streets, drains, and utility connections. Under state laws modeled on the 1927 Standard Planning Enabling Act, the layout of a proposed subdivision and its infrastructure plan must be reviewed and approved by the local municipality before any lots may be sold and/or construction of houses may begin (Figure 9–1).

The regulation of land subdivision is thus a fundamental legal tool for municipal guidance of land development. It applies not only to single-family projects, but also to the development of condominiums (which involve subdivision of a volume of space) and to nonresidential subdivisions such as industrial parks. It does not apply to rental apartments or commercial projects where no division of legal ownership is proposed. (The latter may, however, be subject to an equivalent process of site plan review in some states and municipalities.)

Subdivision regulation supplements but does not modify zoning. The use of land, minimum lot size, and bulk of structures to be built in the subdivision all must conform with applicable zoning (as modified by variances in appropriate cases). Subdivision approval requires that in addition to satisfying all zoning provisions, the proposed development will also meet performance standards for the layout and design of new subdivisions. These standards are set forth in published subdivision regulations adopted by the local planning board or equivalent body.

The community exerts considerable leverage through its subdivision approval function. A subdivision plan must be approved before it may legally be recorded. If

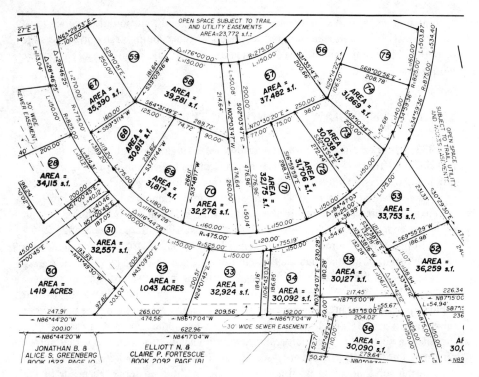

Figure 9.1 Excerpt from a Final Subdivision Plan submitted to the Town of Amherst, Massachusetts, 1987.

it cannot be recorded, the lots shown on it cannot be sold individually since they do not legally exist. Furthermore, lending institutions will not extend credit to purchase or improve land in an unapproved subdivision. Developers are eager to gain approval for their plans, which is prerequisite to starting work on the infrastructure and marketing the lots. The municipality, therefore, may negotiate a variety of concessions from the developer as a condition to subdivision approval.

Since new streets within a subdivision are normally conveyed ("dedicated") by the developer to the municipality or county for future maintenance, the recipient is concerned to ensure that those streets are sufficiently wide, properly paved, and aligned to avoid future maintenance or public safety problems. New rights of way must be mapped on the proposed subdivision plan and detailed engineering specifications must conform with the local requirements for street design (which often are said to be extravagant and costly).

Storm drainage is another subdivision review concern. Conversion of natural land surfaces to impermeable pavement and house sites reduces infiltration of precipitation into ground water. Surface runoff is thereby increased, with possible flooding imposed on land at lower elevation, either within or outside the subdivision boundaries. Subdividers may be required to install drains, retention basins, pumps, or other devices necessary to prevent both on-site and off-site drainage problems.

The subdivision plan must provide for the connection of each house lot to available public services and utilities, including water, sewerage, gas, electricity, telephone, and even cable television. Overhead wires are normally prohibited in new subdivisions. If public water and/or sewer service is not available to the subdivision, each lot must be large enough to accommodate an on-site septic system and/or a well. Where the physical character of the site precludes such on-site facilities, subdivision approval or individual building permits may be denied. State laws usually require percolation tests to be conducted by qualified contractors to determine the septic absorption capability of each lot where public sewerage is not available.

When a subdivision plan is formally approved by the appropriate local body (e.g., the planning board in Massachusetts), it may be filed with the county Registry of Deeds and becomes legally operative. However, much of the work to be performed by the developer remains undone at that time. Normally, the installation of streets, drains, and other infrastructure is phased over time in relation to the sale of lots. Some of the work promised by the developer will not be completed for months or even years after approval of the plan. Thus it is necessary for the community to be able to ensure that the work will be performed as promised in the subdivision plan.

There are two means by which the community may protect itself against default by the developer. One is to require the latter to provide a surety bond to the municipality for an amount of money sufficient to cover the costs of completing the necessary work if the developer should become bankrupt or otherwise fail to satisfy its contractual obligations. The other way is to impose restrictive covenants on each lot which must be released by the community before a lot may be sold. If infrastructure serving a lot or group of lots has not been constructed, the municipality will refuse to release its covenant and the developer cannot sell those parcels.

Private Deed Restrictions

Besides the public controls of zoning and subdivision requirements, a private subdivision is internally regulated through extensive use of private deed restrictions in the form of covenants and easements. It will be recalled that these instruments of private land-use control originated in seventeenth-century England where aristocratic property owners sought to control the use of land in the hands of subsequent buyers or tenants indefinitely. *Tulk* v. *Moxhay* held that a subsequent purchaser with knowledge or notice of a restriction on the use of land was bound to honor that restriction (See Chapter 3). Recording of deeds containing restrictive covenants today provides to purchasers "constructive notice" of all such restrictions. Purchasers are obligated to conform to any restrictions that could be ascertained by reading the subdivision instrument on file at the Registry of Deeds.

Deed restrictions are useful to control many aspects of a private subdivision. Since they arise by voluntary contract between the developer and the lot purchaser, they involve no public regulatory function and thus do not present problems of "reasonableness" or the "taking issue." Deed restrictions may therefore be more restrictive than the public police power. They may, for instance, regulate the external appearance of a dwelling, the kind of landscaping, or the parking of commercial vehicles in driveways. Minor alterations of a premises which would not involve zoning approval may be prohibited by deed restrictions. Condominium-type subdivisions are often especially tightly controlled through private restrictions.

Enforcement of private deed restrictions may take several forms. They may of course be enforced through legal action by the developer/seller against direct purchasers who violate the contractual restrictions. Other property owners, who are subject to similar restrictions, may enforce such provisions against recalcitrant neighbors. Homeowners or condominium associations established to own and maintain common facilities after the developer departs from the scene are also legally vested with power to enforce deed restrictions. Courts will enforce restrictions upon suit by any of the foregoing parties unless it finds the purpose of the restriction to be unenforceable (e.g., a racial restrictive covenant) or it finds that conditions in the area have rendered the restriction moot (e.g., a limitation to single-family homes in a high-density urban district).

Some communities include specific amenities in their subdivision requirements. These may include sidewalks, bike paths, natural buffers between the subdivision and adjoining public roads, and retention of large trees and natural drainage ways. The actual aesthetic appearance of a subdivision is the joint result of public zoning and subdivision regulations, deed restrictions imposed by the developer on the lot buyers, and the individual decisions of the latter regarding planting and landscaping. Depending on the market to which the development is directed, amenity requirements through private deed restrictions may be far more detailed and strict than requirements that the public may constitutionally impose.

Subdivision Exactions and Impact Fees

Residential subdivisions often generate costs to the host municipality which exceed the tax revenue produced by the development. New demands are placed on the community's schools, its water and sewage treatment facilities, its police and fire departments, and its parks and recreation areas. A number of states, led by California, have authorized municipalities to require, as a condition of subdivision plan approval, that the developer "dedicate" (donate) land within the subdivision for necessary park and school sites, in addition to the usual street requirements. If the subdivision is very small or has no land suitable for school or park sites, the developer is required to pay a "fee in lieu" of dedication equal to the value of the land that would otherwise be required. These are "subdivision exactions." (See generally, Smith, 1987.)

Such exactions may not be arbitrary: They must be set according to a local standard applicable to all new subdivisions. The critical constitutional issue, however, is how much land or fees in lieu of land the community may reasonably require. In 1971, the California Supreme Court rejected a developer's claim that exactions must be limited to meeting needs attributable to the subdivision, not reflecting the larger community interest:

> We see no persuasive reason in the face of these urgent needs caused by present and anticipated future population growth on the one hand, and the disappearance of open land on the other to hold that a statute requiring the dedication of land by a subdivider may be justified only upon the ground that the particular subdivider upon whom an exaction has been imposed will, solely by the development of his subdivision, increase the need for recreational facilities to such an extent that additional land for such facilities will be required (*Associated Home Builders* v. *City of Walnut Creek* 484 P.2d 606, at 611).

The court upheld the community requirement of 2.5 acres per 1,000 new residents or a fee equal to the value thereof. The use of such land or fees to remedy community-wide deficiencies, not simply the needs of each new subdivision, was approved in this and many subsequent cases in other states.

A further refinement of subdivision exactions has been to impose "impact fees" on all new development, not simply subdivisions or residential property. Again, the leading state has been California, where Proposition 13 has limited municipal tax revenue in the face of rapid population growth during the 1970s and 1980s. An impact fee is basically a charge levied when a building permit is applied for. Such fees are not limited to subdivisions but may be levied upon offices, shopping centers, or virtually any type of construction. The revenue generated from such fees is devoted to a broad range of municipal purposes including roads, sewer and water facilities, police and fire stations, and even libraries. San Francisco, Boston, and a few other cities have imposed "linkage" fees on new commercial construction to contribute to the cost of affordable housing for the workers to be hired by the building occupants. (Juergensmeyer and Blake, 1984).

In 1984, the California Supreme Court upheld an impact fee (referred to as a "facilities benefit assessment") imposed by the City of San Diego on all new development in designated vacant and developing areas of the city:

> San Diego's general plan is the instrument through which the City seeks to manage an explosive growth with land use controls, development of new and urbanizing communities over a period of years and the financing of public facilities through assessment of benefited property. We recognize the ordinance for what it is—imposing a present lien on undeveloped property to pay in the future an apportioned share of the costs of public facilities required to accommodate the needs of future residents of the properties upon their development (*J. W. Jones Companies* v. *City of San Diego* 203 Cal. Rptr 580, at 587).

The diffusion of subdivision exactions and impact fees around the United States has been uneven. A survey conducted in 1985, which produced 220 responses from 1,000 sample communities, reported that 65.9 percent of those responding have some form of traditional on-site subdivision exactions. Off-site exactions (money to be spent elsewhere in the community on the specified purpose) were imposed by 57.8 percent of respondents, while 45.4 percent utilize impact fees. But impact fees were reported by 81.8 percent of the California respondents, by 45.5 percent of those in other western states, 27.8 percent by southern respondents, 13.3 percent by those in the Mid-Atlantic states, and zero in New England (Bauman and Ethier, 1987, 57–59).

The foregoing suggests that the focus of subdivision regulation with the *physical* layout of each new subdivision has shifted in recent years to a concern with the *fiscal* impact of new development. Off-site subdivision exactions and impact fees have only an indirect relationship to the habitability of the new subdivision per se. Such fees supplement local property tax revenues (which new subdivision residents *also* pay) to fund basic public services within growing communities. Land-use control thereby merges into budget control. Developers seldom challenge impact fees and similar charges since it is cheaper to pay them than to incur much greater costs of delay and legal counsel. The fees are passed along to the ultimate home buyer with the inevitable effect of further increasing housing costs. Like large lot size requirements, high subdivision and impact fees thus serve as de facto exclusionary devices, discouraging the less wealthy from attempting to reside in the community.

How far may communities, even desirable ones, expand the use of impact fees to "privatize" public costs of providing general government services? According to one experienced planning law practitioner, the limit may be in sight:

> It is time for local governments to realize that the exaction "party" is over. The end is near, not because of the courts, but because of a changing economy. In the post–World War II era, an abundance of cheap land, and extraordinary demand for housing, and the consequent suburban sprawl brought about an inflow of housing into previously undeveloped or underdeveloped suburbs that required the construction of substantial amounts of physical infrastructure. The combination of low cost land, high housing demand, and demonstratable public need made subdivision exactions tolerable. Now, however, as the newer suburbs compete for glamorous developments, age overtakes the older suburbs, urban neighborhoods deteriorate and regentrify, and commerce abandons the central city for landscaped office parks, and the leverage of local government over development will decrease. Rehabilitation and redevelopment of urban and older suburban areas is unlikely to take place if it is burdened with exactions that represent a disguised and non-uniform system of taxation. In sum, it is time to work toward a consensus between the public and private sectors on which public facility costs a private developer can reasonably be expected to bear (Smith, 1987, p. 30).

Conclusion

Local control over the subdivision of land has thus expanded since the 1920s from a concern with the layout and paving of streets to a broad range of exactions and fees for both on-site and off-site facilities and services. Traditional subdivision review was primarily a means for regulating the internal planning and habitability of the development and providing for its orderly integration into the physical framework of the community. It still serves this purpose in the majority of states and communities. But impact fees and "linkage" assessments have augmented these public purposes to include reallocation of fiscal burdens from the existing taxpayers to newcomers. This function has little to do with the habitability of the development in question, or with land planning in the community at large. If anything, it may aggravate the tendency to use zoning and other planning tools as a means of optimizing revenue and minimizing public costs. Such goals, which have characterized "fiscal zoning" for many years, are likely to conflict with the explicit purpose of local land-use control: to ensure efficient, equitable, and balanced use of available land in the community.

GROWTH MANAGEMENT

The twin booms of the postwar period, producing quantities of babies and houses, transformed the settlement geography of the United States. Demographic change involved both absolute population growth at various scales and a massive geographic redistribution of households. This redistribution involved both intra- and interregional migration. The former was chiefly characterized by the "white flight" from central cities to suburbs, while the latter took the form of long-distance moves from the "frost belt" of the Northeast and Middle-West census regions to the South and West (Palm, 1981, Ch. 9).

To be more specific, population classified as "metropolitan" by the Bureau of the Census grew between 1960 and 1980 from 118 million to 169 million, with the metropolitan proportion of the U.S. population rising from 65.9 percent to 74.7 percent. But within metropolitan areas, central cities as a class gained only 15.2 percent (many of them experienced net losses), while the balance of the metropolitan area ("suburbs") grew by 71.1 percent (See Figure 1-2 in Chapter 1). Meanwhile, over the same period, the East and Midwest regions grew by only 10.0 percent and 13.9 percent, respectively, but the South and West gained 37.1 percent and 53.94 percent (Table 9-1). By 1984, the latter two regions outnumbered the earlier settled Northeast and Midwest by 18.5 million people. (U.S. Bureau of the Census, 1986, Tables 11, 12, and 20.)

The brunt of metropolitan growth, especially in the newly developing sun belt states has fallen most heavily on small to medium-sized urban places. Population of cities exceeding 1 million remained almost unchanged between 1960 and 1980, and cities of 250,000 to 1 million as a class gained only 4 percent. But urban places of smaller size increased their aggregate populations from a low of 43 percent (size class of 50–100,000) to as high as 58 percent (10–25,000) (Table 9–2).

Smaller urban communities were scarcely prepared for the rapid changes that this population growth signified. Immediate impacts included traffic congestion, demands for new schools and other public services, loss of open space and visual

Table 9-1 U.S. regional population: 1960–1980 (in millions)

Region	1960	1970	1980	% Change 1960–80
Northeast	44.6	49.0	49.1	24.6%
Midwest	51.6	56.5	58.8	32.4%
South	54.9	62.8	75.3	59.8%
West	28.0	34.3	43.1	114.4%

SOURCE: U.S. Bureau of the Census, 1986, Table 11.

amenities—in general an erosion of their cherished "small-town atmosphere." Although most had some form of basic zoning and subdivision regulations, they generally lacked experience in dealing with "big city" developers. Towns under 25,000 very likely had no professional planning staff and depended on their volunteer boards, occasionally assisted by an outside consultant, to cope with the onslaught. It was among these smaller but fast-growing suburban jurisdictions, especially in the sun belt states, that the idea of "growth management" became a virtual religion during the 1970s.

Growth management is not a precise term nor did the idea behind it originate in the late 1960s. Euclidean zoning, at least in theory, "manages" new urban growth through limits on land use, bulk, and density in accordance with comprehensive planning. Subdivision regulation also manages growth by imposing design performance standards and exactions on new subdivisions. But by the mid-1960s, there was a growing consensus in many faster-growing communities that neither zoning nor subdivision regulation could address: (1) the timing or pace at which growth was permitted to occur or (2) the ultimate character of the community when fully developed. "Urban sprawl," a phrase popularized by sociologist William H. Whyte (1958; 1968), included wasteful patterns of land use that encroached on farmland, flood plains, natural areas, and other open space.

Table 9-2 U.S. urban growth by size class of municipalities (in millions)

Size Class	1960	1970	1980	% 1960–80
>1 million	17.4	18.7	17.5	0
250K–1 mill.	22.0	23.3	22.9	4%
100K–250K	11.6	14.2	17.0	46%
50K–100K	13.8	16.7	19.7	43%
25K–50K	14.9	17.8	23.4	57%
10K–25K	17.5	21.4	27.6	58%
Urban Total	125.2	49.3	167.0	33%
U.S. Total	179.3	203.2	226.5	26%

SOURCE: U.S. Bureau of the Census, 1986, Table 17.

Growth management evolved during the early 1970s in a variety of forms reflecting the lack of any dominant technique or approach. Various communities as diverse as Boulder, Colorado; Petaluma, California; Ramapo Township, New York; and Boca Raton, Florida pursued their home-grown strategies with varying success.

A study of 18 local or county programs by Godschalk et al. (1978) disclosed a diversity of both goals and means. The former included such respectable objectives as preserving prime farm land, providing a balanced housing supply, relating new development to public services, and reducing natural hazard vulnerability. More suspect goals included minimizing tax burdens and preserving a "small-town environment." Means used by various communities surveyed included: capital improvement programs, annual permit limits, dwelling unit caps, downzoning (to lower densities), site plan review, fair-share housing plans, and open space acquisition. Most communities were relying on a combination of existing and new techniques:

> Many of the tools have been invented or adapted by individual local governments. Existing approaches are modified in light of experience, changing conditions, or legal challenges. Typical systems consist of scattered parts of policies, amendments to development ordinances, and items in capital programs and budgets. Often the first time a system is completely described and analyzed in writing is by the attorneys who attack and defend it in court (Godschalk, et al., 1978, p. i).

To the extent that it involves the use of the public police power, growth management faces the same constitutional hurdles as traditional zoning: due process, equal protection, and the taking issue. Courts have weighed various growth-management schemes in light of these principles.

The first major growth-management case was decided in 1972 by the New York Court of Appeals (State Supreme Court) in *Golden v. Township of Ramapo* 285 N.E.2d 291. The case involved phasing the tempo of new development in relation to the construction of capital improvements. The defendant town exercised jurisdiction over 60 square miles of unincorporated land, located some 30 miles north of New York City. In response to a doubling of its population in less than a decade, Ramapo in the mid-1960s undertook to prepare a master plan for its future growth (Godschalk, et al., 1978, Ch. 7). This was not unusual: Hundreds of communities prepared such plans during the 1960s with the help of "Section 701" federal planning grants.

What distinguished Ramapo's plan was that it had teeth. It included a capital improvements program to provide the entire town with public services within 18 years. A zoning amendment in 1969 provided that all proposals for residential construction (except individual homes) would be evaluated in terms of the availability of sewers, parks, firehouses, roads, and storm drainage. Points were assigned according to each requirement that was satisfied, and sites lacking at least 15 out of a possible 26 points were not approved (Dawson, 1981, p. 150). In lieu of outright denial of a building permit, however, the town issued a "rain check" in the form of a special permit to become effective in the year when services were scheduled to be provided. The rain check could be sold to other parties to realize some immediate cash return, and property taxes would be reduced during the interim. (Alternatively, the developer could gain points by installing services privately.)

Upon review, the New York Court of Appeals acknowledged that "there is something inherently suspect in a scheme which, apart from its professed purposes, effects a restriction upon the free mobility of a people until sometime in the future when project facilities are available to meet increased demands." (285 N.E.2d, at 300). But it recognized the social and environmental ills associated with rapid growth that gave rise to the plan. The court, in a leap of faith, construed that "phased growth is well within the ambit of existing enabling legislation." It found that a *temporary* delay related to a carefully considered plan for capital improvements and involving no absolute growth cap was reasonable in a constitutional sense:

> We may assume that the present amendments are the product of foresighted planning calculated to promote the welfare of the township. The Town has imposed temporary restrictions upon land use in residential areas while committing itself to a program of development. It has utilized its comprehensive plan to implement its timing controls and has coupled with restrictions provisions for low and moderate income housing on a large scale. Considered as a whole it represents both in its inception and implementation a reasonable attempt to provide for the sequential, orderly development of land. . . (285 N.E.2d, at 303).

The Ramapo plan added "a third dimension to land-use regulation: a temporal aspect to the traditional ones of use and location" and the decision was hailed as ". . . the most significant land-use regulation case, other than *Euclid*" (Callies and Freilich, 1986, p. 834). On the other hand, the Ramapo plan was criticized by Bosselman (1973, p. 249): "The wolf of exclusionary zoning hides under the environmental sheepskin worn by the stop-growth movement." Either way, the approach followed by Ramapo has had few imitators.

The next landmark in judicial review of growth management arose in Petaluma, California, a Victorian-era chicken-farming town 35 miles north of San Francisco. A surge of new building in 1970–1971 alarmed the community into conducting a major planning exercise. The resulting "Petaluma plan" was a complex set of documents, policies, and programs. Unlike Ramapo, Petaluma established a target population for itself considerably below what existing rates of building would indicate—55,000 instead of 77,000. Policies and measures designed to achieve this reduction included an annual limit on building permits, a moratorium on annexation or extension of services into adjacent areas, and deliberate limitation of sewer and water capacity to the needs of 55,000 people (Figure 9–2). The most controversial feature was the limit of building permits to an average of 500 a year (2,000 had been issued in 1970–1971). A point system was introduced by which this quota would be allocated, but no Ramapo-type rain checks were provided.

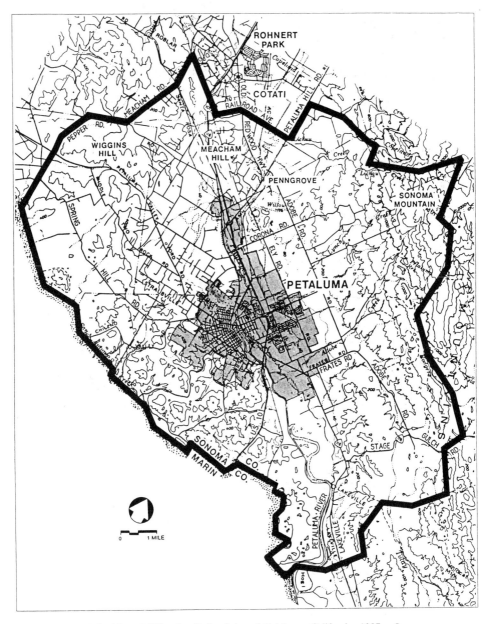

Figure 9.2 Map of "Planning Referral Area," Petaluma, California, 1987. *Source:* City of Petaluma, 1987.

Upon constitutional challenge and appeal by a building trade association, the Federal District Court and the Federal Court of Appeals reached opposite conclusions on the validity of the Petaluma Plan (*Construction Industry Assn. of Sonoma County v. City of Petaluma* 375 F.Supp. 574, 1974 and 522 F.2d 897, 1975, respectively). The contrasting outcomes arguably reflect the importance of judicial perception of geographic context as a factor in determining the reasonableness (i.e., constitutionality) of a land-use regulation.

The District Court held the Petaluma Plan to be invalid on the ground that it inhibited the constitutional right to travel. This conclusion was based on a scholarly discussion of the regional housing market of the Bay Area, including the concepts of "growth poles" and "trickle down" of housing opportunities as new dwellings are added on the periphery. The court concluded that the Petaluma Plan, if widely imitated, would retard new housing construction and thereby restrict housing availability throughout the region.

The Court of Appeals (9th Circuit) saw the issue very differently. It ignored theories of urban geography and planning and even objected to the "oversized" brief submitted by the plaintiffs. In contrast to the regional perspective of the lower court, the appellate court viewed the issue strictly within the geographic compass of Petaluma. The Court of Appeals cited recent decisions of its own and of the U.S. Supreme Court upholding municipal zoning prerogatives where a deliberate intent to exclude was not proven. It found that the plan earmarked 8–12 percent of permits for low- and moderate-income housing (assuming a builder turned up) which it felt nullified any exclusionary motive.

[A careful reading of these two opinions is a valuable lesson in the process of adjudication and legal reasoning. As a class exercise, it is useful to conduct a straw poll as to which opinion was (a) correct in its result and (b) most convincing in its argument.]

Plan or no plan, the growth of Petaluma lagged even behind its planned level. By 1987, the city's population had reached only about 40,000. Petaluma has continued to refine its plan, however, which in 1987 received the Outstanding Planning Award of the California Chapter of the American Planning Association. The following list of chapter topics reflects its broad scope:

- "Community Character,"
- "Land Use and Growth Management,"
- "The Petaluma River,"
- "Open Space, Conservation, and Energy,"
- "Parks, Recreation, Schools, and Child Care,"
- "Local Economy,"
- "Housing,"
- "Transportation," and
- "Community Health and Safety."

SOURCE: City of Petaluma, 1987

The District Court in *Petaluma* warned of a potential diffusion of growth-limitation plans to other developing communities. The voters of Livermore, California in fact had already voted in 1972 to impose a moratorium on any further residential construction pending upgrading of school, sewer, and water services. This differed from Petaluma's plan in utilizing a total ban rather than an annual allotment, and it lacked the timetable for completion of services provided by Ramapo. It was legally a borderline case.

The California Supreme Court in *Associated Home Builders* v. *City of Livermore* 557 P.2d 473, 1975 declined to view the moratorium as inherently exclusionary or unduly vague. But it also avoided the simplistic approach of the appeals court in *Petaluma* that a local town may act as a world unto itself. Instead, the *Livermore* opinion articulated a new constitutional test:

> . . . whether the ordinance reasonably relates to the welfare of those whom it significantly affects. If the impact is limited to the city boundaries, the inquiry may be limited accordingly; if, as alleged here, the ordinance may strongly influence the supply and distribution of housing for an entire metropolitan region, judicial inquiry must consider the welfare of that region (557 P.2d, at 487).

The court specified three steps by which a trial court might apply this test: First, it must "forecast the probable effect and duration of the restriction." Second, it must "identify the competing interests affected by the restriction." The third step is to determine "whether the ordinance, in light of its probable impact, represents a reasonable accommodation of the competing interests" (557 P.2d, at 488). The California Supreme Court then returned the case to the trial court to apply this test.

Livermore thus mandated the weighing of geographical evidence on regional-housing, economic, and demographic trends as a basis for resolving the constitutionality of growth management plans. Like the lower court in *Petaluma,* the California Supreme Court was groping for a rationale for judging the reasonableness of such plans in terms of their real-world impacts.

Livermore, incidentally, included a fervent dissent by Justice Mosk who viewed the moratorium as deliberately exclusionary. Unlike the majority, which passed the buck to the trial court, Mosk applied his own geographical perception to the issue:

> Limitations on growth may be justified in resort communities, beach and lake and mountain sites, and other rural and recreational areas; such restrictions are generally designed to preserve nature's environment for the benefits of all mankind. As Thomas Jefferson wrote, the earth belongs to the living, but in usufruct.
>
> But there is a vast qualitative difference when a suburban community invokes an elitist concept to construct a mythical moat around its perimeter, not for the benefit of mankind but to exclude all but its fortunate current residents (557 F.2d, at 493).

The complex growth-management plans considered so far provoked mixed judicial reactions but were ultimately upheld. More simplistic or blatantly exclusionary measures adopted elsewhere have been emphatically rejected as unconstitutional. A 1979 Florida decision held to be arbitrary and invalid a charter amendment by the City of Boca Raton that attempted to establish an absolute limit of 40,000 dwelling units in the city (*Boca Raton* v. *Boca Villas Corp.* 371 So.2d 254). In Colorado, an attempt by the City of Boulder to withhold water and sewer service from a proposed development outside its corporate limits but adjacent to areas already served was rejected in *Robinson* v. *City of Boulder* 547 P.2d 228, 1976. The court held that the city had assumed the status of a public utility as sole provider of water and sewage treatment in the area and could not therefore deny service on the basis of a growth-limitation plan.

Few cases involving growth management per se have appeared during the 1980s. In part, this reflects a temporary easing of development pressures due to high interest rates early in the decade. Another reason may be that growth management has become a widely accepted practice which, like subdivision approval, generates few constitutional challenges.

Evidence that growth management was alive and well in the 1980s, especially in the sun belt states, was provided in a survey of 24 county and local programs by Bollens and Godschalk (1987). Their study examined the use of computerized land information systems as a basis for administering growth-management programs:

> Jurisdictions that control growth should maintain supplies of available land large enough to allow for a competitive land market. Policies that assure such a land supply can only be formulated if local government and the private sector have up-to-date information on the locations, amounts, and prices of available land supply. . . . The most important advantage of a monitoring system could be its assistance in the design of growth management policies that will balance land supply and demand without contribution to price inflation in fast-growing areas (Bollens and Godschalk, 1987, p. 316).

A further advantage of such geographic information systems is to bolster the constitutionality of growth management against charges of undue discretion and arbitrariness. Also, such systems may facilitate the identification and alleviation of adverse regional effects of local growth management which concerned the California Supreme Court in *Livermore*.

TRANSFER OF DEVELOPMENT RIGHTS

Transfer of development rights (TDR) was widely touted by planners during the 1970s as a means of achieving better regulation of growth while avoiding the problem of the taking issue (e.g., Costonis, 1974; Woodbury, 1975). TDR conceptually involves severing the "right to develop" from a site that the public wishes to have preserved in low-density or open space condition and transferring that right to another site where higher than normal density would be tolerable. Land uses proposed for protection through TDR include agricultural land, wetlands, coastal or riverine flood hazard areas, historical sites, churches, and other privately owned landmarks. Receiving sites for TDRs may include areas where density above the levels allowed in the applicable zoning could be accommodated without doing violence to planning concerns such as overcrowding of population, fire hazards, traffic congestion, or overtaxing of sewer and water services.

TDRs normally require official sanction through amendment of local zoning or other development-control law. This would provide that the local government would award a zoning bonus in the form of increased height or more dwellings per acre to a developer in the receiving area or "growth zone" upon payment by that party to the owner of a site to be preserved. The amount of the payment would be set by mutual agreement as with any real estate transaction. The seller of the development right would record a permanent restriction on the future development, subdivision, or alteration (in the case of historic preservation) of their site. The buyer of the right would then be issued a density bonus usable at the receiving site.

The owner of the preserved site retains existing use rights while receiving compensation for the developmental value foregone. The public ensures the preservation of the site without paying for it. And the buyer of the development right gains legal approval for a more profitable project (Figure 9–3).

Several pitfalls have hampered the utility of TDRs, however. One is the need for a market for TDRs. Smart developers usually can gain extra density through administrative adjustments (e.g., variances) and thus have little incentive to buy TDRs unless the zoning process is relatively inflexible and incorruptible. The neighbors of a receiving site may object that the granting of higher density unfairly harms their property values and is "spot zoning." Alternatively, they may demand to be

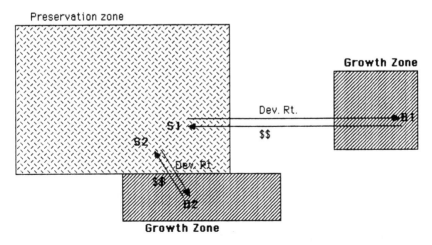

Figure 9.3 Diagram of transfer of development rights.

similarly rezoned, thereby eliminating the incentive for further purchases of TDRs. In the case of wetlands and flood plains, it seems illogical to recognize a "development right" in land that is inherently unsafe to develop. Finally, it is not clear to what extent the uncertain possibility of selling a TDR is adequate to safeguard a restrictive zoning measure against a taking issue challenge.

While not deciding the last question, the U.S. Supreme Court has at least commented favorably on TDRs in *Penn Central Transportation Co.* v. *City of New York* 98 S.Ct. 2646, 1978. The case arose from disapproval by New York's Landmarks Preservation Commission of plaintiff's proposal to build a 55-story office tower above its Grand Central Terminal in midtown Manhattan. Although technically the terminal would survive, the aesthetic effect of surmounting it with a tall modern building would have been profound, and civic leaders, including Jacqueline Kennedy Onassis, mounted a defense of both the landmark and the law that protected it.

The law provided, among other features, that development rights not usable on the preserved landmark site could be transferred to nearby sites under the same ownership. Penn Central owned eight structures in the vicinity, but, instead of exercising the TDR option, it challenged the landmarks law as a taking. The High Court upheld the law on the ground that historic preservation is a valid public purpose and that Penn Central was not denied all economic use of the site, which generates revenue from concessionaires. It thus held the restriction not to be a taking regardless of the TDR option, but bolstered its conclusion with the following observation:

> To the extent appellants have been denied the right to build above the terminal, it is not literally accurate to say that they have been denied all use of even those pre-existing air rights. Their ability to use these rights has not been abrogated; they are made transferrable to at least eight parcels in the vicinity of the terminal, one or two of which have been found suitable for the construction of new office buildings. . . . While these rights may well not have constituted "just compensation" if a "taking" had occurred, the rights nevertheless undoubtedly mitigate whatever financial burdens the law has imposed on appellants and, for that reason, are to be taken into account in considering the impact of regulation (98 S. Ct., at 2666).

Two years earlier and four blocks east, the New York Court of Appeals took a dimmer view of TDRs as a form of compensation. *Fred F. French Investing Co.* v. *City of New York* (50 N.E.2d 381, 1976) involved a law that prohibited building on two privately owned parks in Tudor City, a fashionable older apartment complex near the United Nations. The owner was permitted to sell the development rights anywhere in central Manhattan but had not done so. The court viewed the restriction as a taking because it forced the owner to "provide a benefit to the public without recoupment." It held the TDR option to be inadequate recompense:

> . . . Development rights, disembodied abstractions of man's ingenuity, float in a limbo until restored to reality by reattachment to tangible real property. Put another

way, it is a tolerable abstraction to consider development rights apart from the solid land from which as a matter of zoning law they derive. But severed, the development rights are a double abstraction until they are actually attached to a receiving parcel yet to be identified, acquired, and subject to the contingent future approvals of administrative agencies, . . . By compelling the owner to enter an unpredictable real estate market to find a suitable receiving lot for the rights, . . . the amendment renders uncertain and thus severely impairs the value of the development rights before they were severed (350 N.E.2d, at 388).

According to Dawson (1982, p. 70): "Probably the court was influenced by the fatal words 'public park' since the use of private property as parkland effectively denies the owner any income at all."

An easier situation involves a voluntary agreement to transfer development rights among adjacent parcels of land to which the only objectors may be neighboring property owners. In *Dupont Circle Citizens Association v. District of Columbia Zoning Commission* 355 A.2d 550, 1976, the development rights from a historic mansion and garden were proposed to be transferred to an adjoining site to permit extra height for a new office building. The court held that the preservation of historic structures is a valid regulatory objective that may be achieved by transferring development rights from one parcel to another, provided the aggregate density is within prescribed limits for the combined land area.

Outside cities, various schemes for preserving open land through TDRs have been proposed but few have borne fruit. Perhaps the nation's most successful program has occurred in Montgomery County, Maryland on the outskirts of Washington, D.C. where several hundred parcels of farmland have been protected through TDRs. Faced with intense development pressure on the one-third of its land area that remained in agriculture, the county in 1980 established an "agricultural reserve" of some 73,000 acres. It downzoned this area from one dwelling per five acres to one per 25. Landowners were allowed to sell one development right for each five acres to developers in the other two-thirds of the county. (In Maryland, counties exercise planning and zoning powers throughout their territories.) To ensure a ready market for development rights, the county established a publicly funded TDR "bank." Its function was to buy and sell rights when private demand languished and thus maintain the confidence of agricultural owners (Tustian, 1984, Ch. 11).

FLOODPLAIN AND WETLAND REGULATION

Introduction

For decades, Euclidean zoning took little notice of physical variation in the land. Just as the rectangular townships of the Federal Land Survey marched relentlessly across prairie, desert, and mountain, the geometry of suburbia, imposed by zoning until the 1970s, ignored natural impediments to building. Those limitations included such geographically specific natural hazards as seismic faults, unstable slopes, coastal and riverine floodplains, and wetlands. As a result of this blindness to physical hazards, much postwar urban development was located "in harm's way" and resulting losses have been costly.

Since the 1950s, many researchers have documented the double jeopardy of such ill-considered building practices: (1) the risk to life and property and (2) the damage to natural values and functions of floodplains and wetlands (e.g., Hoyt and Langbein, 1955; White, 1964, 1975; and Schneider and Goddard, 1973). Geographer Gilbert F. White and his associates (1958) in particular found that flood control projects often lead to increased flood losses by encouraging encroachment in the "protected" floodplain, which may then be inundated by a flood exceeding the project's design capacity.

Such disclosures prompted a reversal of public policies that had long tolerated encroachment on floodplains in reliance upon structural food control projects. Consistent with the model in Chapter 2, new approaches to the management of floodplains resulted from better recognition of the impacts of human activities on the natural environment and vice versa. The federal and state dimensions of this new approach are considered in Chapters 12 and 13. This section examines the evolution of a judicial basis for regulation of floodplains and wetlands at the local level.

Physical Characteristics

Floodplains are natural overflow areas adjoining surface waters, including streams, rivers, lakes, estuaries, bays, and the open ocean (Figure 9–4). Riverine floodplains are formed where a watercourse over time has weathered away bordering uplands and deposited sediment on the resulting level surface adjoining its normal channel. The floodplain is narrow or nonexistent in steep and less erodible terrain, and tends to be broader in more erodible lowlands. The latter are characterized by deposits of fertile alluvial soils ideal for agriculture. But in regions such as Appalachia, floodplains may be the only level land available for urban development and communications linkages such as highways, railroads, power lines, and airports. Agriculture is reasonably compatible with occasional flooding, especially that which occurs in the early spring in northern climes, the "spring freshet." But urban development, unless suitably designed and/or fortified by protective measures, is inherently at risk in times of flood.

Where the onset of flooding is very swift ("flash flooding"), threat to human life is of paramount concern. The Rapid City, South Dakota flash flood of 1972 took 238 lives as well as destroyed the city's business district (White, 1975, Ch. V). The 1976 Big Thompson Canyon flash flood in Colorado killed 143 while scouring most structures out of that narrow Rocky Mountain canyon (Gruntfest, 1987).

Floodplains are not uniformly risk prone. Levels of hazard are distinguished in terms of estimated frequency of inundation for specific elevations of land within

Floodplain and Wetland Regulation

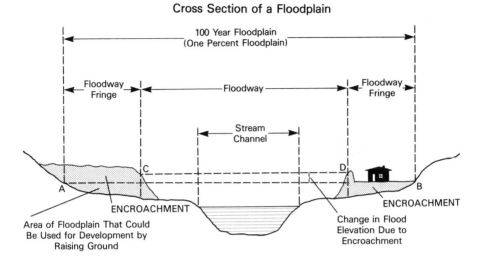

Figure 9.4 Diagram of a riverine floodplain in cross-section.

the overall floodplain. Thus, low ground close to the river may be flooded nearly every year while slightly higher ground further from the channel may be flooded on the average only every 50 to 100 years. (The close-in areas of course are flooded more deeply during those events.) The "100 year floodplain," which has a probability of 1 percent of being flooded in any given year, is the geographic area regulated in most public floodplain management programs in the United States. Within this regulatory floodplain, different levels of control over building are applied to the high-risk "floodway" and the less hazardous "floodway fringe" (Figure 9–4).

Wetlands are natural depressions where the groundwater table is normally at or just below the land surface. (Where the water table is normally above the surface, a pond is formed.) The wetness of wetlands and their special biological characteristics provide nutrients to support a rich diversity of flora and fauna. The ecology of wetlands varies according to their location, soil chemistry, hydrology, elevation, and size (Figure 9–5). Coastal wetlands include salt marshes, mudflats, and other estuarine wetlands subject to tidal influence and some degree of salinity. Inland or freshwater wetlands are of many types including the familiar cattail swamp, riparian habitat along stream banks, isolated bogs, prairie potholes, and bottomland hardwoods of the lower Mississippi Valley (Tiner, 1984; Kusler, 1983). Wetlands of all types are estimated to cover slightly more than 3 percent of the continental U.S. land area—about 70 million acres (U.S. Council on Environmental Quality, 1981).

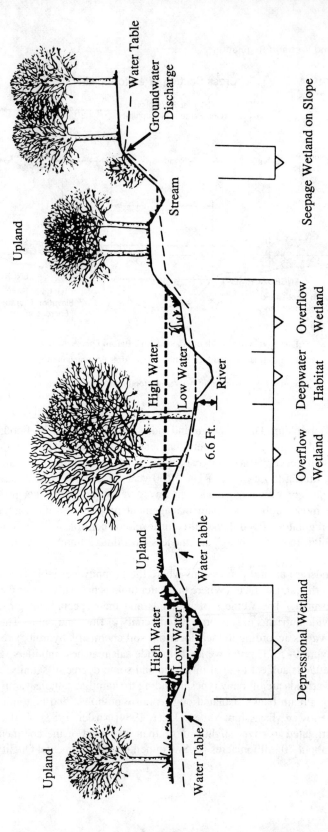

Figure 9.5 Cross-section of various types of freshwater wetlands. *Source*: U.S. Fish and Wildlife Service.

Depending on their location and ecological type, wetlands provide a number of "values" or functions in their natural state:

> These serve as food producers, spawning sites, and sanctuaries for many forms of fish and bird life. They produce timber, peat moss, and crops such as wild rice and cranberries. They serve as storage areas for storm and flood waters, reduce erosion, provide for groundwater retention, and at times purify polluted waters (U.S. Council on Environmental Quality, 1978).

Wetlands and floodplains frequently coincide geographically and, as just stated, wetlands help to alleviate downstream flooding in watersheds and damage to landward property in coastal areas. But, the management of floodplains and wetlands pursues somewhat different goals. In floodplains, the public seeks to reduce hazards to life and property from flooding, while in wetlands, the public seeks to preserve the site in its natural condition or to restore it to that condition. While a properly designed building can be tolerated in floodway fringe areas, any construction in wetlands is usually harmful to its natural functions.

Public programs for managing floodplains and wetlands have evolved independently at the federal and state levels. The federal government addresses flooding primarily through the National Flood Insurance Program administered by the Federal Emergency Management Agency. Federal wetlands regulations are based on Section 404 of the Clean Water Act and are jointly administered by the U.S. Army Corps of Engineers and the U.S. Environmental Protection Agency. (The goals and methods of these programs and their state analogues differ substantially and are discussed in Chapters 12 and 13.)

At the local level, floodplain and wetland management tend to be more synonymous. Both present similar issues of regulating land alteration and building practices on the basis of imperfect, sometimes crude, scientific data and maps. Until the late 1960s, courts balked at the rare attempts by municipalities to restrain building in either floodplains or wetlands. By the 1970s, however, the pendulum swung in the other direction and courts regularly began to uphold such measures. The judicial "watershed" between these two positions is clearly reflected in the following trilogy of cases, one from New Jersey in 1963 and the other two from Massachusetts and Wisconsin, both in 1972.

The Preenvironmental Period

Public efforts to regulate floodplains and wetlands were rare before 1960 and so were judicial decisions reviewing them. In the absence of precedents, Dunham (1959) in a seminal law review article stated an *a priori* argument to support floodplain regulations. He urged that such measures are valid to (1) protect unwary individuals from investing or dwelling in hazardous locations; (2) protect riparian landowners from higher flood levels due to ill-considered encroachment on floodplains by their neighbors; and (3) protect the community from the costs of rescue and disaster assistance. This argument was to bear fruit in later cases, notably in *Turnpike Realty* in Massachusetts. But it did not influence the first landmark decision on floodplain/wetland regulation: *Morris County Land Improvement Co. v. Parsippany-Troy Hills Twp.* 192 A.2d 232 (N.J., 1963).

Morris County involved a 66-acre fragment of a 1,500-acre freshwater marsh known as the Troy Meadows. The Meadows is located in the headwaters of the Passaic River, a sluggish and polluted stream that drains a heavily developed region of northern New Jersey. About three-quarters of Troy Meadows was owned by a private conservation organization, Wildlife Preserves, Inc.

The plaintiff wished to excavate sand and gravel from its site and then fill it with other material to eventually be sold as a building site. Upon complaint by Wildlife, the township rezoned the area as a "Meadows Development Zone." The only permitted uses within this zone were agriculture, horticulture, recreation, and similar open space uses. Plaintiff was prohibited from proceeding with its plans. It challenged the rezoning as a taking of its property value in that no remunerative use was allowed. The trial court upheld the ordinance.

On appeal, the New Jersey Supreme Court took note that Troy Meadows provides both flood reduction and wetland benefits in its natural state. It described the area as:

> . . . a natural detention basin for flood water in times of very heavy rainfall which would otherwise run off more quickly and aggravate damaging flood conditions occurring with some frequency in municipalities further down the Passaic River valley (192 N.E.2d, at 233).

This function would be viewed in later cases as a valid and commendable use of the police power to "prevent public harm." In *Morris County,* it was viewed as "providing a public benefit," which is constitutionally fatal. The court found that two forms of public benefit were sought by the ordinance: (1) flood control and (2) preservation of the land in its natural state to augment the wildlife refuge. The court was thus simultaneously grappling with two novel public purposes, flood-loss mitigation and wetland preservation. Having no precedent on either issue, the court reverted to the analysis of Justice Holmes in *Pennsylvania Coal Co. v. Mahon* and concluded that the regulation "went too far":

> These are laudable public purposes and we do not doubt the high-mindedness of their motivation. But such factors cannot cure basic unconstitutionality . . . Public acquisition rather than regulation is required (192 N.E.2d, at 238).

In the court's defense, the unusual geographic context of this case perhaps explains the result. The site in question was in the headwaters of the drainage system and thus would not be threatened by flooding from upstream. Provision of compensatory detention could have mitigated adverse effects downstream from filling the site. Thus, the flood issue was perhaps offered as a "cover" for the real goal of maintaining the site in its natural state. And the township did not specify the public harm to be prevented by so preserving it.

In any event, *Morris County* was to become the leading decision adverse to floodplain and wetland regulation and as such has been widely cited by landowners. The case later proved an embarrassment to the New Jersey court and particularly to Justice Hall (of *Mount Laurel* fame) who wrote the opinion. In 1975, Hall offered the following revealing comment in *A.M.G. Associates* v. *Springfield Township* 319 A.2d 705:

> It is to be emphasized that we deal in this case only with the split lot situation where there is a deprivation of all practical use of the smaller portion thereof. The approach to the taking problem and the result may be different where *vital ecological and environmental considerations of recent cognizance* have brought about rather drastic land-use restrictions in furtherance of a policy designed to protect important public interests wide in scope and territory; as for example the coastal wetlands act . . . and various kinds of floodplain use regulation. Cases arising in such a context may properly call for a reexamination of some of the statements made 10 years ago in the largely locally limited *Morris County Land* case . . . (319 A.2d, at 711). (Emphasis added)

The flexibility of the law to respond to improved perception of human impacts on the physical world has seldom been better stated.

The Postenvironmental Period

For a decade, *Morris County* cast its shadow over floodplain and wetland regulations, and several northeastern states followed its lead. But in 1972, the same year as the Federal Water Quality Amendments and the Coastal Zone Management Act, the Supreme Courts of Massachusetts and Wisconsin provided a fresh perspective. Landmark decisions in those states held such regulations to be valid. These laid a foundation for the subsequent approval of such measures in nearly every jurisdiction that has considered them.

Turnpike Realty Co. v. *Town of Dedham* 284 N.E.2d 891 (Mass., 1972) turned the tide in the Northeast. This is probably the most quotable and instructive decision by a state court on floodplain/wetland regulations. The case involved a 61-acre parcel located in a wetland adjoining the Charles River in eastern Massachusetts. Like the Boston Marathon, the Charles flows from exurban Hopkinton to Boston. On its route to Boston Harbor, it passes through increasingly built-up communities, which it has flooded from time to time.

The plaintiff's site also adjoined Route 1 in Dedham, a major artery lined with shopping centers, franchise fast-food outlets, and subdivisions, many of them built on filled wetlands. Plaintiff wished to continue this trend. Dedham, however, in 1963 had amended its zoning to establish a "Flood Plain District" that included this site. Filling of plaintiff's site was prohibited.

Greater care was exercised by Dedham than by the New Jersey township in articulating the purposes of the regulations. These included (1) protection of the ground water table; (2) protection of public health and safety against floods; (3) avoidance of community costs due to unwise construction in wetlands and floodplains; and (4) conservation of "natural conditions, wild life, and open spaces for the education, recreation and general welfare of the public." (284 N.E.2d. at 894). The last, of course, raised the specter of "public benefit" that had tainted the *Morris County* ordinance. The *Turnpike Realty* court, however, perceived the principal objectives to be the first three, which it accepted as valid police power purposes. It held that the suspect fourth goal would not justify the measure on its own, but was not fatal since "the by-law is fully supported by other valid considerations of public welfare." Furthermore, the court declared that "the validity of this by-law does not hinge on the motives of its supporters." (284 N.E.2d, at 895).

The court addressed the flood hazard issue carefully. Although plaintiff claimed that the site was only flooded due to improper operation of a floodgate, the court found the site to be a part of the natural floodplain of the Charles River. It cited testimony as to recent severe flooding of the site. It distinguished *Morris County* as not applicable to a site in a "lower reach" of a stream as was the case in *Turnpike Realty*. Relying on the Dunham rationale stated earlier, it upheld the measure, declaring: "The general necessity of flood plain zoning to reduce the damage to life and property caused by flooding is unquestionable." (284 N.E.2d, at 899).

Turnpike Realty thus based its decision on flood hazards rather than wetlands issues, although the site was definitely a wetland. It left dangling the question of whether a wetland regulation per se could be upheld in the absence of flood risk. That issue was tackled in the extraordinary Wisconsin decision: *Just v. Marinette County* 201 N.W.2d 761, 1972.

Mr. and Mrs. Just owned 36 acres bordering Lake Noquebay, one of thousands of "glacial potholes" strewn across Wisconsin and Minnesota. Like most of that region, their land was a boggy wetland, whose only economic use was to be filled for summer cottages. Marinette County, however, had adopted a shoreland zoning ordinance pursuant to state law that required a permit to be obtained for filling of wetlands within 300 feet of navigable waters. The Just property was covered by this requirement, but they proceeded to fill without getting a permit and carried their defiance as far as the Wisconsin Supreme Court. Their defeat in that forum has resounded through land-use law ever since.

The *Just* decision rested, not on flood hazards, but on water quality. Wisconsin is a "public trust" jurisdiction in which the state "owns" navigable waters and land beneath them in trust for the people. The court viewed the purpose of the shoreland zoning ordinance as protecting: ". . . navigable waters and the public rights therein from the degradation and deterioration which results from uncontrolled use and development of shorelands (201 N.W.2d, at 765). This is reasonable in light of the problem of septic seepage from shoreline cottages contaminating lakes of this type.

The court could thus easily have upheld the ordinance as simply giving the

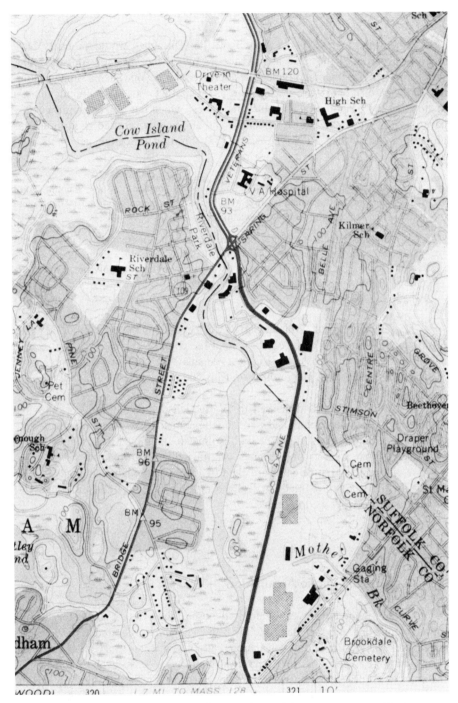

Figure 9.6 Site of the *Turnpike Realty Co.* v. *Town of Dedham* case between Charles River and Route 1 (heavy dark line).

county the right to review plans for new fill. The Justs would probably have received a permit if they had applied for it. But in the spirit of Aldo Leopold, Wisconsin's environmental prophet, the court went far beyond such a simple resolution of the case. It chose to view the issue as "a conflict between the public interest in stopping the despoliation of natural resources, which our citizens until recently have taken as inevitable and for granted, and an owner's asserted right to use his property as he wishes." (201 N.W.2d, 767). Shifting into high gear, the court declared:

> It seems to us that filling a swamp not otherwise commercially usable is not in and of itself an existing use, which is prevented, but rather is the preparation for some future use which is not indigenous to a swamp. . . .
>
> The shoreland zoning ordinance preserves nature, the environment, and natural resources as they were created and to which the people have a present right. The ordinance does not create or improve the public condition but only preserves nature from the despoilage [sic] and harm resulting from the unrestricted activities of humans (201 N.W.2d, at 770-1).

This environmental "Magna Carta" was beyond the call of duty or demands of this modest case. To other courts around the country, it was probably seen as beyond the Constitution as well. But *Just* remains "the law" in Wisconsin and its rhetorical flourishes continue to thrill students and practitioners of natural resource management.

The trilogy of cases discussed in this section delimit the range of possible judicial response to floodplain and wetland regulations. *Morris County* reflected the cautious preenvironmental approach with Holmes looking over the shoulder. *Turnpike Realty* broke new ground through careful judicial craftmanship. *Just* was a glimpse of utopia. All three cases are cited frequently, although challenges to local floodplain and wetland regulations declined in the 1980s. The constitutionality of such measures is now well accepted.

PUBLIC ACQUISITION

Reasons for Public Acquisition

Certain public objectives regarding the use of land cannot adequately be met through regulation of private land. Public land ownership is required to provide sites for schools, parks, fire and police stations, airports, sewerage and water treatment plants, as well as corridors for streets and highways. Where actual *public use* is involved, the Fifth Amendment requires payment of "just compensation" to the owner. Police power regulations such as zoning may be used to protect the "public health, safety, and welfare" but not to provide outright public use. As in *French*

Investing, discussed earlier in this chapter, an attempt to impose public use through regulation is likely to be struck down as a "taking" without compensation.

Beyond sites for public facilities, public ownership may be desired to maintain tight and long-term control over certain types of land by removing them from the uncertainties of the private market. Thus, areas such as wetlands; coastal barriers; wild lands; or sites of scenic, historic, or cultural importance may best be protected through public ownership. Although some of those areas may arguably be protected through the regulatory power (provided public access is not involved), public acquisition provides greater security against subsequent political change and eliminates worry about the taking issue.

Each level of government—municipal, county, state, and federal—may acquire and own land. Public entities differ, however, in their objectives, fiscal resources, and management policies regarding land. Some local governments diligently pursue long-term open space acquisition programs. Others acquire little or no land for "amenity" reasons but instead may acquire land for resale or lease to promote economic development. What government may acquire, it usually may also sell. Favoritism and abuse sometimes debase the buying and selling of public land.

Means of Land Acquisition

The usual means of acquiring public ownership of real property, as depicted earlier in Figure 6–8, are the following:

- negotiated purchase,
- eminent domain,
- gift or dedication, and
- tax default.

Negotiated purchase is the most common way to acquire land for public use. Once authorized by legislation, a governmental agency may approach a private owner like any other prospective buyer. However, a public buyer must obtain one or more independent appraisals of the value of the property and cannot normally pay more than the appraised value so established. (A slight increase may be allowed to avoid condemnation proceedings.) If the offer is accepted, a deed is exchanged for payment and the land becomes publicly owned. No court action is required, the process is reasonably speedy, and everyone is presumably satisfied. If it is impossible to reach agreement on a price within the limit of appraised value, the effort must either be abandoned or pursued through condemnation.

Eminent domain (also known as condemnation) is an inherent power of government in the United States. It is the compulsory purchase of land from a private owner in exchange for "just compensation" as established by a jury. It is a constitutional prerogative as long as the purpose and value paid are appropriate. (Condemnation should be distinguished from "confiscation," which is an illegal taking without compensation.)

Eminent domain pursuant to legislation appeared in London as early as John Nash's Regent Street project in the 1820s and was used on a larger scale in Haussmann's Paris boulevards. Land was "taken" for New York's Central Park in 1856 and eminent domain gained increasing use by municipalities thereafter. Its use by states and the federal government accompanied the expansion of their respective activities in the twentieth century. The New Deal involved a vast increase in public works projects such as highways, parks, dams, and airports, which required eminent domain on a large scale. The use of the power by all levels of government peaked in the 1950s and 1960s.

The widening use of eminent domain in programs funded by the federal government was facilitated by judicial expansion of the meaning of "public use." Early in the twentieth century, it was generally believed that the Fifth Amendment required literal use by the public to justify condemnation of private land. In 1936, the use of the power to acquire slum property for clearance and residential redevelopment was upheld by the New York Court of Appeals in *New York City Housing Authority* v. *Muller,* 1 N.E.2d 153.

The concept that "public use" might include a "public purpose" without actual use by the general public was confirmed in the landmark 1954 U.S. Supreme Court decision in *Berman* v. *Parker* 75 S.Ct. 98. Berman challenged the condemnation of his store by the Urban Renewal Authority of the District of Columbia to be cleared and resold to another private enterprise. He argued that the premises was not "blighted" and the resulting use would benefit another businessman, not the public. Nevertheless, Justice William Douglas in a much-quoted opinion upheld the taking as necessary to the achievement of the urban renewal project which was a legitimate public activity. (His opinion is quoted at page [8-46] supra.) The uses and abuses of eminent domain by the federal government are discussed further in Chapter 11.

One potential use of eminent domain that is not generally sanctioned in the United States is the taking of more land than is needed for a project to lease or resell later at a higher price. This practice, known as "excess condemnation" has been used in Europe, beginning with Haussmann's rebuilding of Paris. But it has been frowned upon by U.S. courts. While, arguably, it is a "public purpose" for government to recoup some of the increment of value that its programs bestow on adjoining land, it has not been viewed as a proper use of eminent domain simply to speculate in the land market.

In one case to the contrary, the New York Court of Appeals in 1963 upheld the taking of a private business for an economic redevelopment project (the World Trade Center) although the site was not functionally related to the project but was taken "solely for the production of incidental revenue . . ." (*Courtesy Sandwich Shop Inc.* v. *Port of New York Authority* 190 N.E.2d 402). The court held that the taking was only an incidental extension of a site required for public use, and was related to the overall purpose of bolstering the city's economic development.

While the use of eminent domain to speculate in the land market is suspect, the public may engage in "land banking" through negotiated purchase. Land may thereby be acquired either for future public use or for eventual lease or resale. This concept was widely advocated during the 1960s but, like transfer of development rights, has been little utilized.

Eminent domain remains a fundamental power of government in the United States but is used more sparingly than in the 1950s and 1960s. Legally, the only major issue is usually that of value, which is referred to a jury (with sometimes extravagant results). Eminent domain is a politically unpopular technique used generally as a last resort when negotiation fails. Sometimes, however, eminent domain is mutually agreeable as a means of efficiently clearing title to private land where the owner is willing to sell.

Gifts of land to public entities are an important source of open space and natural areas. Many national, state, and local parks originated in a philanthropic gift. Examples include portions of Acadia, Great Smokies, and Grand Teton National Parks donated to the National Park Service by the Rockefeller family and the Harriman State Park along the Hudson River Palisades opposite New York City. The writer's community of Northampton, Massachusetts is graced with Look Park and Childs Park, both donated to the city decades ago together with an endowment toward their upkeep.

Gifts of real property to public or nonprofit organizations are usually recognized in federal tax law as charitable contributions. Donors can take the current appraised value of the donated land as a deduction from their taxable income. *Dedication* of land is a "gift" by a developer to the municipality of streets, school sites, or park areas. See discussion of "subdivision exactions" at pp. 224–7 *supra*. This tax subsidy for land donations applies also to gifts of interests less than fee (e.g., a conservation restriction that severs the development rights). It also applies to gifts to nonprofit charitable organizations such as colleges, churches, and eligible conservation organizations, and land trusts.

Tax default is another means of "acquiring" public land at the local level, although it usually is not needed or wanted for public purposes. If the owner fails to pay property taxes on real property, the site is subject to a court lien and possible sale. Occasionally, a tax-delinquent tract may be retained for public purposes, but the former owner may redeem the property on payment of back taxes and interest within a specified period of time.

Costs of Public Acquisition

Public land acquisition through negotiated purchase or condemnation is expensive, especially in the face of metropolitan development. Costs per acre in 1989 may range from about $2,000 for raw farmland not yet ripe for development to more than $100,000 for sites snatched from the bulldozer in prime locations. Clearly, early action on land needed for public purposes may save much tax money. Beyond a certain point, public acquisition becomes prohibitively expensive. Furthermore, the direct cost paid to the seller is only part of the total cost of acquisition. The following are some types of costs of public land acquisition.

1. *Direct cost* of the purchase price reached through negotiation with the owner or through court determination in eminent domain. "Just compensation" implies full market value as estimated by independent professional appraisers. *Appraised* value is usually established by comparison with recent private sales of similar property in the vicinity. The *assessed* value of a parcel of land listed in local tax records is usually not accepted as evidence of market value since assessments tend to be lower than fair market value (even in states that require "full value" assessment).

2. *Legal and administrative costs* associated with the purchase. These include the cost of negotiation, appraisal fees, land survey, title search, and preparation of legal documents. Where federal or state funding is sought, preparation of the necessary paperwork is another project expense.

3. *Financing cost* of interest on money borrowed to pay the purchase price. Like private buyers, governments often borrow to pay for large capital outlays. Financial instruments for this purpose include general obligation bonds (repayable from general tax revenues) and revenue bonds (repayable out of user fees, if any, generated by the facility funded by such bonds). Even where a government can pay the full purchase price up front, there is an implicit loss of future interest on such capital if it were otherwise invested (referred to by economists as "opportunity cost").

4. *Cost of taxes foregone.* Public land is "off the tax roll" and is exempt from local property taxation. This may be a large loss to local governments if the site might have been developed privately as a high-revenue-earning facility such as a shopping center or office park. The cost of tax revenue foregone naturally falls on the local taxing jurisdiction (municipal or county) regardless of what level of government acquires the land. Local governments, therefore, sometimes oppose the acquisition by the state or the federal government of developable land within their jurisdictions. Also, a facility such as a state park or a military base may inflict burdens on the locality in terms of demands for police, schools, water, and sewage treatment. States and the federal government offset lost tax revenue and the fiscal impacts of their facilities by making "payments in lieu of taxes" to local governments. Such payments then add to the costs of ownership by the higher unit of government.

5. *Costs of improvement and maintenance.* Land acquired for public use must be improved to serve its purpose. A park, for instance, requires installation of roads, walks, parking, restrooms, sports facilities, playgrounds, nature interpretation centers, and so forth. The installation of these is a major initial cost; maintenance is an annual expense.

6. *Liability insurance.* Lawsuits against public land-management agencies are increasingly common and costly. Formerly, government was protected from suit by the doctrine of "sovereign immunity" but that has withered away in most states. The federal government may be sued under the Federal Tort Claims Act and when it loses a case it pays out of general revenues, i.e., it is a self-insurer. Many states and local governments also self-insure but it is increasingly common for such units to purchase liability insurance like private landowners. A related trend is that parks and playgrounds are increasingly designed for safety, at the expense of excitement. Some facilities have been closed due to the high costs of maintaining and insuring them.

Counterbalancing the many costs of public land are the direct and indirect benefits that they provide. Some of these are direct economic returns in the form of user fees. Other, and more important, benefits are not easily quantified, such as recreation, scenic amenity, biodiversity, environmental quality, and urban separation. Resource economists such as Marion Clawson have attempted to assign monetary values to such intangible but valuable functions of recreation land. Secondary economic benefits are also important, including money spent by vacationers for motels, food, and supplies in "gateway communities" near public parks.

Intergovernmental Cooperation

Given the fragmented political geography of U.S. metropolitan areas described in Chapter 6, it is clearly impractical for each municipality to act independently to meet all of its needs for public land. Land needed for a highway corridor or to protect riparian habitat along a stream may cross local, county, and even state boundaries. Also, the benefits of public open space usually extend beyond the political jurisdiction(s) in which it is situated. Indeed the local government that has tax jurisdiction over a site may prefer to have it privately developed, while neighboring units with no tax interest prefer to have it preserved. Thus, the spatial distribution of the costs and benefits of public open space may be out of sync with each other (Platt, 1972).

Intergovernmental cooperation is therefore needed to acquire and manage land that lies within or affects multiple political units. Regional plans for land acquisition are prerequisite to such cooperation. Such plans may be prepared by a state planning office, a regional planning agency, a council of governments, or by a private civic organization. (The Commercial Club of Chicago, for instance, commissioned Burnham's Chicago Plan of 1909.) Figure 9–7 depicts a more recent open space plan for the Chicago region, prepared by the Northeastern Illinois Planning Commission in 1969. Although somewhat crude in detail, it identifies the approximate locations of land recommended for acquisition, including potential urban trail routes.

Accomplishing such a plan is not simple. Each tract or cluster of tracts

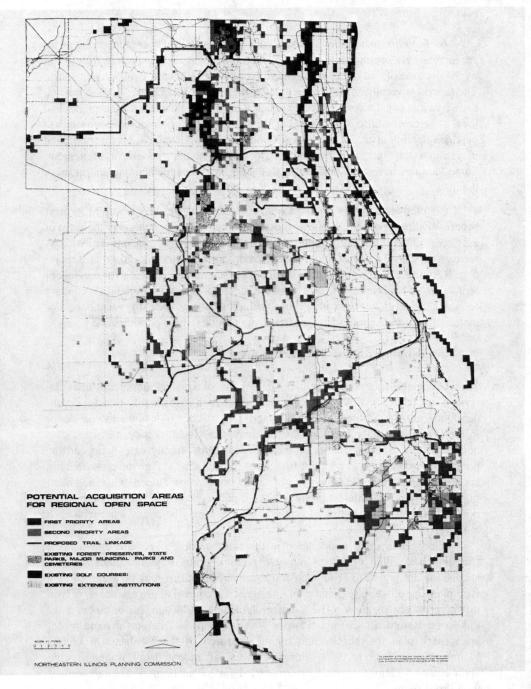

Figure 9.7 Potential acquisition areas for public open space, northeastern Illinois, 1969. *Source:* Northeastern Illinois Planning Commission.

requires an individual strategy, often involving more than one unit or level of government. *Horizontal coordination* involves agreement between neighboring governments that share jurisdiction over a specific resource area. *Vertical coordination* involves agreement between local, state, and federal levels regarding the sharing of costs. The Federal Land and Water Conservation Fund (LAWCON) pays half the cost of eligible open space projects; many states contribute part of the balance.

Such complex arrangements take time. But the private land market does not stand still waiting for government to assemble funds. To rescue key parcels of open land from the bulldozer, intervention by a private conservation organization may be needed. Such organizations play critical roles in many public acquisition programs.

Private Conservation Organizations

There are many kinds of private organizations in the United States that supplement public land acquisition and management efforts. National organizations include The Nature Conservancy (TNC), the National Audubon Society (NAS), and the Trust for Public Lands (TPL). They and others acquire land through purchase and gift but of course they do not have a power of eminent domain. TNC and NAS own substantial reserves of natural land which they manage primarily for ecological protection, education, and research. TPL serves largely as a conduit for open lands which it acquires and then resells to public agencies when possible. TPL sells portions of its holdings for development to defray its operating costs.

Similar functions are performed by a variety of organizations operating at the levels of states, regions, or local communities. The Massachusetts Audubon Society operates a statewide system of sanctuaries which it augments through aggressive acquisition of threatened sites. Sanctuary management balances natural protection, education, and recreation.

Several hundred community or regional land trusts have appeared across the nation since 1970. These may hold land for preservation or serve as agents for public land-acquisition programs. In the latter capacity, a land trust purchases key parcels of open land scheduled for public ownership and holds them until a public agency can muster the funds to buy them from the trust. Many land trusts have "revolving funds" or lines of credit from banks which enable them to move quickly to protect land, pending action by slower-paced agencies.

REFERENCES

BABCOCK, R. F. 1960. "The Chaos of Zoning Administration." *Zoning Digest* 12:1ff.

BABCOCK, R. F. and C. L. SIEMON. 1985. *The Zoning Game Revisited*. Boston: Oelgeschlager, Gunn & Hain.

BAUMAN, G. and W. H. ETHIER. 1987. "Development Exactions and Impact Fees: Survey of American Practices." *Law and Contemporary Problems* 50(1): 51–62.

BOLLENS, S. A. and D. R. GODSCHALK. 1987. "Tracking Land Supply for Growth Management." *Journal of the American Planning Association* 53(3): 315–327.

BOSSELMAN, F. P. 1973. "Can the Town of Ramapo Pass a Law to Bind the Rights of the Whole World?" *Florida State University Law Review* 1(1): 234ff.

CALLIES, D. L. and R. H. FREILICH. 1986. *Cases and Materials on Land Use*. St. Paul: West Publishing Co.

CITY OF PETALUMA, 1987. *Petaluma General Plan: 1987–2005*. Petaluma: The City.

COSTONIS, J. J. 1974. *Space Adrift: Saving Landmarks Through the Chicago Plan*. Urbana: University of Illinois Press.

DAWSON, A. D. 1981. "Management of Residential Growth." *1981 Zoning and Planning Handbook*. Chicago: Clark, Boardman.

———. 1982. *Land-Use Planning and the Law*. Hadley, MA: The Author.

DELOGU, O. E. 1986. "Local Land Use Controls: An Idea Whose Time Has Passed." in *1986 Zoning and Planning Law Handbook* (F. A. Strom, ed.). New York: Clark Boardman.

DUNHAM, A. 1959. "Flood Control Via the Police Power." *University of Pennsylvania Law Review* 107: 1098–1132.

GODSCHALK, D. R., D. J. BROWER, D. C. HERR, and B. A. VESTAL. 1978. *Responsible Growth Management: Cases and Materials*. Chapel Hill, NC: University of North Carolina Center for Urban and Regional Studies.

GRUNTFEST, E. C., ed., 1987. *What We Have Learned Since the Big Thompson Flood*. Special Pub. 16. Boulder: University of Colorado Natural Hazards Research Center.

HOYT, W. G. and W. B. LANGBEIN. 1955. *Floods*. Princeton: Princeton University Press.

JUERGENSMEYER, J. C. and R. M. BLAKE. 1984. "Impact Fees: An Answer to Local Governments' Capital Funding Dilemma." in *1984 Zoning and Planning Law Handbook* (J. B. Gailey, ed.). New York: Clark Boardman.

KRASNOWIECKI. J. 1980. "Abolish Zoning." 31 *Syracuse Law Review* 719–731.

KUSLER, J. A. 1983. *Our National Wetland Heritage: A Protection Guidebook*. Washington, D.C.: Environmental Law Institute.

LISTOKIN, D. (ed.). 1974. *Land Use Controls: Present Problems and Future Reform*. New Brunswick, NJ: Rutgers Center for Urban Policy Research.

PALM, R. 1981. *The Geography of American Cities*. New York: Oxford University Press.

PLATT, R. H. 1972. *The Open Space Decision Process: Spatial Allocation of Costs and Benefits*. Research Paper no. 132. Chicago: University of Chicago Department of Geography.

REPS, J. 1964. "Requiem for Zoning". *Planning—1964*. Chicago: American Society of Planning Officials.

SCHNEIDER, W. J. and J. E. GODDARD. 1974. *Extent and Development of Urban Flood Plains*. U.S. Geological Survey Circular 601-J. Reston, VA: U.S. Geological Survey.

SMITH, R. M. 1987. "From Subdivision Improvement Requirements to Community Benefit Assessments and Linkage Payments: A Brief History of Land Development Exactions." *Law and Contemporary Problems* 50(1): 5–30.

TINER, R. W., JR. 1984. *Wetlands of the United States: Current Status and Recent Trends*. Washington, D.C.: U.S. Government Printing Office.

TUSTIAN, R. E. 1984. "TDRs in Practice: A Case Study of Agriculture Preservation in Montgomery County, Maryland" in *1984 Zoning and Planning Law Handbook* (J. B. Gailey, ed.). New York: Clark Boardman.

U.S. BUREAU OF THE CENSUS. 1986. *Statistical Abstract of the U.S.—1986.* Washington, D.C.: U.S. Government Printing Office.

U.S. COUNCIL ON ENVIRONMENTAL QUALITY. 1981. *Environmental Trends.* Washington, D.C.: U.S. Government Printing Office.

———. 1978. *Our Nation's Wetlands.* Washington, D.C.: U.S. Government Printing Office.

WHITE, G. F. 1964. *Choice of Adjustment to Floods.* Research Paper no. 93. Chicago: University of Chicago Department of Geography.

———. 1975. *Flood Hazard in the United States: A Research Assessment.* Boulder: University of Colorado Institute of Behavioral Studies.

WHITE, G. F. et al. 1958. *Changes in Urban Occupance of Flood Plains in the United States.* Research Paper no. 57. University of Chicago Department of Geography.

WHYTE, W. H., JR. 1958. "Urban Sprawl" in *The Exploding Metropolis* (Editors of *Fortune*). Garden City, NY: Doubleday Anchor.

———. 1968. *The Last Landscape.* Garden City. NY: Doubleday & Co.

WOODBURY, S. R. 1975. "Transfer of Development Rights: A New Tool for Planners." *Journal of American Institute of Planners* 38(1): 3–14.

10

MANAGEMENT OF THE FEDERAL LANDS

INTRODUCTION

The federal government of the United States owns outright about 730 million acres, or about one-third of the nation's total land and water area. This vast expanse, about the size of India, is unevenly distributed spatially. Alaska alone accounts for about 313 million acres; the balance is predominantly located in the sparsely settled mountain and arid regions of the western states. Several states are substantially owned by the federal government, for example, Alaska (96 percent), Nevada (86 percent), Utah (66 percent), Idaho (63 percent), and Oregon (52 percent). Federal ownership east of the Rocky Mountains is much sparser, although most land beyond the Appalachians was at one time federally owned. This chapter reviews the history and policy conflicts that have characterized management of the federal lands since the nation's founding.

THE PUBLIC DOMAIN

The vast majority of federal lands, past and present, belong to what is known as the *public domain*. The public domain consists of lands that came into federal ownership in massive transfers by virtue of cession or purchase from a previous sovereign owner. Excluded from the public domain are federal lands acquired by purchase or gift from private owners, as has occurred in many locations since 1960. The distinction between public domain and other federal land lies in legal technicalities: For the sake of simplicity, this discussion uses the terms *public domain* and *federal lands* interchangeably.

At the time of its independence, the United States extended from the Atlantic to the Mississippi River, excluding Florida and the Gulf Coast, which still belonged to Spain and France. Several of the newly formed states claimed territories extending westward to the Mississippi. But in settlement of Revolutionary-War debts, seven states between 1784 and 1786 ceded their claims to land beyond the Appalachians to the national government. Thus began the public domain and presented the nation's founders with a literal embarrassment of (land) riches. According to Treat (1962, p. 7):

> It seems paradoxical on the face of it that a Congress too poor to own and maintain a capital, too weak to protect itself from the insults of a band of ragged mutineers, should yet be concerned with the disposal of a vast domain of over 220,000 [actually 370,000] square miles of the richest of virgin soil.

As if this were not enough, Thomas Jefferson's Louisiana Purchase in 1803 added another 831,000 square miles, extending to the Continental Divide (western boundary of the Mississippi-Missouri drainage system). This acquisition doubled the territory of the nation and tripled the public domain (Clawson, 1983, p. 17). Later territorial cessions and purchases further expanded the public domain, which peaked in 1867 with the acquisition of Alaska. At one time or another, 1,800 million acres (2.8 million square miles) have belonged to the public domain, of which 730 million acres remain today. The disposal of nearly 1,100 million acres clearly was the dominant feature of federal land management until the late nineteenth century (Figure 10–1).

Disposal

The fate of the vast empire of unsettled public land was substantially influenced by a law adopted before the U.S. Constitution was adopted. The Land Ordinance of 1785 was the most important act of the short-lived Confederation. Its chief contribution was to establish the rectangular Federal Land Survey as a means of organizing the management, disposition, and use of the public domain. From its "point of beginning" in 1785 where the Ohio River crosses the western boundary of Pennsylvania, the federal survey has imposed its grid on the nation's hinterland to the Pacific Ocean, as far west as the Aleutian Islands and as far south as Key West. It ultimately covered all of the continental U.S. except for the original thirteen states, three states carved out of them (West Virginia, Kentucky, Tennessee), Texas, and parts of Ohio. (See technical description of the federal survey in Chapter 5 supra.)

The Land Ordinance and its survey thus prepared the way for disposition and settlement of the federal lands, initially the Northwest Territories of Ohio, Indiana,

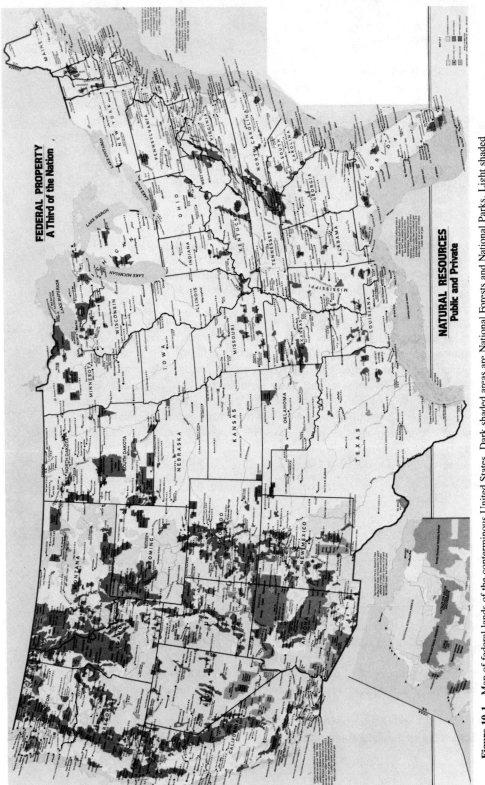

Figure 10.1 Map of federal lands of the conterminous United States. Dark shaded areas are National Forests and National Parks. Light shaded areas in western states are administered by the Bureau of Land Management. *Source*: Public Land Law Review Commission.

Illinois, Michigan, Wisconsin, and part of Minnesota. But the federal objectives in disposing of its land were by no means settled. Competing policies advocated by Thomas Jefferson and Alexander Hamilton, respectively, proposed: (1) disposal free of charge or for a modest fee to promote agrarian settlement and a buffer against Indian or foreign attack; and (2) disposal at a significant price to augment the nation's meager cash reserves. The debate was further complicated by a schism within the "settlement school" between those advocating the New England pattern whereby towns were settled incrementally on the edge of the already consolidated territory, versus the plantation model of extensive settlement of large units, widely spaced from each other (Treat, 1962).

The Land Ordinance compromised between the Jefferson and Hamilton proposals. Settlement, on the New England plan, was served by the requirement that land be precisely surveyed prior to sale and that it be conveyed in parcels of 640 acres (one section) or fractions thereof, reasonably suited to agricultural settlement, at least in the Midwest. The revenue objective was served by a requirement that land be sold at public auction with "substantial" minimum price of $1 per acre (raised in 1796 to $2!).

The disposition issue was not, however, settled by this apparently equitable solution. According to Hibbard (1965), the land disposal situation in 1800 was complicated by several factors: (1) granting of free lands to veterans of the Revolutionary War; (2) conveyance of large tracts to private companies for resale to settlers; (3) speculation by persons who bought land cheaply from either of the two preceding classes of recipients and resold it below the prevailing federal market price. These factors undermined the revenue objective since relatively little land was purchased directly from the federal government.

An additional complication was the prevalence of illegal squatters who entered and settled federal land without any legal title. Technically law breakers, these pioneers epitomized the westward movement. During the early decades of the nineteenth century, support increased among northern and "western" states to grant a *"right of preemption"* to such settlers, to allow them to buy the land for a token price before speculators could bid. The southern states through their spokesman Henry Clay opposed preemption on the ground that it would reward illegal conduct (and incidentally foster settlement from nonslave parts of the country). The Preemption Act of 1841 awarded the right to purchase 160 acres at $1.25 per acre without competition. While technically still a revenue measure, it further undermined the principle of sale to the highest bidder (Hibbard, 1965).

The revenue policy was further eroded by wholesale Congressional land grants to many kinds of recipients during the mid-nineteenth century, for example, public improvement companies, states, and land grant schools (Table 10–1). Grants for public improvements, including canals, national roads, and railroads totaled 125 million acres, and those to states for various purposes eventually totalled 328 million acres. The railroad grants, which foreshadowed contemporary grants in aid for economic development, conveyed land in a checkerboard of square-mile sections. Alternate sections were conveyed and retained within several miles of the

Table 10-1 Disposition of public lands

Total Public Domain (1781–1867)	1,837 million acres
Minus Grants to states	328 million acres
Grants to railroads	94 million acres
Grants to homesteaders	287 million acres
Other	435 million acres
Total Dispositions (1803–1975)	*1,144 million acres*
Remaining Public Domain (1975)	*693 million acres*

SOURCE: Bureau of Land Management, *Public Land Statistics: 1975*, Tables 2 and 3.

proposed right of way. It was expected that the retained sections would become more valuable due to demand generated by access to the railroad. But most settlers bought land (often to their grief) from the railroad itself and once again, hopes of revenue proved illusory.

The threadbare revenue objective was largely abandoned with the passage of the Homestead Act in 1862. After decades of controversy, the Act was finally adopted upon the secession of the southern states from the Union. It provided that settlers would receive clear title to 160 acres upon filing a claim and living on the land for a period of five years. The new policy was claimed as a victory by the western states who foresaw a massive influx of new settlers. Statistically, it appeared to have this effect: During the decade 1870–1880, some 140,000 claims were filed involving 16 million acres of land (Hibbard, 1965, p. 396). Altogether, nearly 300 million acres have been conveyed under the Homestead Act.

But the results were often tragic. Gates (1962) notes that during the decade following adoption of the Homestead Act, Congress granted 127 million acres to railroads, five times the total of the previous 12 years. This land, by definition, was the most accessible to rail service; potential homesteading sites were perforce isolated and often of lesser quality than valley lands favored by the railroads. Furthermore, 160 acres was simply insufficient to support a family in the arid west without access to irrigation. The homily that "rainfall follows the plow" proved to be a public relations myth perpetrated by railroads and land speculators. Homesteaders either had to obtain additional land, legally or illegally, or starve. No one has described the plight of the 2 million homesteaders better than Wallace Stegner (1953/1982, pp. 220–1):

> Suppose a pioneer tried. Suppose he did (most couldn't) get together enough money to bring his family out to Dakota or Nebraska, or Kansas or Colorado. Suppose he did (most couldn't) get a loan big enough to let him build the dwelling demanded by the [Homestead] law, buy a team and a sodbuster plow. . . . Suppose he and his family endured the sun and glare on their treeless prairie, and were not demolished by the cyclones that swept across the plains like great scythes. Suppose they found fuel in a fuelless country . . . and sat out the blizzards and the loneliness of their tundra-like home. Suppose they resisted cabin fever, and their family affection withstood the hard

fare and the isolation, and suppose they emerged into spring again. It would be like emerging from a cave. Spring would enchant them with crocus and primrose and prairies green as meadows. It might also break their hearts and spirits if it browned into summer drouth.

Land continued to move out of the public domain at an incredible pace during the late nineteenth and early twentieth century. Of 1.3 billion acres in the public domain in 1850, about half were distributed to individuals by 1909 either for a price or as a free grant. Another 11 percent were granted to states for various purposes (Hibbard, 1965, p. 529). Accompanying this massive change of ownership was much fraud and misuse of lands obtained for particular purposes. For instance, lands conveyed on condition of physical improvement by the recipient in the form of drainage, irrigation, or timber planting were frequently never used as specified. But according to Clawson (1983, p. 25):

> The process of land disposal was a lusty affair—a headlong, even precipitous process, full of frauds and deceits, but one which transformed a great deal of land into valuable private property—and one which built a nation. Historians, political scientists, conservationists and others have frequently criticized the speed, disorder, fraud, and other undesirable features of the long disposal process. If the process had been slower, with more opportunity to learn from experience, some mistakes possibly could have been avoided. But would even worse consequences have ensued?

Retention

At the very peak of the disposal binge during the last third of the nineteenth century, a new conflict regarding the public domain was emerging, namely, between *disposal* and *retention* of federal lands. Prior to the Civil War, there seemed to be consensus at least that federal land should be conveyed, not retained. But by the 1860s the effects of profligate squandering of the nation's natural resources were becoming inescapable. According to Steward Udall (1963, p. 66): "It was the intoxicating profusion of the American continent which induced a state of mind that made waste and plunder inevitable." The lumber industry moved rapidly westward from Maine to Minnesota, and on to the Pacific Northwest, stripping forests and leaving a wasteland. This in turn caused soil erosion which clogged streams and lakes. In the Rocky Mountains and California, mining fever drew thousands of fortune seekers from the East. Like the lumber industry, mining severely damaged the natural environment and left behind a swath of abandoned "boom towns," denuded hill slopes, and polluted streams. The pursuit of oil and gas in the late nineteenth century caused equivalent devastation to many areas in the Southwest.

In 1864, George Perkins Marsh, an inquisitive Vermonter, published a seminal treatise: *Man and Nature Or, Physical Geography as Modified by Human Action*. In company with Chadwick, Olmsted, and (later) Howard, Marsh was a versatile amateur. Lacking formal training in natural science or geography, Marsh produced the first and one of the most celebrated studies of human impacts on the natural environment. According to Lowenthal (1965, p. ix):

> Few books have had more impact on the way men view and use land. Appearing at the peak of American confidence in the inexhaustibility of resources, it was the first book to controvert the myth of superabundance and to spell out the need for reform. . . . *Man and Nature* was indeed "the fountainhead of the conservation movement."

Marsh was inspired by the contemporary German geographers Alexander von Humboldt and Carl Ritter whose research on human alteration of natural regions underlay the subsequent development of geographical inquiry in Europe and the United States (Hartshorne, 1939). Marsh traveled and read widely, recording evidence, particularly from European history, of deforestation, soil erosion, loss of biological diversity, and hydrologic alteration. He related his findings with a normative, even moralistic tone that anticipated the rhetoric of the environmental movement a century later:

> Man has too long forgotten that the earth was given to him for usufruct alone, not for consumption, still less for profligate waste . . . Man everywhere is a disturbing agent. Wherever he plants his food, the harmonies of nature are turned to discords. The proportions and accommodations which insured the stability of existing arrangements are overthrown. Indigenous vegetable and animal species are extirpated, and supplanted by others of foreign origin, . . . Of all organic beings, man alone is to be regarded as essentially a destructive power . . . [against which] nature . . . is wholly impotent (Lowenthal, ed., 1965, p. 36).

Marsh's plea joined a growing chorus of lamentation against the despoliation of the nation's natural resources. It served as scientific counterpoint to the romantic yearning for a vanishing America expressed in the great Hudson River School paintings of Thomas Cole, Frederick Church, and Albert Bierstadt; the bird paintings of John James Audubon; the poetry of William Cullen Bryant, Emily Dickenson, and Henry Wadsworth Longfellow; the essays of Henry David Thoreau and Ralph Waldo Emerson, and the novels of Nathanial Hawthorne and James Fenimore Cooper (White and White, 1964; Nash, 1982).

In the 1880s, John Muir, the wilderness mystic of the Sierra Nevada began to fan the flames of public indignation. From his base in San Francisco, he divided his time between treks into the mountains (where he allegedly relished climbing tall trees during thunderstorms) and courting the eastern establishment through essays, poems, and lectures (Fox, 1951). In 1891 he founded the Sierra Club, and his influence peaked in the Hetch Hetchy dispute two decades later.

A very different voice from the West was that of Major John Wesley Powell, one-armed geologist, geographer, ethnologist, and exemplar of scientist in service to government. Powell gained fame for his explorations of the Grand Canyon of the Colorado River in 1869 and 1871. These were followed by survey expeditions through the Colorado Plateau and the arid regions of the Columbia, Rio Grande, and Missouri River basins. His 1878 *Report on the Lands of the Arid Region* and subsequent reports of the U.S. Geological Survey (U.S.G.S.), (of which he was the second director and, in effect, founder) documented the unworkability of existing national policy toward disposition of the public domain in arid areas (Stegner 1953/1982). Powell advocated the need to organize human use of western land in relation to its physical limitations, especially water. He urged that allocation of public land be based, not on political determinations in Washington, but rather on scientific appraisal of the physical resources of the area in question. The U.S.G.S. topographic map system was initiated under his direction.

Powell's advocacy of large-scale dams and irrigation projects, which served as a blueprint for the future work of the Bureau of Reclamation and the Army Corps of Engineers in the West, earned Powell the contempt of today's environmentalists (e.g., Reisner, 1986). But his criticism of prevailing national policies lent authority to arguments against heedless disposal of the public domain. And his insistence on a scientific basis for government decisions regarding land anticipated the modern practice of environmental impact assessment.

Gradually, the diverse voices of Marsh, Muir, Powell, and others collectively eroded the political viability of disposal per se as the public land policy. As perception of the impact of human actions on natural resources changed, so too, eventually, did the laws and policies governing such actions.

The beginning of a new retention philosophy was marked in the establishment of Yellowstone Park by Congress in 1872. The spectacular scenery, lakes, waterfalls, wilderness, hot springs, and geysers of this 2-million-acre preserve, carved out of the public domain, remain one of the nation's leading national parks. It was to be an exemplar for the national parks of the United States and the world (Nash, 1982, p. 108). This was followed in 1891 by federal legislation establishing "forest reserves," under which 40 million acres were withdrawn from the public domain. This was the beginning of the national forests, which today total 187 million acres. Also in 1891, New York State established its Adirondack Forest Preserve to be "kept forever as wild forest lands" (Nash 1982). (Much of the 6-million-acre Adirondack Park today remains in private ownership, however, with considerable development in progress (Kunstler, 1989).

Reformist fervor in the early twentieth century yielded two additional federal land-preservation programs. In 1903, the National Wildlife Refuge System originated with the designation of Pelican Island in Florida as a refuge by President Theodore Roosevelt. By 1984 the system had grown to include 427 units totaling more than 90 million acres. In 1916, the National Park Service was established by Congress. Its holdings now total about 75 million acres, of which 40 million are in Alaska.

CONFLICT AND MANAGEMENT IN THE TWENTIETH CENTURY

At the opening of the twentieth century, disposal and retention were thus competing as alternative public land policies. But soon to emerge was still another round in the succession of conflicts that have characterized United States public land history.

This involved a schism among the retentionists themselves, and, specifically, between the wilderness ethic of John Muir and his admirers and the philosophy of "rational use of public lands" advocated by the first director of the U.S. Forest Service, Gifford Pinchot.

The venue of the conflict was the pristine Hetch Hetchy Valley on the west slope of the Sierra Nevada adjoining Yosemite National Park (which had been established in 1890). The 1906 San Francisco earthquake and fire had forcefully demonstrated the inadequacy of that city's local water supply. The quest for a reliable external water source focused on Hetch Hetchy, which was proposed to be dammed and flooded. The valley, which was designated as "wilderness preserve" by Congress, would remain public land but would be physically transformed. Pinchot, a Yale-educated forester and Theodore Roosevelt's chief conservation advisor, supported the project. John Muir passionately opposed it, for example:

> Dam Hetch Hetchy! As well dam for water-tanks the people's cathedrals and churches, for no holier temple has ever been consecrated by the heart of man (quoted in Nash, 1982, p. 168).

The conflict which raged from San Francisco to Washington crystallized the dilemma in the management of retained public lands between preservation in their natural state *versus* human use. Muir's position, being absolute, was simpler in concept: "wilderness is wilderness"—it cannot be tampered with. Pinchot represented the orthodox "conservation" position as the term was then used, namely that natural resources should be managed rationally to promote the public welfare. The handwriting of Powell in this doctrine is manifest, (although Powell himself took little interest in forests and died in 1902) (Stegner, 1953/1982, p. 354).

When Hetch Hetchy erupted in 1908, President Theodore Roosevelt was caught in the middle between his trust in Pinchot and his admiration for Muir (Nash, 1982, pp. 162–4). Ultimately the decision to dam the "cathedral" was made by President Woodrow Wilson in 1913. But in the words of Nash (p. 180): "The preservationists had lost the fight for the valley, but they had gained much ground in the larger war for the existence of wilderness."

This point was exemplified in the Echo Park Dam controversy 50 years after Hetch Hetchy. In that case, opponents led by the Sierra Club successfully blocked construction of a Bureau of Reclamation dam in a valley within Dinosaur National Park in Utah and Colorado. In the 1960s, the Bureau constructed the Glen Canyon Dam on the middle reach of the Colorado despite similar opposition. But this project, whose 186-mile-long impoundment is named Lake Powell after the bureau's patron saint, was one of the last major dams to be constructed in the United States, for both economic and environmental reasons. A subsequent proposal by the bureau to dam a portion of Grand Canyon was handily demolished by outraged environmentalists in 1966. The ill-advised Teton Dam built by the bureau in Idaho collapsed on June 5, 1976 before its completion (Reisner, 1986, Chs. 8 and 11).

Multiple Use

Not every tract of public land is a Hetch Hetchy or a Grand Canyon. Hundreds of millions of federal acres consist of undistinguished forests, grasslands, desert, tundra, and wetlands. (Table 10–2.) While most of this is "wilderness" in the sense of absence of visible human presence, even the most ardent preservationist has not called for the nonuse of *all* federal land. On the other hand, total laissez faire of the nineteenth century variety is scarcely a viable political position today. The era of the public domain as an unregulated commons ended with the Taylor Grazing Act of 1934 which established grazing districts in place of the chaotic and mutually destructive practice of unlimited grazing of private livestock on federal grasslands. The "open range" thus gave way to range management, as has the been the case with minerals, timber, water, recreation, and other valuable resources of the federal lands.

Thus a compromise between total control and no control is required for federal land not designated as wilderness or in the National Park System. Powell and Pinchot have prevailed and resource management, as their "conservation" is now called, has been national policy since the 1920s.

But this scarcely has led to a quieting of the passions that the public domain has generated since the nation's founding. To the contrary, management of public lands has been if anything more contentious than non-management. Marion Clawson (1983, pp. 39–59) refers to the period since 1960 as one of "consultation and confrontation."

The central concept guiding public lands management since 1960 has been "multiple use." This concept was declared national policy in two Congressional acts one concerning the Forest Service (USFS) in 1960 and the other, the Bureau of Land Management (BLM) in 1964. Together these agencies account for nearly 600 million of the 730 million acres of retained federal land. (Table 10–2). Both acts were replaced by the Federal Lands Policy and Management Act of 1976 that further broadened federal land planning objectives. Thus, instead of listening chiefly to its

Table 10–2 Holdings of major federal land-management agencies (1984)

Department of the Interior	
Bureau of Land Management (BLM)	398 million acres
National Park Service (NPS)	75 million acres[a]
Fish and Wildlife Service (FWS)	43 million acres
Department of Agriculture	
U.S. Forest Service (USFS)	188 million acres
Other	41 million acres
Total Federal Land Ownership	*745 million acres*

SOURCE: Clawson, 1983, Table 2–2

[a]U.S. Bureau of the Census, 1985, Table 380

timber company constituents (who lease timber rights), U.S.F.S. is required to incorporate watershed management, wildlife habitat protection, and recreation into its management plans for each national forest. Similarly, BLM must weigh those needs alongside its traditional grazing, mining, and timber interests (Cutter, Renwick, and Renwick, 1985, pp. 176–7). But as Clawson (1983, p. 137) suggests:

> The meaning [of multiple use] implied by those acts and in today's popular usage expresses a desire for a kind of management that is defined in the minds of the users rather than in specific instructions to the agencies. Almost everyone supports the general idea; it is its translation into practice that produces controversies.

Public participation is fundamental to planning procedures by federal land agencies. Draft plans are published and available for public comment, which is often negative. Public hearings are required at each step in the development of plans. Environmental impact statements under the National Environmental Policy Act of 1969 are required for major plans and policy initiatives, and these generate another layer of public involvement and contention. Finally, lawsuits by public interest organizations or special user groups frequently challenge and delay the implementation of proposed actions on federal lands.

An issue of particular interest to western states is that expressed in the "Sagebrush Rebellion" of the early 1980s, namely, whether states may assume control and even ownership of federal lands within their territories. Nevada in 1979 adopted a law making such a claim. While partly symbolic, this and equivalent statements reflect a frustration by states and private economic interests with federal control over the water, timber, grasslands, minerals, and wildlife located on the public domain. Closely allied to this view is the question of outright privatization of certain federal lands or resources. This idea was supported rhetorically by Secretary of the Interior James Watt during the first Reagan Administration, but with little substantive result.

National Forests

Although they include only about 29 percent of total commercial timberlands, the national forests include over 40 percent of the timber volume and account for a large proportion of mature, higher-quality trees, particularly western conifers (Clawson, 1963, p. 93). Western forests, mostly in federal ownership, are estimated to contain about 50 million acres of old-growth sawtimber, of which 40 percent is unexploited virgin forest—a resource virtually unknown in the eastern United States (Landsberg, 1964, p. 162). Such ancient forests possess great attraction in terms of public recreation and wildlife habitat in their natural state and therefore cutting is widely opposed by environmental organizations such as the Sierra Club and Friends of the Earth. On the other hand, foresters argue for increased cutting of these old stands since they grow very slowly in comparison with younger trees that would replace them. The Forest Service is thus caught in a perpetual dilemma regarding the management of its older western forests (Clawson, 1983, p. 82).

Cutting of timber in the national forests is conducted by private companies that buy timber rights from the U.S. Forest Service. Between 1975 and 1984, sales of timber rights in national forests exceeded the volume of timber actually harvested by the buyers. The average annual volume sold during the decade was 10.8 billion board feet (U.S. Council on Environmental Quality, 1984, p. 267). Thus, recent experience indicates that timber companies are in no hurry to harvest the timber that they have purchased as they wait for the market price to rise.

Meanwhile, foreign sources of timber products, especially Canada, continue to increase their share of U.S. domestic consumption. Domestic production and consumption of forest products each rose by about 37 percent between 1960 and 1983, but imports of these products increased by 120 percent, rising from 16 percent to 26 percent of total domestic requirements (U.S. Bureau of the Census, 1985, p. 675). U.S. exports have risen substantially since 1960 but most of that growth occurred before 1970. Exports increased by only 3 percent per year during the 1970s and amounted to only about half of total imports in 1983. The U.S. Forest Service believes that this country will remain a net importer of forest products (Conservation Foundation, 1984, p. 166).

The foregoing trends signify that U.S. forest resources are not unduly stressed. In fact, growing stock of timber on both public and private lands has increased by 18 percent since 1952. This reflects a growth of 56 percent in the East and 48 percent in the South, in contrast to a loss of 5 percent in the more massive but slower growing timber stock of the West (U.S. Bureau of the Census, 1985, p. 672).

National Parks

Within the broad spectrum of federal lands conflicts since 1960, some of the most publicized and bitter disputes have involved the National Park System. In terms of acreage, national parks account for a modest portion of the total federal lands: about 75 million acres in 1984. Forty million acres of this total in Alaska were transferred to the National Park Service (NPS) by Congress for safekeeping in 1978 and are essentially inaccessible. The real park system consists of 35 million acres scattered throughout the 50 states, an increase of ten million acres (40 percent) since 1960.

The National Park System is best known for its "crown jewels": the great western parks of Yosemite, Yellowstone, Rocky Mountain, Grand Canyon, Zion, Bryce Canyon, Grand Teton, Glacier, and Olympic. In the East, Acadia in Maine attracts 4.5 million visitors a year and Shenandoah receives 2 million.

Since 1960 these traditional scenic marvels have been supplemented by a variety of new kinds of NPS facilities established by Congress. Beginning with Cape Cod in 1961, NPS now operates 10 national seashores (nine on the Atlantic and Gulf and one, Point Reyes, on the Pacific) and four national lakeshores on the

Great Lakes. Seventeen national recreation areas have been established in or near urban areas, including Gateway in New York City, Golden Gate in San Francisco, and Santa Monica in Los Angeles. NPS also operates many smaller facilities including historic sites and battlefields, national monuments (including the venerable Muir Woods north of San Francisco), and miscellaneous special facilities such as an industrial museum park at Lowell, Massachusetts (Conservation Foundation, 1985, Ch. 2).

This varied assortment of public sites, both traditional and nontraditional, has provoked endless possibilities for conflicts over management policies of the NPS. Some place-specific and generic issues have involved:

- water supply for the Everglades,
- off-road vehicles at Cape Cod and elsewhere,
- mining in Death Valley,
- whether to move the Cape Hatteras Lighthouse,
- clearcutting timber adjacent to national parks,
- controlling forest fires in Yellowstone and elsewhere,
- airplane flights through the Grand Canyon,
- nude bathing,
- quotas for wilderness backpacking,
- traffic and parking,
- concessions,
- land-use planning in "gateway communities,"
- wildlife management,
- erosion control,
- rights of inholders and enclave communities, and
- sewage and solid waste.

In charting its course through these and other minefields of public contention, the NPS has little statutory guidance from Congress. The 1916 National Park Service Organic Act charged NPS:

> . . . to conserve the scenery and the natural and historic objects and the wildlife therein and to provide for the enjoyment of the same in such manner and by such means as will leave them unimpaired for the enjoyment of future generations (16 U.S.C. Sec. 1).

This mandate thus posed the eternal dilemma for NPS as to how it may both preserve the resources entrusted to it while facilitating public enjoyment of them. Further guidance is provided in authorizing legislation for particular facilities such as the Cape Cod National Seashore Act of 1961. And of course, NPS is subject to the National Environmental Policy Act of 1969 and its own planning regulations.

Conflict and Management in the Twentieth Century

Much of the controversy swirling around the National Park System involves long-standing issues pertaining to the internal management of parks: recreation, traffic, wildlife, primitive areas, off-road vehicles, and so forth. Such issues relate to the essence of the parks, their very raison d'etre—scenery, water resources, biodiversity, quiet, and solitude. Protection of these qualities arouses passionate advocacy even among members of the public and interest groups far from the scene, who may never have visited the site in question. This tradition dates back at least to Hetch Hetchy.

But the users of parks have been equally outspoken on behalf of more facilities for the public, more concessions, parking, campgrounds, and opportunities for specialized pastimes. Visitorship to the National Park System quadrupled from 79 million in 1960 to 332 million in 1984. Increasing mobility of middle-class Americans has facilitated access to and pressure on the "crown jewels." Meanwhile, the newer urban-oriented facilities have attracted visitation far disproportionate to the relatively small acreage that they represent. New recreational pastimes and inventions such as snowmobiles, jet-skis, high-tech rafting, and hang-gliding create demands for special-use areas or privileges within parks. Cultural and ethnic diversity of users (a contrast to the recent past) create additional needs (e.g., bilingual signs). Handicapped and elderly users require specially designed trails and other facilities.

A separate cluster of management issues stems from the intermingling of public and private ownership in many facilities. The great western parks consist of large tracts of NPS land surrounded by other federal land managed by USFS or BLM. Private ownership adjacent to or within these parks consists largely of agricultural or timber holdings. There are few local governments to deal with except for gateway communities like Estes Park at the entrance to Rocky Mountain National Park in Colorado.

A different legal geography applies to NPS facilities elsewhere, especially the national seashores and lakeshores. Those sites were not carved out of the public domain like the western parks but instead have been acquired through negotiated purchase, condemnation, or gift from previous owners. Typically, the authorized areas of these facilities are much larger than the land actually acquired to date, since Congress almost never appropriates enough money to buy the entire park outright. The result is a hodge-podge of NPS land scattered throughout the authorized area, interspersed with private holdings and tracts owned by other federal, state, county, and municipal governments.

This crazy quilt of ownership and jurisdiction presents challenging problems for NPS management. In coastal areas, passions are especially aroused due to the high property values and scarcity of access to the shore. NPS is a mixed blessing to pre-existing communities such as those affected by the Cape Cod, Fire Island, and Cape Hatteras national seashores. Federal acquisition preserves key areas from development, such as the Sunken Forest on Fire Island, a renowned and also well-known ecological preserve. But the NPS mission to facilitate public usage brings traffic, crowds, pollution, and ethnic diversity to previously aloof summer enclaves.

Under such circumstances, NPS is thrust reluctantly into the local and regional planning process. At Cape Cod and Fire Island, Congress authorized NPS to intervene in local development decisions within the authorized park boundaries. Elsewhere, NPS exercises at least the same rights as any other landowner to testify in zoning proceedings and otherwise make its views known. Subdivisions, condominiums, shopping centers, and amusement parks, which are often attracted to the vicinity of national parks, may detract from the quality of the park environment. Adverse impacts include visual blight, traffic congestion, water quality degradation, littering, and reduction of natural wildlife habitat.

Sticky questions also arise where growth-inducing facilities such as roads, water lines, sewers, and electrical lines must be routed across NPS land to serve private inholdings, as at Cape Hatteras. The leverage this provides to NPS is tempered by its desire to promote harmonious relations with its nonfederal neighbors. Park management in these situations involves as much diplomacy as technical expertise on the part of park superintendents and their staffs. Cape Cod National Seashore, the archetype of the new generation of national parks, has gained nearly three decades of experience with a formal advisory committee that represents the various public and private interests of that complex region (Foster, 1987).

Overriding the myriad specific disputes that influence NPS management is the paramount philosophical concern as to what is the purpose of the national parks. Ultimately, should they cater to the desires of the user public (however those may be defined) or should they provide a "park experience" that uplifts and refreshes those who choose to experience it (Conservation Foundation, 1985, Ch. 7). This question has been most cogently addressed by Joseph Sax (1980) who strongly advocates the latter option. Drawing on Thoreau and Olmsted, Sax argues that the park must offer "contemplative recreation" in contrast to "artificial recreation" (e.g., scenic vistas, not water slides):

> Engagement with nature provides an opportunity for detachment from the submissiveness, conformity, and mass behavior that dog us in our daily lives; it offers a chance to express distinctiveness and to explore our deeper longings. . . .
>
> From this perspective, what distinguishes a national park idea from a merely generalized interest in nature may be the special role that the nature park plays as an institution within a developed and industrialized society, in contrast to those traditions in which nature is offered as an alternative to society (Sax, 1980, p. 42).

CONCLUSION

This chapter has briefly surveyed the history and present status of federal land management. The major eras may be identified as: (1) disposal—1785 to about World War I; (2) retention—1872 to the present; and (3) multiple-use management and advocacy planning—1960 to the present. Throughout the nation's history, federal land policy has generated conflicts—revenue versus free disposal, retention

Conclusion

versus disposal, preservation versus "conservation," single use versus multiple use, and "artificial recreation" versus "contemplative recreation." Classes of participants in federal land-decision making include politicians, professional land managers, private economic interests (e.g. mining, timber, water users, grazing interests, recreation providers), adjacent and inholder property owners, the environmental community, and the recreation-seeking public. Each of these major classes is in turn subdivided into subgroups of more specialized interests (e.g., off-road vehicle users versus backpackers).

From the beginning of the public domain in the 1780s, revenue to the public treasury has fared poorly as an object of federal land management. The shift from

Table 10-3 Significant events in the management of the federal lands

1784–6	Cession of "Northwest Territories" to national government
1785	Land Ordinance—beginning of Federal Land Survey
1787	Northwest Ordinance—organization of territorial governments
1803	Louisiana Purchase—doubled size of U.S.
1823–68	Road and canal land grants
1841	General Improvement Act—land grants to ten states
1849–50	Swampland Acts—further grants to certain states
1850–72	Land grants to railroads
1862	Homestead Act—free land grants to settlers
1872	Yellowstone Park established—first national park
1878	Powell Report on the arid lands
1879	U.S. Geological Survey established
1891	Forest Reserve Act—beginning of National Forest System
1902	Bureau of Reclamation established
1905	U.S. Forest Service established
1908–13	Hetch Hetchy controversy—first "environmental" battle
1911	Weeks Act—to purchase private land for national forests
1916	National Park Service established
1920	Mineral Leasing Act
1933–40	New Deal Programs, e.g., TVA, SCS, CCC
1934	Taylor Grazing Act—closing of the "open range"
1953	Submerged Lands Act
1960	Multiple Use and Sustained Yield Act—U.S. Forest Service
1961	Cape Cod National Seashore Act—first national seashore
1964	Land and Water Conservation Fund Act Wilderness Act Classification and Multiple Use Act—BLM
1968	National Wild and Scenic Rivers Act National Trails Act
1969	National Environmental Policy Act
1970	Public Land Law Review Commission Report
1976	Federal Land Policy and Management Act

sale to free grant marked by the 1862 Homestead Act and its sequels marked the first retreat from a revenue policy. But land retained in federal ownership, except for national parks and designated wilderness areas, has been eligible for lease of timber, mining, grazing, and other economic rights to private interests. Clawson (1983), however, reports that federal leasing policies in general fail even to cover the administrative costs of the management agencies (USFS, BLM). The major exception is oil and gas leasing from submerged lands of the Outer Continental Shelf (OCS) which, since the late 1960s, "has produced about two-thirds of the BLM's receipts, completely overwhelming the onshore oil and gas, the timber, and especially the grazing receipts" (Clawson, 1983, p. 91). (Much of the OCS revenue is earmarked for the Land and Water Conservation Fund.)

Recreation remains essentially a free public good with gate fees at NPS and other federal facilities either nonexistent or nominal. Long lines of expensive recreation vehicles, laden with all kinds of costly high-tech recreation gear, are admitted to the nation's parks at a cost of a few dollars per day. Concessionnaires in those parks charge substantial prices for their services and facilities with comparatively little return to the responsible federal agency (Conservation Foundation, 1985, Ch. 5).

With the privatization agenda of James Watt now a footnote in federal lands history, there is little likelihood of a change in the foregoing policies. But the pressures for multiple usage and particularly multiple recreational use of public lands continue to mount, fueled in part by the low cost of entering federal facilities. There will certainly be much debate ahead as to the necessity of limiting access to certain facilities, through quotas, higher fees, or other means, in order to protect the quality of the park experience (as is already the case in wilderness areas). In the future, this issue, among many others, may be expected to provide much grist for continued confrontation and dispute regarding federal lands.

REFERENCES

CLAWSON, M. 1963. *Land for Americans: Trends, Prospects, and Problems*. Chicago: Rand, McNally.

———. 1983. *The Federal Lands Revisited*. Baltimore: Johns Hopkins University Press for Resources for the Future.

CUTTER, S. L., H. L. RENWICK, and W. H. RENWICK. 1985. *Exploitation, Conservation, Preservation: A Geographic Perspective on Natural Resource Use*. Totowa, NJ: Rowman and Allanheld.

CONSERVATION FOUNDATION. 1984. *State of the Environment: An Assessment at Mid-Decade*. Washington: The Foundation.

———. 1985. *National Parks for a New Generation*. Washington: The Foundation.

FOSTER, C. H. W. 1987. *The Cape Cod National Seashore: A Landmark Alliance*. Hanover, NH: University Press of New England.

FOX, S. 1981. *John Muir and His Legacy*. Boston: Little Brown.

GATES, P. W. 1962. "The Homestead Law in an Incongruous Land System." In *The Public Lands,* (V. Carstenson, ed.). Madison: University of Wisconsin Press.

HARTSHORNE, R. 1939. *The Nature of Geography.* Lancaster, PA: The Assn. of American Geographers.

HIBBARD, B. H. 1965. *A History of the Public Land Policies.* Madison: University of Wisconsin Press.

KUNSTLER, J. H. 1989. "For Sale" in *The New York Times Magazine* (June 18), pp. 23 ff.

LANDSBERG, H. H. 1964. *Natural Resources for U.S. Growth.* Baltimore: Johns Hopkins University Press for Resources for the Future.

LOWENTHAL, D., ed. 1975. See Marsh, G. P. 1864/1975.

MARSH, G. P. 1864/1975. *Man and Nature Or, Physical Geography as Modified by Human Action.* (D. Lowenthal, ed.) Cambridge: Belknap Press of Harvard University Press.

NASH, R. F. 1982. *Wilderness and the American Mind.* New Haven: Yale University Press.

———. 1990. *American Environmentalism: Readings in Conservation History* (3rd ed.). New York: McGraw-Hill.

REISNER, M. 1986. *Cadillac Desert: The American West and Its Disappearing Water.* New York: Penguin Books.

SAX, J. L. 1980. *Mountains Without Handrails: Reflections on the National Parks.* Ann Arbor: University of Michigan Press.

STEGNER, W. 1953/1982. *Beyond the Hundredth Meridian: John Wesley Powell and the Second Opening of the West.* Lincoln: University of Nebraska Press.

TREAT, J. P. 1962. "Origin of the National Land System under the Confederation" in *The Public Lands,* (V. Carstenson, ed.). Madison: University of Wisconsin Press.

UDALL, S. L. 1963. *The Quiet Crisis.* New York: Avon Books.

U.S. COUNCIL ON ENVIRONMENTAL QUALITY. 1984. *Environmental Quality–1984.* Washington: U.S. Government Printing Office.

U.S. BUREAU OF THE CENSUS. 1985. *Statistical Abstract of the United States—1985–86.* Washington: U.S. Government Printing Office.

WHITE, M. and W. WHITE. 1962. *The Intellectual Versus the City.* Cambridge: Harvard and M.I.T. University Presses.

11

FEDERAL LAND USE PROGRAMS BEFORE 1970

INTRODUCTION

Twice during the twentieth century, demographic thresholds signaled important innovations in land use control practices. The 1920 Census was the first to report that population of urban places outnumber rural residents. In 1960, the Census found that suburbanites exceeded central city inhabitants for the first time. In both cases, the shift in the demographic center of balance—from rural to urban in 1920 and from city to suburb in 1960—closely coincided with a radical change in the structure of public authority over land use. In the earlier case, the advent of Euclidean zoning marked the assertion of municipal power in relation to private property rights. In the 1960s, the federal government began to reassert the powers with which it had experimented during the New Deal, and to play an increasingly significant, albeit indirect, role in shaping the contemporary American metropolitan region.

Two factors may be cited to account for the 1960 "threshold." First, suburban growth involved a proliferation of increasingly fragmented local governments. While these clung tenaciously to their land use prerogatives, it was gradually recognized that local units are inadequate—spatially, fiscally, and philosophically—to address metropolitan, state, and national needs. Second, spatial redistribution of population was accompanied by a shift in political power in state legislatures and in Congress. This was the result of the 1963 Supreme Court decision in *Baker* v. *Carr* (369 U.S. 186) which declared that electoral districts must be redrawn to reflect demographic shifts to ensure the principle of "one man, one vote."

Under the U.S. Constitution, the federal government lacks authority to directly regulate the use of nonfederal land. According to the Tenth Amendment:

> The powers not delegated to the United States by the Constitution, nor prohibited by it to the states, are reserved to the states respectively, or to the people.

Thus in theory, public power over private land use is vested in the sovereign states, not the federal government. And the states, as discussed in earlier chapters, have long delegated most of their authority to local municipalities and counties. The mighty federal government of the United States is powerless to interfere in a local zoning determination. States for their part have the legal power but often lack the political will to influence local actions, and they generally ignored land use entirely until the 1970s.

Between the 1960s and the late 1980s, however, the relative balance of local government versus state and federal influence over land use decisions shifted markedly toward the latter. As Popper (1988, p. 291) has written:

> Two decades ago, American land use regulation consisted almost entirely of local zoning: it no longer does. Instead it has become increasingly centralized—that is, more likely to originate with regional, state and federal agencies rather than with local ones.

This chapter traces the unfolding of this transition that led up to the Federal Environmental Decade of the 1970s.

NATIONAL PLANNING IN THE NEW DEAL

Federal land use planning, while alien to the Eisenhower Administration in 1960, was not without precedent. President Franklin D. Roosevelt's New Deal during the 1930s had briefly interjected the federal government into the planning and development of the nation's natural resources on a massive scale. Within months after taking office in 1933, Roosevelt established the National Planning Board under the direction of the Public Works Administrator, Harold L. Ickes. During the balance of the 1930s, this Board metamorphized into the National Resources Board (1934–1935), the National Resources Committee (1935–1939), and finally into the National Resources Planning Board (1939–1943). Despite the changes in name, these Boards comprised a continuum of national planning activity with many of the same personnel involved throughout (Clawson, 1983, p. 273). They will be referred to collectively as the "Board."

The Board exemplified the "brain trust" approach to national problems under the New Deal. The Board's volunteer expert committees addressed a dazzling array of planning issues, for example, water resources, agriculture, forestry, soil erosion, economic planning, flood control, industrial siting, and urban and rural land-use planning. The committees were assisted by a small staff based in the Old Executive Office Building adjoining the White House. (Geographer Gilbert F. White, who served on this staff while completing his doctoral degree in geography at the University of Chicago, recalls exchanging poems with FDR in the margins of staff memoranda) (Platt, 1986, p. 63).

In 1933, the Board grouped its activities under four headings:

1. Planning and programming of public works
2. Stimulation of city, state, and regional planning
3. Coordination of federal planning activities
4. Research

Achievement of this ambitious agenda was facilitated by studies conducted in 1929 by President Hoover's Research Committee on Social Trends. In 1933 that committee was, however, "totally out of touch with social and political reality" (Clawson, 1981, p. 32). Within six months after its formation, the Board produced a "Final Report" containing remarkably detailed analyses of land, water, and regional economic issues (National Planning Board, 1934).

The Board advanced the art of regional planning in the United States in two principal sectors: river basin management and regional land use planning. The 1930s experienced not only economic depression but also severe floods and drought. President Roosevelt proposed a massive public works program to construct flood and erosion control projects, while alleviating unemployment. In response, Congress appropriated billions of dollars for river basin projects and other public works.

To counteract the allocation of this money by the usual Congressional practice of "pork barrel," President Roosevelt charged his Planning Board with the development of river basin plans and criteria for water resource project approval. The Water Resource Committee of the Board between 1933 and 1939 prepared a series of reports on water resources, some dealing with specific river basins in detail while others dealt with substantive problems of pollution, hydrology, soil erosion, and small watershed management (Clawson, 1981, Appendix B.)

The Tennessee Valley Authority (TVA) was established by Congress in 1933 as a laboratory for federal involvement with water and land management. Proposals to establish federal authorities for river basins elsewhere met with political opposition and were never carried forward. However, the Board, through its many reports, contributed greatly to the recognition of river basins as logical units for land and water management. It promoted the concept of "multiple objectives" in the comprehensive management of river basins. This would be soon joined by the concept of "multiple means" for achieving a specific objective such as flood loss reduction (White, 1969). Federally sponsored water resource planning revived a generation later under the Water Resources Planning Act of 1965 (until terminated by the Reagan Administration in 1981).

The Board also promoted state and regional comprehensive planning. According to Clawson (1981, p. 167):

> During the period of its existence the NRPB was in the forefront of promoting and developing regional planning. Among the difficult problems, never fully resolved then or later, were how to define a 'region'—by natural features, such as river systems, or by economic interests; whether to follow or to disregard state boundaries; whether to have regions for all purposes follow the same boundaries, and other problems. In its own work, the NRPB used 'flexible' boundaries, the exact contours depending upon different activities.

Generally, the concept of region as used by the Board entailed interstate or subnational units. Regional planning commissions were established for groups of states such as New England and the Pacific Northwest. A principal function of these commissions was to plan and coordinate the flow of federal funds and activities within the regions in question. (In the 1960s this role was assigned to "regional clearing houses" by the Bureau of the Budget.) A closely related NRPB initiative was the formation of state planning boards in nearly every state between 1933 and 1936. Some of these bodies continue to the present while others have been abolished. The Board itself was abolished in 1943 by Congress, which was frustrated by years of the Executive Branch meddling with its prerogatives (Platt, 1986, p. 47).

Various other New Deal programs introduced land planning in one form or another. TVA was the exemplar of federal planning for the betterment of an impoverished region: the drainage basin of the Tennessee River. TVA is best known for its series of main stem dams that harnessed the river for power, navigation, recreation, and flood control. But TVA also developed pioneering programs in soil erosion management, reforestation, economic development, and improvement of housing, medical care, schools, and recreation.

Both TVA and another New Deal program, the Resettlement Administration, dabbled with the planning and construction of small new towns. The latter agency hoped to establish a series of "greenbelt towns" nationally, following the precepts of Ebenezer Howard's Garden City Movement. But only three such towns were actually built: Greenbelt, Maryland; Green Hills, Ohio; and Greendale, Wisconsin (Parsons, 1990).

The Soil Conservation Service (SCS) was created by Congress in 1935 in response to a national soil erosion crisis. SCS, an agency of the U.S. Department of Agriculture, fostered the establishment of soil and water conservation districts in most of the nation's counties. With SCS funding and guidance, the districts assist private landowners to improve land management and development practices.

The New Deal also fostered a new and ultimately pervasive federal impact on the location, design, and occupancy of new residential construction. Kenneth T. Jackson in his book *Crabgrass Frontier* (1985, Ch. 11) attributes the origins of socioeconomic "redlining" of neighborhoods and communities to the appraisal practices of the Home Owners Loan Corporation (HOLC) established by Congress in 1933. HOLC's mission was to provide federal mortgage assistance at low interest rates to forestall owner default on home loans during the Depression. To promote the reliability of lending practices, it conducted surveys of residential neighborhoods. These studies codified prevailing real estate assumptions regarding the effects of race, religion, and wealth on residential property values. Detailed "residential security maps" prepared by HOLC for many cities and suburbs throughout the nation influenced the lending practices of financial institutions and thus became self-fulfilling prophecies. Furthermore, the home loan guarantee programs of the Federal Housing Administration (FHA) beginning in 1934, and the Veterans Administration (VA) after 1944, reflected for many years the racial and economic assumptions regarding neighborhood quality and housing type fostered by the HOLC. The post-war white middle-class, single-family suburb was the result (Jackson, 1985, pp. 206–9).

POST-WAR FEDERAL PROGRAMS

Post-war America was much more interested in building than in planning. Single-family residential subdivisions proliferated with federal low interest loans and mortgage guarantees from the FHA and VA. In 1956, Congress launched the 42,500-mile Interstate Highway System that further promoted development at the fringes of metropolitan areas. These programs produced the suburban population surge noted in the 1960 Census. For central cities, Congress established the Urban Renewal Program in the Housing Acts of 1949 and 1954 to acquire, clear, and redevelop "blighted" urban land.

These federal programs and policies profoundly changed the face of metropolitan America. While technically they did not violate the doctrine that land use is a nonfederal concern, they demonstrated the immense capability of the federal government to indirectly influence—through spending, tax incentives, and technical guidelines—the use of private land. FHA regulations literally specified suburban single-family homes as the approved style of housing to be constructed with its assistance (Fried, 1971, pp. 66–70). Tying strings to federal benefits was thus a means of exerting federal influence over the form of urban development in the 1950s, whether or not so recognized at the time.

Although planning was sometimes equated with "communism" (Cassella, 1983, p. 20), Congress paradoxically at the height of the McCarthy Era enacted Section 701 of the Housing Act of 1954, the nation's first comprehensive planning assistance law. Section 701 authorized "planning grants to state, metropolitan, and other regional planning agencies . . . to encourage comprehensive planning, including transportation planning, for states, cities, counties, metropolitan areas and urban regions, and the establishment and development of the organizational units needed therefor." The 701 program nurtured a generation of plans and planners:

During the 27 years of the program's existence, annual federal appropriations for local planning assistance increased from $1 million to $100 million; by the time the program terminated in 1981, it had allocated more than $1 billion to local planning (Feiss, 1985, p. 175).

To the extent that Section 701 supported local planning, it reinforced rather than resisted the dominance of localism in land use control. But it was supplemented by the Housing Act of 1959 which supported the preparation of comprehensive plans at the metropolitan, regional, state, and interstate levels (Brooks, 1988, p. 243). Federal-level planning is conspicuously missing from this list. But the precedent was established for federal support of nonfederal planning at various scales that would later characterize the Coastal Zone Management Program and other federal initiatives of the 1970s.

Another pre-1960 federal contribution to regional planning was the beginning of a data base of metropolitan statistics. The Metropolitan Statistical Area (MSA) was first applied in the 1950 Census to aggregate demographic, housing, and other census data at a regional scale. The number and boundaries of MSAs have been revised frequently since 1950, as have the range of topics for which regional data are collected. If care is utilized in making inter-census comparisons, the MSA data is an invaluable tool for analysis of metropolitan development (see Chapter 1).

RUMBLINGS OF DISSENT

Two periods of intellectual and public outcry against prevailing land use practices were discussed in earlier chapters. The mid-nineteenth century protest against the waste of natural resources that began with Marsh's *Man and Nature* reversed federal land policy to favor retention and conservation over disposal. The progressive crusade against urban congestion of the early 1900s in its turn prompted the advent of city planning and zoning. The 1960s were to experience a third major wave of dissent and reaction to prevailing land use policies that would later be known as the conservation movement. This in turn would merge into the environmental movement of the 1970s. Contemporaneous with these, but very different in its implications for urban policies, was the Civil Rights Movement.

All this ferment was scarcely foreseen during the complacent 1950s. An important exception, however, was the 1955 Wenner-Gren Conference at Princeton on *Man's Role in Changing the Face of the Earth*. Together with its 1,193 page proceedings volume (Thomas, ed., 1956), this symposium marked a scholarly watershed as significant as that defined by George Perkins Marsh a century earlier (to whom *Man's Role* was dedicated). Seventy-six distinguished scholars described "the multiple impacts of human beings as agents of vast and often fearsome change in the world" (Kates, 1987, p. 529). Like Marsh, *Man's Role* focused on degraded "landscapes" of the world: soils, forests, water, biotic species, with new attention to cities, climate, and wastes (Kates, 1987).

Lewis Mumford (1956), one of the conference convenors, sounded a tocsin of the coming environmentalism, noting that modern metropolitan development tends:

> . . . to loosen the bonds that connect [the city's] inhabitants with nature and to transform, eliminate, or replace its earth-bound aspects, covering the natural site with an artificial environment that enhances the dominance of man and encourages an illusion of complete independence from nature (Mumford, 1956, p. 386).

And pursuing one of his lifelong themes:

> Within a century, the economy of the Western world has shifted from a rural base harboring a few big cities and thousands of villages and small towns, to a metropolitan base whose urban spread . . . is fast absorbing the rural hinterland and threatening to wipe out many of the natural elements favorable to life which in earlier stages balanced off against the depletions of the urban environment (Mumford, 1956, p. 395).

Mumford's concern was geographically cosmic and not based on study of any particular urban region (although he had a love-hate relationship with New York City). By contrast, the French geographer Jean Gottmann in 1961 published the classic study of the urbanized northeastern seaboard of the United States which he named *Megalopolis*. This region was loosely defined by Gottmann as a series of connected metropolitan areas extending from southern Maine to Virginia. He included all adjacent counties having a minimum average density of 600 people per square mile, which included a lot of still-rural land. It also included several of the nation's leading cities: Boston, New York, Philadelphia, Baltimore, and Washington, D.C. (The region is sometimes called "BosNyWash".) Contrary to Mumford, Gottmann extolled the idea of the megalopolis:

> Because of its concentration of people, wealth, and economic activities, Megalopolis stands out on the map of the present world as a stupendous monument erected by titanic efforts. (Gottmann, 1961, p. 23).

Gottmann's 800-page study examined the history, geographical functions, linkages, and land-use patterns that characterized this vast region. His findings were more sanguine than most later metropolitan studies: poverty, pollution, and politics were not his primary concerns. But with respect to the "symbiosis of urban and rural," his findings reinforced more polemical writers:

> In Megalopolis the fully urbanized and built-up sectors are many and of impressive size, but there still remains a great deal of thinly occupied space devoted to woods, fields, and pasture . . .

On closer examination, however, we shall find that present and future use of these green spaces within Megalopolis is completely dependent on the march of urbanization. We shall also discover that, while the actual crowding is still localized and open land is available on a much larger scale than is usually recognized, present trends indicate an urgent need for new policies if Megalopolitan populations are not to find themselves even more fenced in than are the people in other highly urbanized regions of the world (Gottmann, 1961, p. 218).

Gottmann's study strengthened perception of Megalopolis as a new kind of region characterized by a high degree of functional interdependence, yet fragmented by political boundaries. His penultimate chapter, "Sharing a Partitioned Land," was devoted to the need for intergovernmental cooperation in solving the problems of Megalopolis and its analogues elsewhere. (For a comparison of Megalopolis with its British counterpart, see Clawson and Hall, 1973.)

The manifesto of the 1960s Conservation Movement was a short, punchy essay entitled "Urban Sprawl" by journalist William H. Whyte, Jr. (1958). Whyte, an acute observer and critic of American land development, argued that zoning had failed to promote orderly and efficient use of land. He portrayed the results of recent land development as wasteful of land, costly to provide with public services, and aesthetically a disaster as open space vanished from the reach of metropolitan residents. Whyte for three decades has promoted new techniques to improve land-development practices and save open space (e.g., Whyte, 1959, 1968, 1989).

The perception of loss of open space prompted an outpouring of scholarly and civic reports by many respected organizations, for example, the Committee for Economic Development (1960), the Urban Land Institute (1959), the New York Regional Plan Association (Siegel, 1960), and Resources for the Future (Wingo, ed., 1963). "Open space" was to the conservationists of the Kennedy/Johnson era what "congestion" was to the progressives of Theodore Roosevelt's time.

Increasingly, the concern was directed above the municipal level to states and the federal government:

Open space has become the subject of a remarkable new interest. The words are echoing even in the halls of Congress and state legislatures. And voices behind the words are those of *urban* senators, *urban* congressmen, and *urban* legislators. This is no faddist, back-to-nature movement; it is a direct expression of concern about the present and future use of urban space." (Tankel, 1963, p. 57. [Emphasis in original]).

The earliest Congressional response to this growing perception was the creation in 1958 of the Outdoor Recreation Resources Review Commission (ORRRC). That act and its legacy, the Land and Water Conservation Fund, anticipated an expansion of federal involvement in land use and urban development during the 1960s (Table 11–1).

Table 11-1 Major federal laws of the 1960s concerning cities and land use

1961—Cape Cod National Seashore Act (PL 87-126)
1963—Outdoor Recreation Act (PL 88-29)
1964—Wilderness Act (PL 88-577)
Civil Rights Act (PL 88-352)
1965—Appalachian Regional Development Act (PL 89-4)
Land and Water Conservation Fund Act (PL 88-578)
Department of Housing and Urban Development Act (PL 89-174)
Highway Beautification Act (PL 89-285)
Water Resources Planning Act (PL 89-80)
1966—Historic Preservation Act (PL 89-665)
Department of Transportation Act (PL 89-670)
Demonstration Cities and Metropolitan Development Act (PL 89-754)
1968—Civil Rights Act (PL 90-284)
Fair Housing Act (Title VIII)
Housing and Urban Development Act (PL 90-448)
Urban Mass Transportation (Title VII)
New Communities Act (Title IV)
National Flood Insurance Act (Title XIII)
Interstate Land Sales Act (Title XIV)
Wild and Scenic Rivers Act (PL 90-542)
National Trails System Act (PL 90-543)

OUTDOOR RECREATION

ORRRC was the first in a series of blue ribbon commissions established by Congress or the White House that played major roles in promoting new federal laws during the 1960s. Chaired by Laurance S. Rockefeller and composed of members of Congress and private citizens, ORRRC was well constituted to attract political support. Its legislative mandate was to:

(a) . . . set in motion a nation-wide inventory and evaluation of outdoor recreation resources and opportunities, . . . [and]

(b) . . . compile such data and . . . determine the amount, kind, quality, and location of such outdoor recreation resources and opportunities as will be required by the year 1976 and the year 2000 and shall recommend what policies should best be adopted . . . (P.L. 85-470, Sec. 6).

ORRRC's mission was predominantly nonurban: "outdoor recreation" was defined by Congress to *exclude:*

. . . recreation facilities, programs, and opportunities usually associated with urban development such as playgrounds, stadia, golf courses, city parks, and zoos (P.L. 85-470, Sec. 2).

The 27 working papers and final report published by ORRRC over the next four years reflected the best professional thinking of the time on recreation, open space, and urban planning. ORRRC found that "public areas designated for outdoor recreation include one-eighth of the total land of the country" but "the problem is not one of number of acres but of effective acres—acres of land and water available to the public and usable for specific types of recreation." (Outdoor Recreation Resources Review Commission, 1962, p. 49.)

The final report discussed statistically, graphically, and in prose the existing and the anticipated status of outdoor recreation opportunities, with the notable exception of central cities. The report estimated that outdoor recreation provided a $20-billion-a-year market for goods and services, as well as nonquantifiable social benefits to the nation (Beard, 1970, p. 97).

Today, the report is remarkably dated in its nonrecognition of minorities, handicapped persons, and the elderly. Its photographs uniformly depict white, middle-class families enjoying themselves in bucolic surroundings. Hugh Davis, a former ORRRC staff planner, recalls that the Izaak Walton League and other hinterland interest groups prevailed over YMCA-type interests who sought more attention to cities and athletics.

Few national commission reports have had faster and more lasting impact. Within three months after its release, Secretary of the Interior Stewart Udall established a Bureau of Outdoor Recreation (BOR) by administrative action: This was ratified by Congress in the Outdoor Recreation Act of 1963. The ORRRC recommendation to establish a federal land-acquisition fund was immediately endorsed by President Kennedy and ultimately adopted in the Land and Water Conservation Fund Act of 1965 (P.L. 88-588).

This act (known as LAWCON) has been the primary source of federal matching grants for acquisition and improvement of open space by all levels of government. Despite the shortcomings of the ORRRC report, it addressed urban as well as hinterland recreation needs, and both active and passive forms of recreation. LAWCON is funded through earmarked revenue from (1) proceeds from the sale of surplus federal property, (2) motorboat fuel taxes, (3) entrance fees to federal parks, and, under a 1968 amendment (4) off-shore oil and gas lease proceeds. However, these funds are not automatically available: Congress must appropriate money annually for LAWCON and program outlays have accordingly fluctuated over the years. (Table 11-2).

Sixty percent of appropriated LAWCON funds is allocated for matching grants to states with the balance available for federal land acquisition. The share of LAWCON each state receives is established by a Congressional formula. But a prerequisite to such outlays is the preparation of a State Comprehensive Outdoor Recreation Plan (SCORP) acceptable to the administering agency (as of 1989, the National Park Service). A SCORP identifies how the federal money will be allocated by the state to remedy its recreational needs. Thus, the Land and Water Conservation Fund Act built on the precedent of Section 701 in providing federal assistance to nonfederal planning programs.

Table 11–2 Land and Water conservation Fund: Annual appropriations to states and federal agencies FY 1965–1988 ($ millions)

Year	$ millions
1965	$ 15.2
66	105.0
67	79.0
68	96.5
69	150.4
70	116.0
71	308.4
72	327.3
73	263.3
74	71.6
75	265.9
76	267.9
TQ*	64.6
77	434.7
78	680.8
79	629.9
80	471.5
81	242.7
82	133.4
83	236.5
84	203.6
85	169.1
86	93.6
87	110.1
88	60.7
Total 1965–88	$5,598.9

*TQ = Transitional Quarter
SOURCE: National Park Service Unpublished Data.

LAWCON provides 50 percent of eligible costs of acquiring and/or improving open space at the state or local level. Many states have created their own funds to contribute an additional 25 percent of local project costs, leaving only 25 percent to be covered by a local or county government. Even this cost is an obstacle to central cities where land is costly and budgets are tight. As of the mid-1970s, the benefits of LAWCON heavily favored suburbs, especially affluent ones (Platt, 1976, Ch. 9). Another federal open space program administered by the Department of Housing and Urban Development was more oriented to cities but lasted only from 1961 until 1972. It spent $350 million in matching grants to acquire 350,000 acres of urban open land by 1971 (U.S. Council on Environmental Quality, 1970, p. 196).

URBAN HOUSING

While the FHA and VA underwrote suburban sprawl and ORRRC studied ways to provide recreation for the white middle class, federal policies toward central cities were increasingly criticized by advocates of civil rights and urban housing. Before 1960, federal programs for central cities took two forms: public housing and urban renewal. Both were to be challenged in the 1960s as ineffective and counterproductive.

Direct federal assistance to local public housing programs began with the Public Works Emergency Housing Corporation in 1933 (Feiss, 1985, p. 176). In 1938, Congress established the Slum Clearance Program for replacement of tenements with publicly owned housing projects. By 1970 this program, as modified in various later housing acts, had resulted in the construction of about 870,000 units of low-rent public housing. If fully occupied, these could accommodate about 3 million people or 1.5 percent of the nation's population—compared with 25 million people (13 percent of the U.S. population) who were below the federally established poverty level in 1970 (Downs, 1973, p. 48).

Although "shamefully small in relation to the nation's housing needs" (Fried, 1971, p. 73), the actual picture was even worse. Many public housing units were uninhabitable by the mid-1960s due to inappropriate design, isolated location, occupancy policies, and lack of upkeep. Most were in large high-rise projects that lacked convenient access to jobs, decent schools, social services, and physical security for their inhabitants. Rife with crime and drug problems, much of the public housing built with federal assistance has been abandoned. The infamous Pruitt-Igoe Project in St. Louis, for example, consisted of 43 eleven-story buildings. By 1970, 26 of these were boarded up, and in 1972 all were demolished by the city's housing authority (Newman, 1972).

Aside from poor design, a fundamental objection of civil rights advocates to federal public housing policy was the practice of locating most projects (except elderly housing) in black ghetto areas, thus reinforcing patterns of racial segregation since most occupants were nonwhite. In a suit filed by the NAACP, the federal district court in Chicago held that this practice violated the Fourteenth Amendment and ordered new public housing to be scattered in small clusters throughout the city, including white neighborhoods (*Gautreaux* v. *Chicago Housing Authority* 296 F. Supp. 907, 1969 and 304 F. Supp. 736, 1970). The subsequent history of this case, which yielded some 20 further opinions, including one by the U.S. Supreme Court, has been vividly recounted by Babcock and Siemon (1985, Ch. 9).

Little public housing has been built in the United States since the 1960s. As of 1984, the nation had only 950,000 occupied units of family public housing and 350,000 units of elderly housing. In the same year, 33.7 million people were below the poverty level, 9 million more than in 1970 (of whom 23 million were white, 9.5 million black, and 4.9 million Hispanic) (U.S. Bureau of the Census, 1986, Tables 766 and 1319). Federal public housing policy since the early 1970s has emphasized housing allowances to eligible households rather than construction of additional public units. The record of federal involvement in housing for the poor, however, has generally been shameful in comparison with the many forms of federal subsidy to the habitability of middle- and upper-class suburbs.

The federal Urban Renewal Program was also widely discredited in the 1960s. Under the Housing Acts of 1949 and 1954, urban renewal funded local authorities to plan, acquire, clear, and redevelop designated areas of "urban blight." With the approval of the U.S. Supreme Court in *Berman* v. *Parker* in 1954 (see Chapter 9), the program cleared thousands of acres of inner-city tenements and displaced tens of thousands of low-income households and small businesses. Aside from areas rebuilt with public facilities such as schools and parks, most urban renewal land was sold at a subsidized price to private redevelopers to be reused according to the urban renewal plan. This resulted in the construction of new office buildings, hotels, shopping malls, and medium- to high-cost dwelling units in place of the former tenements. Some sites were never redeveloped, leaving pockets of litter-strewn vacant land in many inner-city neighborhoods.

Urban renewal was attacked from several standpoints. Housing advocates complained that poor people displaced by the projects could not afford the new units built on the same sites and were given little help in finding alternative housing. More than 400,000 lower-income units were demolished by urban renewal during its first 18 years, while new housing constructed on those sites included only 10 percent of that number of units for low- and moderate-income households (Fried, 1971, p. 86).

Others criticized the disruption of ethnically diverse urban neighborhoods and their replacement by sterile, unsafe high-rise developments (Gans, 1962; Jacobs, 1961; Newman, 1972). Another criticism was that federal subsidies were enriching private redevelopers by clearing sites in favorable locations that were not necessarily blighted (Abrams, 1965). The lack of compensation by urban renewal authorities to displaced tenants and small businesses was another issue (Berry, Parsons, and Platt, 1968).

Other forms of discrimination, however, preceded urban housing on the civil rights agenda. The Civil Rights Act of 1964, inspired by the Freedom Riders and the 1963 March on Washington, D.C. led by Dr. Martin Luther King, Jr., prohibited racial discrimination in voting, public accommodations, public education, and employment but did not reach the issue of housing. In 1965, Congress created the Department of Housing and Urban Development (HUD) as recommended by President Johnson's Task Force on Metropolitan and Urban Problems to serve as an umbrella for federal housing programs (Weaver, 1985).

The next year, Congress adopted the Demonstration Cities and Metropolitan Development Act of 1966 pursuant to another task force on urban problems, chaired by Robert C. Wood. This launched the "Model Cities" program, which sought to upgrade community facilities, housing, jobs, and social programs in some 150 cities. "Model Cities" was viewed by its proponents as an alternative to urban renewal and a comprehensive approach to making inner-city neighborhoods more habitable.

For middle-class black families, however, the prospect of remaining in the inner city, even if better lighted, policed, and rehabilitated, was unacceptable. Efforts to open up exclusive white suburbs to minority households led to adoption of the Fair Housing Act (Title VIII of the Civil Rights Act of 1968), which prohibited racial discrimination by lenders, realtors, and other parties to real estate transactions. Within three years, this Act covered 80 percent of the nation's housing supply (Weaver, 1985, p. 466). It also provided a legal basis for challenging exclusionary zoning barriers in the federal courts (e.g., *Metropolitan Housing Development Co. v. Village of Arlington Heights* 469 F. Supp. 836, 1979). (See discussion of state court decisions on exclusionary zoning in Chapter 8.)

In the same year, Congress in the Housing and Urban Development Act of 1968 (P.L. 90-448) reaffirmed the goal of the 1949 housing act to provide: "a decent home and a suitable living environment for every American family." It called for the construction or rehabilitation by 1978 of 26 million dwelling units, including 6 million for low- and moderate-income households (President's Commission on Urban Housing, 1968). The Act established new federal subsidies for "lower-income" rental and ownership housing, revised the public housing and urban renewal programs, and launched new programs in urban mass transportation, "new communities," flood insurance, and interstate land sales. This centerpiece (and finale) of President Johnson's War on Poverty marked the greatest expansion of federal involvement in urban development since the New Deal. The vast concrete headquarters of HUD and the Department of Transportation, which occupy adjacent sites of former urban renewal land in the southwest quadrant of Washington, D.C.,

Table 11–3 Rise and fall of two federal housing subsidy programs (dollar obligations in thousands)

Year	Sec. 235[a]	Sec. 236[b]
1969	$ 11,083	$11,090
1970	78,412	93,162
1971	129,731	91,355
1972	100,059	83,096
1973	44,701	54,759
1974	4,882	23,385
1975	797	14,984
1976	69	11,275

[a]Home Ownership Program established by 1968 Housing and Urban Development Act
[b]Rental Housing Program under same act

SOURCE: U.S. Department of Housing and Urban Development, 1976. Tables 5 and 6.

are monuments to this brief but exuberant era of domestic concern (sadly overshadowed by the other war in Vietnam).

To enact programs is one thing, to achieve results is another. Robert Weaver (1985), the first Secretary of HUD, writes that federal housing policy has been stymied by its stop-go nature. Programs were piled on top of other programs—public housing, urban renewal, housing subsidies, rent supplements, model cities, new communities, and, in the 1970s, housing vouchers. Strategies to "gild the ghetto" conflicted with efforts to open the suburbs, and programs to house lower-income families competed for resources with housing for the very poor, who ended up as "the homeless" of the 1980s.

The Nixon Administration in the 1970s ended the War on Poverty, converting many grant and subsidy programs to "revenue sharing," or eliminating them entirely (Table 11-3). That marked an ideological reversion to localism, in contrast to the strong federal initiatives of the 1960s. In the face of federal funding reductions, continued suburban exclusionary zoning, the *Gautreaux* decision, and other fiscal and legal obstacles, federal housing policy was virtually a dead letter. The Reagan Administration deplored the plight of the homeless while permitting HUD to enrich its bureaucrats and consultants in achieving little, if anything, to improve urban housing.

METROPOLITAN AND REGIONAL PLANNING

Before the 1960s, metropolitan and regional planning was generally a nongovernmental activity. Two leading examples included Daniel Burnham's 1909 "Plan for Chicago" sponsored by the Commercial Club of Chicago and the 1929 plan for Greater New York prepared by the Regional Plan Association. (The latter organization continues to sponsor regional planning studies funded largely by corporate contributions.)

Since World War II, regional and metropolitan planning has increasingly become a governmental function whose vigor has varied directly with the rise and fall of federal support and sponsorship (McDowell, 1983). The immediate postwar objective was the elimination of slums and redevelopment of the nation's central cities. Officially, under the Housing Act of 1949, these efforts took the form of a federal-local partnership with no regional dimension. However, Section 701 of the Housing Act of 1954 for the first time authorized planning grants to state, metropolitan, and other regional planning agencies. This section prompted the establishment of regional planning agencies for most of the nation's metropolitan areas.

Regionalization was further encouraged by Congress in the Demonstration Cities and Metropolitan Development Act of 1966, the preamble of which stated:

> The Congress hereby finds that the welfare of the Nation and of its people is directly dependent upon the sound and orderly development and the effective organization and functioning of the metropolitan areas in which two-thirds of its people live and

work . . . it further finds that metropolitan areas are especially handicapped in this task . . . by the complexity and scope of governmental services acquired in such rapidly growing areas, the multiplicity of political jurisdictions and administrative arrangements available for cooperation among them. It further finds that present requirements for area-wide planning and programming in connection with various federal programs have materially assisted in the solution of metropolitan problems, but that greater coordination of federal programs and additional participation and cooperation are needed from the states and localities in perfecting and carrying out such efforts. . . . (P.L. 89-754, sec. 201-A).

The A-95 Clearinghouse Process

In furtherance of this objective, Congress in Section 204 of the same Act established the procedure for "regional clearing house" review of federal spending actions. This required that requests from local governments and other applicants for federal assistance under several dozen funding programs must be reviewed by "an areawide agency which is designated to perform metropolitan regional planning for the area . . ." Each state was to designate an area-wide agency within each metropolitan region. By late 1967, regional planning agencies had been created and designated for 203 of the nation's 231 SMSAs (Commonwealth of Massachusetts, 1970, p. 68). These were of two principal types: (1) regional planning agencies (RPAs) and (2) councils of governments (COGs).

RPAs are established under state law as quasi-governmental entities that may hire staff, enter into contracts, and advise governments within their planning area on a variety of issues, for example, transportation, housing, land use, open space, sewer and water services, education, and elderly services. Local governments appoint representatives to serve on the RPA advisory body. By contrast, COGs are actually assemblies of elected officials (or their representatives) from constituent local and county governments. COGs perform the same planning functions as RPAs but may exert more influence over the actions of local governments due to the direct involvement of elected officials. Neither type of planning organization normally has taxation or land acquisition powers. Their funds are derived from constituent governments, and from federal, state, and private support.

The regional clearinghouse function, during the late 1960s and 1970s, provided both RPAs and COGs with a modest degree of political leverage. As clearinghouse, the regional agency would evaluate local requests for federal assistance in terms of compatibility with the agency's regional comprehensive plan. A negative recommendation from the regional agency was not binding on the federal government but could be used to achieve modification or even cancellation of local projects. This function became known as the "A-95 Areawide Review Process" after the number of the directive of the federal Office of Management and Budget, which administered the process until its termination by the Reagan Administration in 1981.

Regional planning agencies were not funded to conduct the A-95 review but have been supported under various federal programs concerned with specific topics such as regional transportation, area-wide water quality planning, air quality, and housing. By 1979, some 39 distinct federal programs were financially supporting and/or requiring regional planning (McDowell, 1983, p. 31). In 1980, regional planning organizations served about 99 percent of the counties in the nation with about three-fourths of their collective budgets provided by the federal government (McDowell, 1984, p. 131). Most of these have been organized as "councils of government." By the late 1980s, budgets for many of these agencies have been drastically reduced due to withdrawal of federal funding. Also, elimination of the A-95 Clearinghouse responsibility, except in states that have preserved it, diminishes the overall clout of such agencies. Many survive on the basis of funding for special-issue studies such as solid and hazardous waste management. While this function is needed, the effectiveness of RPAs and COGs as voices for comprehensive regional planning has unquestionably been muted during the 1980s.

Appalachian Regional Commission

The Appalachian Regional Commission (ARC) was established by the Appalachian Redevelopment Act of 1965, an element of President Lyndon Johnson's War on Poverty. The geographic area defined as the *Appalachian Region* encompasses all of the State of West Virginia and portions of 12 other states extending from New York to Mississippi. (See map, Figure 11–1). This area covers 397 counties, with a total land area of 195,000 square miles. The population of the region in 1986 was 20.6 million people.

Physical conditions and resource-management issues that led to the designation of the Appalachian Region include its steep, rugged terrain, poor soil, widespread erosion, flash flood hazards, physical isolation, water pollution, mine surface subsidence, acid mine drainage, and unstable mining waste tips. Social problems as of 1960 included the following:

- one-third of families below poverty level;
- average per capita income one-third below national level;
- unemployment 40 percent above national level;
- 32 out of 100 adults have completed high school;
- 5 out of 100 adults have completed college;
- substandard housing twice the national average;
- infant mortality one-third higher than national average;
- two-thirds the number of doctors as compared with the nation as a whole;
- infectious disease mortality 33 percent above national average;
- widespread incidence of black lung disease;
- net loss of population from the region due to out-migration;
- erratic economic swings;
- stagnant and dying industries;

- exportation of capital to nonresident owners accompanied by nonreinvestment in the region.

The Appalachian Regional Commission is a unique federal-state entity, not a federal agency. Its executive board consists of the governors of the 13 Appalachian states, plus a federal co-chairperson. Decisions of the commission must be approved by unanimous vote of the foregoing. The commission has an executive director and a small staff of about 55 employees. Its office is located in Washington, D.C. Its administrative budget is shared half by the federal government and half by the 13 states.

The raison d'etre of the ARC is to focus federal and state funds upon the

The Appalachian Region

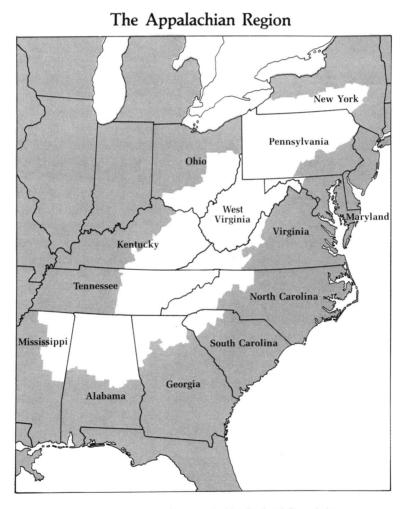

Figure 11.1 Territory of the Appalachian Regional Commission.

economic, social, and environmental upgrading of the Appalachian Region. The keystone of its program has been the planning and partial construction of some 4,200 miles of new four-lane highways. These were needed both to reduce internal isolation within the region and to facilitate the movement of goods and services to and from the region. In 1960, only a single major highway, the Pennsylvania Turnpike, traversed the Appalachian Region from east to west. Two-thirds of the planned Appalachian Highway System has now been completed, at a cost of over $5 billion (75 percent federal; 25 percent state).

Aside from highway construction, the ARC has promoted diverse objectives in the region, for example:

- *public health*—local health clinics, hospital improvements, expanding medical staffs;
- *education*—upgrading local schools, colleges, and vocational schools;
- *environment*—reduction of air and water pollution, acid mine drainage, solid waste management;
- *natural hazards*—mitigation of flash flood and landslide hazards;
- *housing*—new construction and renovation of existing housing stock;
- *community development*—better sewer and park systems.

Decision making regarding the location and form of ARC activities is decentralized. Priorities are proposed in annual plans prepared by each state for approval by the commission. The Appalachian portion of each member state is subdivided into "local development districts" consisting of one or more counties. Altogether, there are 69 local development districts designated by states, which cover the entire Appalachian Region. These local districts have planning staff and play a key role in helping to formulate the ARC annual plan within their respective states.

ARC takes credit for promoting a significant upgrading of economic, social, and environmental conditions in the Appalachian Region. Some 430,000 new jobs have been created in the region along or near highways, of which 214,000 jobs are in manufacturing and the rest in service industries, transportation, or commerce. Sixty percent of new manufacturing plants in the region have located within 30 minute of existing or proposed new highways.

The commission has promoted the conversion of cutover woodlands and marginal croplands to pasture and has cooperated with the Soil Conservation Service in promoting land stabilization and erosion control, and reforestation. It has initiated a mine-related hazards program and helped to win support for adoption of the 1977 Federal Surface Mine Reclamation Act. It has reduced flood hazards through installation of automated flash flood warning systems and channel clearance projects and relocation of some chronically flood-prone properties. (Ninety-nine ARC counties are also within the jurisdiction of the Tennessee Valley Authority, which administers a local flood-plain management assistance program.) ARC has also promoted the establishment of new industrial parks, has encouraged an inflow of venture capital and modernization of existing manufacturing and mining facilities. It has facilitated the development of new water, sewer, and solid waste facilities in mining communities, rapidly changing communities, and backward rural areas.

RIVER BASIN COMMISSIONS

A different experiment with federally sponsored regional planning was conducted under authority of the Water Resources Planning Act of 1965. This Act established the U.S. Water Resources Council (WRC) as a central clearinghouse in Washington, D.C. on which all federal water management agencies were represented. WRC between 1966 and its termination in 1981 made significant contributions to the methodology of water resource planning in the United States by government agencies. In particular, it designated the "100-year flood" (or 1 percent flood) as the standard "design event" for purposes of the National Flood Insurance Program. (See Chapter 12.) WRC also promulgated "Principles and Standards" for water resources planning which, for the first time, placed environmental consequences on an equal footing with economic impacts as criteria for the approval of federal water resources projects.

Title II of the Water Resources Planning Act authorized the creation of federal-state river basin commissions on the initiative of the states sharing jurisdiction over specific river basins. Six commissions were established under this authority: (1) New England; (2) Ohio River; (3) Great Lakes; (4) Upper Mississippi River; (5) Missouri River, and (6) Pacific-Northwest river. (These did not include other river basin commissions established under interstate compacts, for example, the Delaware, the Susquehanna, and the Colorado.)

The primary purpose of the Title II river basin commissions was to prepare plans at various levels of detail for the management of the water resources and associated lands within their respective regions. The problems studied by the commissions varied according to the hydrology and settlement characteristics of their respective basins. The New England River Basins Commission devoted much effort to studies of nonstructural floodplain management in the Connecticut River, to reduction of coastal storm hazards, and to low-head hydropower development. The Great Lakes Basin Commission was more concerned with water quality and shoreline erosion, while the Missouri RBC addressed issues relating to irrigation and navigability. Studies prepared by each commission were funded jointly by federal and state governments and were overseen by the Water Resources Council. Unlike the Tennessee Valley Authority, the RBCs were given no power to implement their proposals. Upon their termination in 1981, the policy recommendations that they made were left to the federal, state, and local governments to pursue or to ignore.

CONCLUSION

The decade of the 1960s, largely coinciding with the Democratic administrations of Presidents Kennedy and Johnson, marked a revival of the New Deal vision of the federal government as a stimulus to the planning and management of land use. This extended from central cities to their suburbs and hinterlands, to distressed rural areas, to river basins, and to federal lands. Instruments of federal policy included financial assistance to planning programs at all levels, subsidies for housing, open land acquisition, and community development; tax incentives (another form of subsidy) to encourage private actions in furtherance of public goals; and technical assistance. In the latter category, government agencies directly and through grants and contracts with universities and consultants sponsored an avalanche of publications on community planning, conservation, and resource management.

Prior to 1970, however, Congress remained wary of direct environmental regulation other than on federally owned lands. The use and abuse of land, air, and water remained fundamentally under the jurisdiction of the states with the federal

---

SOME COMMON FEDERAL ACRONYMS AND ABBREVIATIONS

CEQ	U.S. Council on Environmental Quality
CFR	Code of Federal Regulations
COE	U.S. Army Corps of Engineers
CZM	Coastal Zone Management
EIS	Environmental Impact Statement
EPA	U.S. Environmental Protection Agency
FEMA	Federal Emergency Management Agency
FIRM	Flood Insurance Rate Map
FWS	U.S. Fish and Wildlife Service
LWCF	Land and Water Conservation Fund
NALS	National Agricultural Land Study
NEPA	National Environmental Policy Act
NFIP	National Flood Insurance Program
NPS	National Park Service
NRI	National Resources Inventory
NWI	National Wetlands Inventory
OCRM	Ocean and Coastal Resource Management
PCS	Potential Cropland Study
SCS	Soil Conservation Service
TVA	Tennessee Valley Authority
USC	U.S. Code
USCA	U.S. Code Annotated
USDA	U.S. Department of Agriculture
USGS	U.S. Geological Survey

government serving as a sanctimonious, and often hypocritical, source of unsought advice. Direct federal regulation of environmental conditions, with the important exception of private land use, was to be the central theme of the 1970s.

Paradoxically, the federal government itself during the 1960s was also a major contributor to the problems that its advice and incentives sought to mitigate. As funder of the Interstate Highway System; as builder of dams and other water development projects; as reclaimer of arid lands for crops; as sponsor of environmentally unsafe military facilities; licensor of nuclear power plants; and lessor of rights to exploit offshore oil, timber, and minerals on federal lands, the federal contribution to environmental degradation was profound. The adoption of the National Environmental Policy Act of 1969 (signed by President Nixon on January 1, 1970) marked a critical threshold in federal policy. For the first time, Congress acknowledged its own culpability in the befouling of the national nest.

REFERENCES

ABRAMS, C. 1965. *The City is the Frontier.* New York: Harper and Row.

BABCOCK, R. F. and C. L. SIEMON. 1985. *The Zoning Game Revisited.* Boston: Oelgeschlager, Gunn, and Hain.

BEARD, D. P. 1970. "Meeting the Costs of a Quality Environment: The Land and Water Conservation Fund Act." In *Congress and the Environment* (R. A. Cooley and G. Wandesforde-Smith, eds.). Seattle: University of Washington Press.

BERRY, B. J. L., S. J. PARSONS, and R. H. PLATT. 1968. *The Impact of Urban Renewal on Small Business.* Chicago: University of Chicago Center for Urban Studies.

BROOKS, M. P. 1988. "Four Critical Junctures in the History of the Urban Planning Profession." *Journal of the American Planning Association* 54(2): 241–8.

CASSELLA, W. N., JR. 1983. "Regional Planning and Governance: History and Politics." In *Regional Planning: Evolution, Crisis and Prospects* (G. C. Lim, ed.). Totowa, NJ: Allanheld, Osmun.

CLAWSON, M. 1981. *New Deal Planning: The NRPB.* Baltimore: The Johns Hopkins University Press for Resources for the Future.

———, 1983. "Land and Water Planning in the New Deal." In *Beyond the Urban Fringe: Land Use Issues of Nonmetropolitan America,* (R. H. Platt and G. Macinko, eds.). Minneapolis: University of Minnesota Press.

CLAWSON, M. AND P. HALL. 1973. *Planning and Urban Growth: An Anglo-American Comparison.* Baltimore: The Johns Hopkins University Press for Resources for the Future.

COMMITTEE FOR ECONOMIC DEVELOPMENT. 1960. *Guiding Metropolitan Growth.* Washington, D.C.: The Committee.

COMMONWEALTH OF MASSACHUSETTS. 1970. *Report Relative to Regional Government.* Boston: Legislative Research Council.

DOWNS, A. 1973. *Opening Up the Suburbs: An Urban Strategy for America.* New Haven: Yale University Press.

FEISS, C. 1985. "The Foundations of Federal Planning Assistance." *Journal of the American Planning Association* 51 (2): pp. 175–184.

FRIED, J. P. 1971. *Housing Crisis U.S.A.* New York: Praeger.

GANS, H. 1962. *The Urban Villagers*. New York: Free Press of Glencoe.

GOTTMANN, J. 1961. *Megalopolis*. Cambridge: The M.I.T. Press.

JACOBS, J. 1961. *The Death and Life of Great American Cities*. New York: Random House.

JACKSON, K. T. 1985. *Crabgrass Frontier: The Suburbanization of the United States*. New York: Oxford University Press.

KATES, R. W. 1987. "The Human Environment: The Road Not Taken, The Road Still Beckoning." *Annals of the Association of American Geographers* 77(4): pp. 525–534.

MCDOWELL, B. D. 1983. "The Federal Role in Regional Planning: Past and Present." In *Regional Planning: Evolution, Crisis, and Prospects* (G. C. Lim, ed.). Totowa, NJ: Allanheld, Osmun.

_____. 1984. "Regions at the Crossroads." *Journal of the American Planning Association* 50(1): 131–132.

MUMFORD, L. 1956. "The Natural History of Urbanization." In *Man's Role in Changing the Face of the Earth* (W. L. Thomas, Jr., ed.). Chicago: University of Chicago Press.

NATIONAL PLANNING BOARD. 1934. *Final Report*. Washington, D.C.: U.S. Government Printing Office.

NEWMAN, O. 1972. *Defensible Space: Crime Prevention Through Urban Design*. New York: Macmillan.

PLATT, R. H. 1976. "The Federal Open Space Programs: Impacts and Imperatives." In *Urban Policymaking and Metropolitan Dynamics: A Comparative Geographical Analysis* (J. S. Adams, ed.). Cambridge: Ballinger.

_____. 1986. "Floods and Man: A Geographer's Agenda." In *Geography, Resources, and Environment*, Vol. II: *Themes from the Work of Gilbert F. White*. (R. W. Kates and I. Burton, eds.). Chicago: University of Chicago Press.

OUTDOOR RECREATION RESOURCES REVIEW COMMISSION. 1962. *Outdoor Recreation for America*. Washington, D.C.: U.S. Government Printing Office.

PARSONS, K. C. 1990. "Clarence Stein and the Greenbelt Towns: Settling for Less." *Journal of the American Planning Association* 56(2): 161–183.

POPPER, F. J. 1988. "Understanding American Land Use Regulation Since 1970: A Revisionist Interpretation." *Journal of the American Planning Association*. 54(3): 291–301.

PRESIDENT'S COMMITTEE ON URBAN HOUSING. 1968. *A Decent Home*. Washington, D.C.: U.S. Government Printing Office.

SIEGEL, S. A., 1960. *The Law of Open Space*. New York: Regional Plan Association.

TANKEL, S. B. 1963. "The Importance of Open Space in the Urban Pattern." In *Cities and Space: The Future Use of Urban Land* (L. Wingo, ed.). Baltimore: The Johns Hopkins Press for Resources for the Future.

THOMAS, W. L., JR. 1956. *Man's Role in Changing the Face of the Earth*. Chicago: University of Chicago Press.

U.S. BUREAU OF THE CENSUS. 1986. *Statistical Abstract of the United States*. Washington, D.C.: U.S. Government Printing Office.

U.S. COUNCIL ON ENVIRONMENTAL QUALITY. 1970. *Environmental Quality*. Washington, D.C.: U.S. Government Printing Office.

U.S. DEPARTMENT OF HOUSING AND URBAN DEVELOPMENT. 1976. *1976 Statistical Yearbook*. Washington, D.C.: U.S. Government Printing Office.

WEAVER, R. C. 1985. "The First Twenty Years of HUD." *Journal of the American Planning Association* 51(4): 463–474.

WHITE, G. F. 1969. *Strategies of American Water Management*. Ann Arbor: University of Michigan Press.

WHYTE, W. H., JR. 1958. "Urban Sprawl." In *The Exploding Metropolis* (Editors of Fortune, eds.). Garden City: Doubleday Anchor.

———. 1959. *Securing Open Space for Urban America: Conservation Easements* (Technical Bulletin no. 36). Washington, D.C.: Urban Land Institute.

———. 1968. *The Last Landscape*. Garden City, NY: Doubleday.

———. 1989. *City: Rediscovering the Center*. Garden City, NY: Doubleday.

WINGO, L., JR., ed. 1963. *Cities and Space: The Future Use of Urban Land*. Baltimore: The Johns Hopkins Press for Resources for the Future.

12

THE FEDERAL ENVIRONMENTAL DECADE OF THE 1970s AND BEYOND

INTRODUCTION

The cornucopia of new federal laws during the 1960s concerning open space and urban development (listed in Table 11–1 in the last chapter) appeared to some observers to have exhausted the potential interest of Congress in the quality of the environment. In a modern counterpart to the proposal in the late nineteenth century to close the Patent Office because "everything has been invented," a preface to *Congress and the Environment* written in 1969 speculated:

> And what about the future? There is some indication that the close of the Ninetieth Congress late in 1968 may mark the end of this remarkable period in the history of conservation politics in the United States. Some members of Congress, as well as of the new Republican Administration, have suggested that we are reaching the end of a long wave of significant and highly visible progress, and that the widely hailed "environmental crisis" has, in a certain sense, passed the peak of critical national interest and public concern (Cooley and Wandesforde-Smith, eds., 1970, p. xvi).

But in the clear light of hindsight we can now view the conservation movement of the 1960s as merely a prelude to the environmental revolution of the 1970s. The former was the larval stage, so to speak; and the latter, the butterfly.

The "federal environmental decade" was inaugurated symbolically by the first Earth Day, April 15, 1969, and by the signing of the National Environmental Policy Act on January 1, 1970. Also in 1970, the creation of the U.S. Environmental Protection Agency (EPA) consolidated several existing federal environmental

Introduction

Table 12–1 Selected federal environmental laws since 1970

1970—National Environmental Policy Act (PL 91-190)
Environmental Quality Improvement Act (PL 91-224)
Clean Air Act Amendments (PL 91-604)
Resources Recovery Act (PL 91-512)
Occupational Health and Safety Act (PL 91-596)

1972—Federal Water Pollution Control Act Amendments (PL 92-500)
Noise Control Act (PL 92-574)
Coastal Zone Management Act (PL 92-583)
Federal Environmental Pesticide Control Act (PL 92-516)
Marine Protection, Research, and Sanctuaries Act (PL 92-532)

1973—Flood Disaster Protection Act (PL 93-234)
Endangered Species Act (PL 93-205)
Safe Drinking Water Act (PL 193-523)

1976—Resource Conservation and Recovery Act (PL 94-580)
Federal Land Policy and Management Act (PL 94-579)
Surface Mining Control and Reclamation Act (PL 95-87)
Toxic Substances Control Act (PL 94-469)

1977—Soil and Water Resources Conservation Act (PL 95-102)

1980—Comprehensive Environmental Response Compensation and Liability Act (PL 96-510) ("Superfund")

1982—Coastal Barrier Resources Act (PL 97-348)

1984—Hazardous and Solid Waste Amendments (PL 98-616)

1985—Food Security Act (PL 99-198)

1986—Superfund Amendments and Reauthorization Act (PL 99-499)

(*Caveat*: Subsequent amendments to these acts are not listed. Consult U.S. code annotated for current version).

programs and provided an administrative umbrella for many new initiatives soon to be legislated by Congress.

The new environmental laws of the 1970s (Table 12–1) did not merely refine the conservation measures of the 1960s. They enlarged both the range of problems addressed through federal programs and the spectrum of means by which such problems were to be attacked. In air and water pollution, the federal role shifted from a timid reliance on state programs to direct setting and enforcement of national standards. Other new federal laws addressed such problems as noise, pesticides, solid and hazardous wastes, flood plain management, wetlands, mining reclamation, drinking water, occupational safety, ocean dumping, oil spills, and coastal management.

Constitutionally, these new programs relied on a broadened interpretation of federal powers, particularly the commerce clause, which grants Congress the power to "regulate commerce . . . among the several states" (U.S. Constitution, Art. I, Sec. 8). The emission of pollution by vehicles or in economic production of goods for sale in interstate commerce has been held in many court decisions to justify federal pollution regulations. Also, the movement of air and water across state lines itself gives rise to a federal interest in controlling the pollution that they convey (Dolgin and Guilbert, 1974, pp. 22–27).

Thus, the environmental reforms of the 1970s redefined the meaning of federalism—the relative balance of power between the federal government on the one hand and states and local governments on the other. Like the New Deal in the 1930s, this marked an irreversible enlargement of the federal role. (Even the antiregulatory ideology of the Reagan Administration failed to dismantle the environmental programs started during the 1970s, although it did damage some of them through neglect and mismanagement.) In a parallel development, certain states redefined their relationships with local governments during the 1970s as well. (See Chapter 13.)

The intellectual roots of the environmental movement reached back to George Perkins Marsh, whose 1864 treatise *Man and Nature* examined the impacts of human activities on the natural world. The science of ecology was "born" in the pioneering studies of plant succession in the Indiana Dunes by Henry C. Cowles (Engel, 1985) and was later refined in the work of Paul Sears and Eugene and William Odum. The genre of natural history writing for the nonscientist was pioneered by Henry Beston in his 1927 classic *The Outermost House*. This tradition was continued by Edwin Way Teasle, Joseph Wood Krutch, and this writer's father, Rutherford Platt (1942/1988; 1947; 1966).

Aldo Leopold in *A Sand County Almanac* (1949/1966) synthesized Marsh's view of man as an agent of environmental disturbance with the growing recognition of ecological succession and process. Leopold proposed a "land ethic" to guide human use of natural resources. This urged that humans live in harmony rather than in conflict with the biosphere of which they are a part. *Sand County Almanac*, republished in 1966 and again in 1989, became holy scripture of environmentalism.

The book that finally ignited the environmental movement was Rachel Carson's *Silent Spring* (1962). She warned against the pernicious and irreversible effects of introducing DDT and other hazardous chemicals into the environment and thus the food chain:

> The most alarming of all man's assaults upon the environment is the contamination of air, rivers, and sea with dangerous and even lethal materials. This pollution is for the most part irrecoverable; the chain of evil it initiates not only in the world that must support life but in living tissues is for the most part irreversible. In this now universal contamination of the environment, chemicals are the sinister and little-recognized partners of radiation in changing the very nature of the world—the very nature of its life" (Carson, 1962, p. 6).

This powerful book marked the advent of the politics of ecology on the national scene. Subsequent contributions included Barry Commoner's *The Closing Circle* (1970), the Teals' *Life and Death of the Salt Marsh* (1969), Garret Hardin's "The Tragedy of the Commons" (1968), Edward Abbey's *Desert Solitaire* (1968),

Introduction

Ian McHarg's *Design with Nature* (1968), John McPhee's *Encounters with the Archdruid* (1971), and the photographic essays of Eliot Porter.

Coinciding with the "Age of Aquarius" and the Antiwar Movement, the environmental movement attracted an unlikely alliance of "flower children," concerned scientists, journalists, public interest lawyers, ladies in tennis shoes, and Richard M. Nixon. Its troubadour was Pete Seeger, whose sloop "Clearwater" became both a symbol and an organization for environmental protest. The movement's success was both a cause and effect of the rising influence and affluence of national environmental organizations such as the Sierra Club, National Wildlife Federation, Friends of the Earth, and the National Resources Defense Council. New journals were started, such as *Environment, Environmental Action,* and *Environmental Management.* Conferences were held by the dozens, and media reporters and environmental activists avidly courted each other.

The shift in rhetoric from "urban sprawl" in the 1960s to "environmental deterioration" in the 1970s signified a subtle change in political and geographical emphasis. While cities had occupied central stage in the earlier wave of new federal programs, the environmental movement expanded the area of concern to the entire nation with a consequent downgrading of strictly urban concerns such as housing and transportation. (This geographical enlargement would extend to the entire planet in the 1980s, with reference to global warming, statospheric ozone depletion, acid rain, and tropical deforestation.) With the inauguration of the Republican presidency of Richard Nixon in January, 1969, the urban and community development agenda of the Democrats under Presidents Kennedy and Johnson was substantially dismantled.

In the process, political clout shifted from urban planners in the 1960s to ecologists, chemists, and climatologists in the next two decades. A foretaste of the imminent shift in geographic and disciplinary emphasis was provided in two major publications that appeared in 1965. The first was a special issue of *Scientific American,* "Cities," featuring articles by some of the leading urban scholars of the day. The second was a volume entitled *Future Environments of North America* (Darling, ed., 1965). The latter, the proceedings of a conference dominated by natural scientists, entirely ignored cities, even though they would be the "future environment" for over three-quarters of North Americans.

The cover of *Time* magazine for February 2, 1970 was emblazoned: "Environment: Nixon's New Issue." In cooperation with Congress, the Nixon Administration presided over adoption of the National Environmental Policy Act, the far-reaching amendments to the federal air and water quality acts, and the creation of the Environmental Protection Agency. Even a national land-use policy act was nearly achieved: the federal Coastal Zone Management Act of 1972 was the survivor of many bills filed on that topic. The laws passed between 1970 and 1976 have served the nation well in many respects over the past two decades. Collectively, they comprised a massive and creative legal response to the findings of environmental scientists and the common sense of the average voter.

THE NATIONAL ENVIRONMENTAL POLICY ACT

The National Environmental Policy Act (NEPA) was the keystone of federal environmental reforms of the 1970s. NEPA united both a statutory declaration of national commitment to a safer, healthier environment with a new decision-making procedure applicable to all federal agencies. It also created a new agency, the U.S. Council on Environmental Quality, to administer the new policy and procedures established by the Act.

NEPA reflected a perception that the federal government should get its own house in order before, or at least while, it sought improvement in nonfederal activities affecting the environment. Federally sponsored domestic and military construction programs of the 1950s and 1960s were accompanied by widespread land degradation, air and water pollution, habitat destruction, and aesthetic blight. Also, federal licensing and regulatory authorities were deemed to be administered in disregard of environmental consequences of proposed actions. According to the first Annual Report of the U.S. Council on Environmental Quality (1970, p. 191), harmful environmental impacts arise from:

> A myriad of federal loans, grants, projects, and other programs enacted for specific public purposes.... The most significant federal activities include the highway, airport, and mass transit programs, the sewer and water grant programs, ... the location of Federal facilities, and water resource projects.

In addition to such spending programs, environmental neglect was charged in the administration of diverse federal licensing and regulatory activities involving, for example, pesticide usage, offshore oil and gas leasing, nuclear and fossil fuel power plant siting and design, discharges into navigable waters, and federal land management. Some notable controversies involving federal actions during the 1960s included:

- the proposal by the Bureau of Reclamation to dam portions of the Grand Canyon;
- a proposed 39-square-mile jetport to be built just north of Everglades National Park in Florida;
- the Cross-Florida Barge Canal initiated (but never completed) by the Army Corps of Engineers;
- competing proposals for a national park and a federally funded harbor in the Indiana Dunes on Lake Michigan (Platt, 1972; Engel, 1985).
- the 1969 oil spill disaster in Santa Barbara Channel (Baldwin, 1970);
- innumerable conflicts over the siting and design of Interstate Highways, for example, Franconia Notch, NH; San Francisco's Bay Freeway; New Orleans; Boston; Seattle; and so forth;
- the Rampart Dam proposal for the Yukon River in Alaska; and

- the North American Water and Power Alliance (NAWAPA) proposal to impound massive quantities of water from British Columbia for diversion to arid regions of the United States and Canada (Reisner, 1986, pp. 506–14).

There were precedents for a requirement, as adopted in NEPA, that adverse implications of a proposed action be identified before a federal action is taken. The Fish and Wildlife Coordination Act of 1958 (16 USC Secs. 661-666c) required that any proposal to impound, divert, deepen, or otherwise control or modify any stream or water body under the auspices of a federal project or permit must be reviewed by the U.S. Fish and Wildlife Service (FWS) of the Department of the Interior. The objective of this review is "the conservation of wildlife resources by preventing loss of and damage to such resources." The FWS cannot veto directly a project that would endanger wildlife habitat, but its comments are appended to the report to Congress or other authorizing agency. Disclosure of potentially serious impacts may lead to modification or abandonment of the project by Congress or the agency sponsor. The proposed Rampart Dam in Alaska was an example of a project abandoned due in part to foreseeable impacts on wildlife habitat.

Another precedent was Section 4(f) of the Department of Transportation Act of 1966, which limited the encroachment of DOT projects on preserved open spaces. It prohibited approval of a project that ". . . requires the use of any publicly owned land from a public park, recreation area, or wildlife and waterfowl refuge of national, state or local significance . . . unless (1) there is no feasible and prudent alternative . . . and (2) such program includes all possible planning to minimize harm . . ." to such areas. Together with parallel provisions in the related acts, this doctrine applied to federally assisted highways, mass transportation facilities, and airport development. Section 4(f) provided a basis for several challenges to federal transportation projects prior to NEPA. In *Citizens to Preserve Overton Park* v. *Volpe* 401 U.S. 402, 1971, the U.S. Supreme Court directed DOT to reevaluate a proposed alignment of Interstate-40 through a Memphis park. Although the alternative was to cut through urban neighborhoods, the park was eventually spared.

NEPA, like the Declaration of Independence, is short, clear, and bold. It first states the intent of Congress:

> . . . to foster and promote the general welfare, to create and maintain conditions under which man and nature can exist in productive harmony, and fulfill the social, economic, and other requirements of present and future generations of Americans (PL 91-190, Sec. 101(a)).

It then articulates national goals to:

1. fulfill the responsibilities of each generation as trustee of the environment for succeeding generations;
2. assure for all Americans safe, healthful, productive and esthetically and culturally pleasing surroundings;

3. attain the widest range of beneficial uses of the environment without degradation, risk to health or safety, or other undesirable and unintended consequences;
4. preserve important historical, cultural, and natural aspects of our national heritage . . . ;
5. achieve a balance between population and resource use which will permit high standards of living . . . ;
6. enhance the quality of renewable resources and approach the maximum attainable recycling of depletable resources (Sec. 101(b)).

To add force to this rhetoric, NEPA requires all federal agencies to prepare "detailed statements" disclosing potential environmental consequences of their proposed actions. The requirement applies to any proposed "major federal actions significantly affecting the quality of the human environment" (Sec. 102(c)). This includes (1) direct federal actions such as the siting of federal facilities, (2) funding commitments for nonfederal activities, (3) federal licensing and permits, and (4) proposals for federal legislation (little observed in practice).

Under NEPA, an environmental impact statement (EIS) must consider:

i) The environmental impact of the proposed action;

ii) Adverse environmental effects which cannot be avoided should the proposal be implemented;

iii) Alternatives to the proposed action;

iv) The relationships between local short-term uses of man's environment and the maintenance and enhancement of long-term productivity; and

v) Any irreversible and irretrievable commitments of resources which would be involved in the proposed action should it be implemented (Sec. 102(c)).

Many early EISs were excessively long and detailed, impeding their usefulness. Environmentalists complained that some agencies engaged in overkill to bury opposing views in prolixity. The agencies, on the other hand, responded that they were only trying to avoid being sued for producing an insufficient EIS.

In 1978, CEQ issued new regulations to refine the EIS preparation process (40 CFR Parts 1500-1508). The policy of "scoping" was introduced whereby affected agencies would agree in advance as to environmental issues to be addressed and the level of detail to be devoted to each. Federal agencies were required to develop new procedures for complying with NEPA by ensuring orderly and timely preparation of EISs. The latter must be circulated for comment to all interested public officials and agencies, and to private citizens. The sponsoring agency must consider comments received *prior* to its final decision.

Even with the 1978 refinements, NEPA has generated a mountain of paperwork. Between 1978 and 1984, a total of 6,689 draft and final EISs were filed by federal agencies (U.S. Council on Environmental Quality, 1984, Table A-69). Two-thirds of these were submitted by the major construction agencies: U.S. Department of Transportation—1,474 statements; Army Corps of Engineers—1,062; Department of Housing and Urban Development—843; and Department of Agriculture—816.

The purpose of all of this time-consuming and costly effort has been to "provide full and fair discussion of significant environmental impacts and . . . inform decision-makers and the public of the reasonable alternatives which would avoid or minimize adverse impacts" (40 CFR sec. 1505.1). Unlike Section 4(f) of the DOT Act, NEPA includes no directive to disapprove actions deemed likely to produce avoidable adverse consequences. Rather, its impact has been indirect, through the political process and through adjudication.

Political response to an EIS involves a variety of pressures that may bear on the decision agency. These include formal objections from sister agencies; from members of Congress, governors, mayors, and other elected officials; by environmental organizations and residents of the area to be affected, and adverse publicity in the public media. These various forms of pressure, fueled by the mandatory disclosures of the EIS, may lead to funding cuts or other sanctions by Congress, to a presidential veto, or simply to modification or cancellation of the project by the decision agency. Or they may produce no change at all, leaving the project to proceed as proposed.

Probably the more profound effects of NEPA have occurred through lawsuits challenging particular EISs. Several hundred legal challenges based on NEPA grounds have been filed since the early 1970s by parties objecting to proposed federal actions. The plaintiffs typically have been either major environmental organizations, for example, the National Resources Defense Council, Sierra Club, or National Wildlife Federation or ad hoc organizations formed in opposition to specific proposals. NEPA does not literally prohibit unwise federal actions but requires full disclosure of adverse implications. Therefore, lawsuits generally challenge the sufficiency of such disclosures in an EIS or demand that one be prepared where it has not been.

Challengers have frequently sought to broaden the meaning of "the human environment" while federal agencies and some court opinions have attempted to limit that phrase to mean "impacts on natural systems or to direct human consequences of alterations in natural systems" (Culhane, Friesema, and Beecher, 1987, p. 31).

According to Bear (1989, p. 10067): "The development of NEPA law and its enforcement is closely intertwined with NEPA case litigation." However, the number of NEPA cases filed has declined from a high of 189 in 1974 to 71 cases filed in 1986 (Bear, 1989).

Unquestionably, delay is a common motive in filing a NEPA lawsuit. Years may pass pending the outcome of (1) preliminary hearings, (2) a decision at the trial level, and (3) appeals. The defendant agency is precluded from approving the disputed action until final resolution or settlement is reached. Delay adds to the cost of the project in addition to the costs of the litigation itself and may result in its abandonment or substantial modification.

To take one example among many, the state of Maine in 1978 proposed to construct a small bulk cargo port on Sears Island at the head of scenic Penobscot Bay (Figure 12–1). A federal permit to "dredge and fill" in estuarine wetlands was granted, and certain federal funds were committed to the project. Upon suit by the Sierra Club, the federal appeals court in Boston ordered that an EIS be prepared and

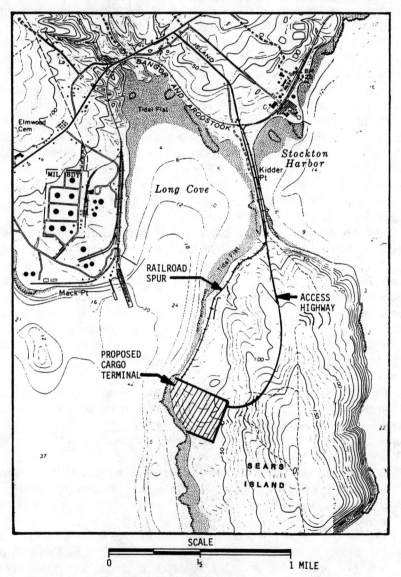

Figure 12.1 Map of Sears Island, Maine: site of proposed bulk cargo terminal.

enjoined the start of construction (*Sierra Club* v. *Marsh* 769 F.2d 868, 1985). A draft and final EIS were duly prepared by the relevant federal and state agencies.

The Sierra Club, with the support of the Boston regional office of EPA, returned to court to contend that the EIS failed to fairly consider the alternative of expanding an existing small port a few hundred yards west of Sears Island. Also, they argued that the EIS narrowly addressed only the direct impacts to the site of the port itself, while ignoring secondary effects on the rest of the still-undeveloped Sears Island and Penobscot Bay. An underlying economic issue, not covered by NEPA, was the lack of any interest by shippers in using the Sears Island facility if it were built, making it likely to be a "white elephant" (Platt, 1987).

As of early 1990, fill for the causeway had been installed but the project was otherwise under court injunction with construction halted. Port advocates blame the Sierra Club and EPA for using NEPA litigation to obstruct development in an economically depressed area of the state. Others, however, suggest that the 900-acre undeveloped Sears Island could better be used as a state park, which would also generate construction and economic benefits.

NEPA is poorly adapted to fostering a broader regional planning perspective and it certainly cannot force states to spend money for parks. But NEPA sometimes may be useful to suspend a project long enough to allow the entire plan to be reconsidered. A recent appraisal by the Environmental Law Institute (1989, p. 10060) celebrates NEPA as "Congress's first modern environmental law [which] has set the tone for the complex superstructure of federal environmental law that was to follow."

On the other hand, Lynton K. Caldwell (1989), who participated in the drafting of the original NEPA in the late 1960s, urges that a constitutional amendment on the environment would offer more reliable protection than the statute alone, whose application has often been swayed by politics.

COASTAL ZONE MANAGEMENT

The tidal and Great Lakes shorelines of the United States extend about 58,000 miles (National Oceanographic and Atmospheric Administration, 1975). These shorelines include many different geomorphic types, of which the following are representative:

> crystalline bedrock (e.g., central and northern Maine)
> eroding bluff (e.g., outer Cape Cod, Great Lakes)
> pocket beach (e.g., southern New England, Pacific Coast)
> strandplain beach (e.g., Myrtle Beach, SC)
> coastal barriers (e.g., Long Island, NY to Texas)
> coral reef and mangrove (e.g., south Florida)
> coastal wetland (e.g., Louisiana)
> (National Academy of Sciences, 1990)

Coasts also vary as to type and intensity of human usage and modification. Coastal settlements include: (1) totally urbanized industrial, port, and resort development; (2) medium-density, older summer colonies, often converted to year-round use and higher-density condominiums; (3) still-picturesque fishing and artist villages becoming "yuppified"; and (4) new megadevelopments like Hilton Head, South Carolina, or Amelia Island Plantation, Florida.

Public or quasi-public agencies own extensive tracts of unspoiled or less-developed coastal lands. The National Park Service operates 10 national seashores, four national lakeshores, and several other coastal recreation facilities. Other undeveloped shorelines are included in units of the National Wildlife Refuge System or are owned by the military. There are 18 National Estuarine Sanctuaries in 15 states under the administration of the National Oceanographic and Atmospheric Administration (NOAA). Some key ecological sites on the coast have been preserved by private organizations such as The Nature Conservancy (e.g., the Coastal Reserve in Virginia) and the various Audubon Societies.

Of course, not all privately owned shorelines are developed. Some remain undeveloped due to isolation, personal preference of the owner, or some form of development restriction. Certain undeveloped and privately owned coastal barriers were placed off limits to flood insurance and other federal growth incentives by the Coastal Barrier Resources Act of 1982. But accessible private shore frontage tends to be extremely valuable for development and usually is densely built-up.

Though limited in geographic extent, coastal areas encompass much of the nation's population growth: Eight of the 10 largest metropolitan areas are situated on tidal waters or the Great Lakes. The aggregate population of counties substantially within 50 miles of coasts grew from 92.7 million in 1960 to 123.4 million in 1984, comprising slightly over half the nation's total population (U.S. Bureau of the Census, 1986, Table 8).

Both within and outside metropolitan areas, coasts are "awash with disputes" (Conservation Foundation, 1977). They are the scene of intense competition between public and private interests, between economic and environmental values, and between diverse land and water uses: residence, business, industry, transportation, recreation, fisheries and natural habitat. The results are sometimes mutually conflicting as in the case of the Indiana Dunes where fifty years of controversy yielded a national lakeshore wrapped around a major industrial complex, to the detriment of both (Platt, 1978).

Vast areas of estuarine wetlands and related natural habitat have been lost to filling, dredging, pollution, and land subsidence. The National Wetlands Inventory estimated that 372,000 acres of "estuarine vegetated wetlands" were inundated or filled between the mid-1950s and the mid-1970s. Of this total, about 5,000 acres were lost to urbanization annually during that period (Frayer et al., 1983, p. 24). Tidal freshwater wetlands and mangrove habitat also have sustained losses at the hands of developers. Coastal wetland losses endanger commercial and sport fisheries, and impair other natural functions of wetlands such as bird habitat, water purification, and protection against wave damage (Mitsch and Gosselink, 1986, Ch. 16).

Another concern is the vulnerability of coastal development, especially on low-lying barrier beaches, to natural hazards including hurricanes, northeasters, erosion, landslides, erosion, and tsunamis. Often the very physical characteristics that attract humans to the shore are directly responsible for potential disaster. Pacific Coast residents seeking ocean views build on unstable slopes that collapse during heavy winter rains. Atlantic and Gulf coast barrier residents in "cities on the beach" may be entirely stranded at times of storm surge, unable to flee to the mainland across an impassable causeway. Cottages on the Great Lakes cling to the rim of eroding bluffs and are undermined during periods of high lake levels. The experience of the "Great 1938 Hurricane" in New England, several others in the 1950s and 1960s, and the catastrophic Hurricane Camille in 1969 signaled even greater losses in the future as the coasts fill up with homes. In 1989, Hurricane Hugo damaged or destroyed nearly all of several hundred shorefront homes lining the coastal barriers near Charleston, South Carolina.

The Federal Coastal Zone Management Act

The federal Coastal Zone Management Act (CZMA) (PL 92-583) arose from the ashes of more than 320 land and water management bills filed in the Ninety-first and Ninety-second Congresses (U.S. Congress, 1973). The rancor surrounding various proposals for a National Land Use Policy Act apparently could not withstand the charm of the seacoast. The CZMA passed the Senate by a vote of 68-0 and the House by 376-6 and has continued to enjoy strong Congressional support even in the face of hostility from the White House.

The CZMA was another product of the wave of federal activism that yielded NEPA and other legislation listed in Table 12–1. The specific idea for a federal CZM program was proposed by the Commission on Marine Science, Engineering, and Resources (1969) (the Stratton Commission). Another influence was the Model Land Development Code drafted by the American Law Institute as a blueprint for an expanded role of states in managing land use. (See Chapter 13.) A few states had adopted their own coastal management laws before CZMA was enacted in 1972, notably California, Washington, and Rhode Island. These provided models for what the federal act might encourage other coastal states to do.

The CZMA is an experiment in creative federalism. It eschews direct federal regulation of the coast for obvious political reasons, but it is not limited to funding endless planning studies. It strikes a balance between these two extremes by supporting *state* planning and management programs subject to *federal* guidelines. It seeks to achieve its objectives by working with and through coastal states and territories and granting much latitude for them to develop programs consistent with their particular physical, settlement, and political characteristics.

But the 1972 act retained considerable influence over the scope and content of such plans, distinguishing between funding for plan preparation (Section 305) and for plan implementation (Section 306). Funding under 305 was limited to a few years. To gain longer-term 306 funding, a state CZM plan had to be approved by NOAA's Office of Coastal Zone Management (now renamed the Office of Ocean and Coastal Resource Management or OCRM). Plan approval required prolonged negotiation between federal and state coastal officials over the scope and operative provisions of each state plan. This process avoided political recrimination and lawsuits and produced generally valuable results (Matuszeski, 1985). As of 1989, 29 of the 35 eligible coastal states and territories were approved to receive Section 306 funds; six other coastal states had dropped out of the program (Table 12–2).

Although the CZM program has established a basis for federal-state cooperation, the task of managing the coast remains formidable. In the first place, the 1972 Act specified a broad range of competing interests to be somehow reconciled in state programs, for example:

> . . . industry, commerce, residential development, recreation, extraction of mineral resources and fossil fuels, transportation and navigation, waste disposal, and harvesting of fish, shellfish, and other living marine resources, wildlife. . . (Sec. 302(c)).

Later sections and amendments added more concerns, for example, public access to beaches and coastal waters, natural hazard reduction, energy development, estuarine research, and protection of cultural and natural landmarks.

An important preliminary step was the designation of a *coastal zone* as the

Table 12–2 Status of state coastal management programs (as of March 1985)

State	Status		State	Status	
Alabama	Approved	1979	Mississippi	Approved	1980
Alaska	Approved	1979	New Hampshire	Approved	1982
American Samoa	Approved	1980	New Jersey	Approved	1978
California	Approved	1978	New York	Approved	1982
Connecticut	Approved	1980	North Carolina	Approved	1978
Delaware	Approved	1979	Northern Mariana Islands	Approved	1980
Florida	Approved	1981	Ohio	Withdrew	1980
Georgia	Not approved	1980	Oregon	Approved	1977
Guam	Approved	1979	Pennsylvania	Approved	1980
Hawaii	Approved	1978	Puerto Rico	Approved	1978
Illinois	Withdrew	1978	Rhode Island	Approved	1978
Indiana	Withdrew	1981	South Carolina	Approved	1979
Louisiana	Approved	1980	Texas	Withdrew	1981
Maine	Approved	1978	Virginia	Pending	1985
Maryland	Approved	1980	Virgin Islands	Approved	1979
Massachusetts	Approved	1978	Washington	Approved	1976
Michigan	Approved	1978	Wisconsin	Approved	1978
Minnesota	Withdrew	1978			

SOURCE: Office of Ocean and Coastal Resource Management

geographic focus of CZM planning in each state. The Act defined the coastal zone to include "coastal waters . . . and the adjacent shorelands . . . strongly influenced by each other and in proximity to the shorelines of the several coastal states." The offshore boundary was set by the Act as the seaward limit of state sovereignty (generally three miles from mean high water along ocean shorelines). The inland boundary was to be designated by each state. A combination of physical and cultural reference lines (e.g., elevation contours and coastal roads) were employed in most cases.

Within its coastal zone, each state was required to identify "permissible land and water uses" and "areas of planning concern" within which special restrictions would apply. All this sounded suspiciously like "state zoning," a notion stoutly opposed by most local governments. States therefore have been compelled to walk a tightrope between adopting significant restrictions on the use of coastal land as specified in federal guidelines while minimizing interference with the prerogatives of local governments and private owners.

The CZMA further required as a condition of plan approval that states exercise some form of control over important coastal land-use decisions through at least one of the following methods:

(A) State establishment of criteria and standards for local implementation, subject to administrative review and enforcement of compliance;

(B) Direct state land- and water-use planning and regulation; and

(C) State administrative review . . . of all development plans, projects, or . . . regulations [with power to disapprove if they are inconsistent with the state CZM Plan] (Sec. 306(e)).

Satisfying this requirement was no simple task: "Arrayed against those options was the nearly overwhelming complexity of 35 sets of state laws." (Matuszeski, 1985, p. 270). Some states offered comprehensive coastal management laws that they had adopted either before or after the federal CZM Act, for example, California, Washington, Rhode Island, North Carolina, and Michigan. Others, however, in the absence of a single coastal law, offered a "network" of statutes, regulations, and executive orders that individually addressed such issues as coastal wetlands, facility siting, beach access, pollution abatement, and hazard protection. In 1978, Massachusetts was the first state to be approved on the basis of "networking." Others followed gradually, after years of negotiation with OCRM over the "sufficiency" of each state's policies and procedures.

Looming over the federal Coastal Zone Management Program like a huge storm cloud has been endless controversy over the leasing and development of offshore oil and gas reserves. The Gulf of Mexico has long been a major source of domestic petroleum production, and extensive fields lie off the California coast and in the Georges Bank and Baltimore Canyon off the Atlantic Coast. Exploitation of offshore energy resources requires extensive land support facilities, which may transform coastal fishing communities into industrial harbors as in Louisiana. Furthermore, undersea drilling poses great risks of leaks and blowouts, which threaten marine habitat and shorelines with massive pollution as in the Santa Barbara Channel in 1969 (Baldwin, 1970). In 1976, the CZM Act was amended to authorize special "energy impact" grants to states facing major impacts from offshore oil development.

Coastal states directly control oil and gas within their own areas of jurisdiction (usually three miles seaward of the mean high water line), and many have refrained from developing such resources. Beyond that limit, seabed resources of the "outer continental shelf" (OCS) are federal property, under the jurisdiction of the Bureau of Land Management (BLM) of the Department of the Interior. Revenue from federal offshore oil and gas development is earmarked for the Land and Water Conservation Fund, as discussed in Chapter 11. Nevertheless, proposed lease sales and further development of OCS resources have been vigorously opposed by environmentalists and certain state governments.

Section 307 of the CZM Act requires that federal actions be "consistent with approved state management programs." State CZM plans may thus theoretically "veto" federal activities, a reversal of the normal federal-state relationship. The U.S. Supreme Court in *Secretary of the Interior* v. *California* 464 U.S. 310, 1984 held that Section 307 did not apply to federal lease sales but would limit subsequent exploration, development, and production of OCS oil and gas resources in accordance with state plans.

The federal CZM program has been popular with recipient states, which have successfully lobbied in Congress for continuation of funding. However, the substantive results of the program are difficult to evaluate. According to Healy and Zinn (1985, p. 308):

> Congress decided, for a variety of reasons, not to prescribe for the nation's coastline substantive federal standards of the type contained in contemporaneous pollution control laws, but rather to encourage each coastal state to set up a rational comprehensive process for making decisions about coastal resources. . . . [B]ecause of this lack of federal standards, it is impossible to evaluate rigorously the federal program's national impact on the protection of the coastal environment or on the rate of coastal development.

The CZM program is also difficult to evaluate due to its multiplicity of objectives—environmental, economic, social—and the complexity of federal-state-local-private interaction from which identifiable results emerge. Thus, the preservation of coastal wetlands, for example, may result from the CZM program, but may equally be attributable to the federal wetlands program (Sec. 404 of the Clean Water Act), to state or local laws, or to private actions. Nevertheless, federal CZM officials are not bashful in claiming credit for a variety of specific improvements in coastal management. A recent agency review of program achievements (Office of Ocean and Coastal Resource Management, 1988) lists a variety of achievements under the following headings:

Coastal Zone Management

- hazards protection,
- natural resource protection,
- natural resource development,
- public access,
- urban waterfront redevelopment,
- ports and marinas, and
- improved government operations.

Apart from individual projects, the CZM program may fairly be credited with fostering a variety of new state laws, regulations, and bond issues concerning

Table 12-3 Selected planning tools used by federally approved state coastal programs

State	Permits	Special review for large projects	Mitigation	Critical areas	Acquisition	Development promotion
Maine	○	●	●	○	●	●
New Hampshire	○	○	●	○	○	●
Massachusetts	○	○	○	●	●	●
Rhode Island	●	○	○	●	●	○
Connecticut	○	○	●	○	○	○
New York	○	○	○	●	●	●
New Jersey	●	○	●	○	○	○
Pennsylvania	○	○	○	○	●	●
Delaware	○	○	○	●	○	●
Maryland	○	●	●	●	●	●
North Carolina	●	●	●	●	●	○
South Carolina	●	●	●	●	●	○
Florida	○	●	●	●	●	●
Alabama	●	●	●	○	●	●
Mississippi	●	○	●	○	○	●
Louisiana	●	○	●	●	○	○
Puerto Rico	○	○	○	●	●	○
Virgin Islands	●	●	○	●	○	○
Michigan	○	○	●	●	●	●
Wisconsin	○	○	●	●	○	●
California	●	○	●	●	●	●
Oregon	○	○	●	●	●	●
Washington	●	○	○	●	●	○
Alaska	○	○	○	○	○	○
Hawaii	○	○	●	●	●	●
Guam	●	○	○	○	○	○
American Samoa	●	○	○	○	○	○
Northern Marianas	●	○	○	●	○	○

Key: ● = tool used; ○ = tool not used
Adapted from: R. G. Healy and J. A. Zinn, 1985, Table 2.

coastal management (Table 12–3). An important example is the adoption of coastal setback requirements for new construction in 13 states, which are cited as a model for a federal erosion management policy (National Academy of Sciences, 1990). Thus the federal CZM program promotes actions by states, which then, in turn, may influence the adoption of a federal standard to cover states that have not yet acted.

Overall, the federal CZM program has facilitated an upgrading of state capabilities to plan and administer coastal planning. The program may sometimes be excessively timid (as in the Sears Island dispute mentioned earlier, in which federal and state CZM officials were conspicuously silent). But, in balance, it has strengthened state skills and confidence in confronting coastal disputes, and may indirectly have contributed to upgrading the management of noncoastal resources as well.

THE NATIONAL FLOOD INSURANCE PROGRAM

Riverine and coastal disasters inflict an ever-rising toll of economic, social, and emotional costs on the United States. Annual loss of life due to floods has declined on average, thanks to improved warning and evacuation capabilities. But the average annual damage to public and private property caused by floods was crudely estimated to be 1 billion dollars in the mid-1960s (U.S. Congress, 1966a, p. 3), $2.2 billion per year in the mid-1970s (in 1967 dollars) (U.S. Water Resources Council, 1977), and about $5 billion in 1985 (current dollars). Hurricane Hugo alone in 1989 caused approximately $7 billion in flood, wind and other damage in Puerto Rico, the Virgin Islands, and the Carolinas.

The geographic area flooded varies according to the magnitude of the storm causing the flood. This presents the issue as to whether public programs should address only the area frequently flooded or should extend more widely to areas flooded only rarely. The federal government is chiefly concerned with hazards within the "100-year" or "1-percent annual chance" floodplain, namely the area with a probability of at least 1 percent of being flooded in any given year. (See Chapter 9 for discussion of the physical nature of floodplains.) This standard represents a compromise between very frequent (e.g., five-year flood events and the rare catastrophic event on the order of the 1927 Lower Mississippi flood or Hurricane Camille in 1969. Although it sounds fairly remote, a 100-year flood has a 26 percent chance of occurring during the lifetime of a 30-year mortgage. And only the outer edges of the 100-year floodplain have a risk as low as 1 percent per year: closer to the stream channel the risk is progressively higher. The total land area within 100-year floodplains is approximately 253,000 square miles (162 million acres) or about seven percent of the nation's land area (Federal Emergency Management Agency).

Ninety percent of these flood hazard areas are rural and thus exposed chiefly to agricultural and natural resource losses (U.S. Department of Agriculture, Soil Conservation Service, 1981, p. 28). But urbanized floodplains, both riverine and coastal, account for most of the human investment at risk, and most of the annual losses caused by floods. In 1975, 6.4 million dwelling units were estimated to be located in flood hazard areas nationally (U.S. Congress, 1975, p. 127). Urban flood-plains also contain industrial and commercial structures, highways, airports, bridges, water and sewer facilities, electrical substations and power lines, and myriad other elements of the urban infrastructure. Thus, the regional effects of a flood may extend much further geographically than just the area inundated, in terms of economic, transportation, and utility disruptions. Conversely, urban flooding often extends beyond the area mapped as the 100-year floodplain. Storm drainage backup accounts for extensive problems in many metropolitan areas such as Chicago and Houston.

Flash floods pose special difficulties for public flood response. Unlike gradual riverine floods that develop slowly and may allow days of notice to evacuate downstream areas, flash floods crest within hours or even minutes after a heavy deluge, thus endangering both lives and property. Flash floods are very destructive in narrow, steep canyons or valleys, as in the Rocky Mountains and Appalachia. The 1976 Big Thompson Canyon flood in Colorado was described as a "wall of water" carrying boulders, trees, cars, building debris, and bodies. (Gruntfest, 1987). Flash floods also strike small watersheds in urban areas where paving and sewering of natural land surfaces convey runoff directly to local streams, converting them into raging and destructive torrents.

Clearly, much of the flood problem stems from human encroachment on natural floodplains. Floods have long been regarded as "acts of God" or, paradoxically, natural phenomena to be controlled by human intervention. Floodplains attract development despite the risk involved because they offer level building sites, favorably located near waterways, highways, and other arteries of commerce. If properly situated and engineered, certain types of buildings may be compatible with occasional flooding. Wise use, rather than nonuse of flood plains has long been a national objective. But too often, encroachment has simply proceeded in ignorance or disregard of flood hazards. Also, the private investor may fairly conclude that if the site is unsafe, "the government" will offer protection against loss.

Structural Flood Control

In the 1936 flood control act, Congress undertook to do exactly that. Motivated by a rash of serious floods in the midst of the Great Depression, the Act was intended to alleviate both problems by authorizing a massive program of federal flood control projects. In portentous language, the Act declared that:

> . . . the federal government should improve or participate in the improvement of navigable waters or their tributaries, including watersheds thereof, for flood-control purposes if the benefits to whomsoever they may accrue are in excess of the estimated costs, and if the lives and social security of people are otherwise adversely affected (33 U.S.C. sec. 701a).

This Act launched a 10-billion dollar, 30-year effort to control flooding through the construction of dams, reservoirs, levees, diversion channels, and coastal protection works. Depending on the location, the major construction agencies have included the U.S. Army Corps of Engineers, the Bureau of Reclamation, and the Tennessee Valley Authority. Since 1954, the Soil Conservation Service of the Department of Agriculture under P.L. 83-566 has supplemented the large dam-building efforts of these agencies with conservation and flood control projects for small rural watersheds.

Flood control thus joined river and harbor improvements as the object of Congressional generosity, and of competition among members of Congress for projects to be located in their districts or states. (The next generation would have its interstate highways and military bases.) Cost sharing in flood control projects by states or local governments was limited to providing land and future maintenance: the cost of construction was entirely federal. The only constraint on the allocation of funds according to pork-barrel politics ("I'll support your project if you support mine") was the requirement in the 1936 Act that projects should be approved only ". . . if the benefits to whomsoever they may accrue are in excess of the estimated costs. . . ." Like NEPA 35 years later, the requirement of a favorable benefit/cost ratio generated elaborate economic evaluation procedures and even more elaborate ways of ensuring that the benefits would indeed exceed the costs. This was sometimes accomplished by assigning extravagant economic value to potential recreation opportunities of proposed reservoirs.

In the case of many projects, especially urban levees, the chief benefit would be the protection of buildings in the floodplain from flooding. Herein lay a fallacy of federal efforts to alleviate flood losses through flood control projects. Such projects made economic sense only if they protected buildings from flooding. The more buildings protected, the more favorable the B/C ratio. Thus, structural flood control implicitly was intended to attract *more building* in "protected" flood plains. Although future construction technically was not counted as a "benefit" in project justification, little effort was made to restrain further building in areas behind levees or downstream from flood control reservoirs. The projects themselves gave an illusion of safety to new investments in floodplains. And the Corps of Engineers in 1975 could claim that the nation had been spared $60 billion in additional losses since the 1930s through structural flood control (Platt, et al., 1980, p. 21).

But flood control projects are designed to a particular level of safety. Floods that exceed that design level can overtop a levee or overwhelm the storage capacity of a reservoir, creating havoc in the "protected" floodplain. A classic example occurred in Jackson, Mississippi where the Corps of Engineers constructed levees and channelized the Pearl River during the 1960s. New commercial and public buildings constructed in the floodplain behind the levee were inundated by up to 14 feet of water when the river overtopped the levee during the "Easter Flood" of April, 1979 (Platt, 1982). Overtopping of levees was blamed for at least 40 percent of the losses from Tropical Storm Agnes in 1972—until Hugo, the largest natural catastrophe in terms of property damage in U.S. history (White, 1975, p. 10).

Earlier studies by White and others (1958, 1964) indicated that average annual flood losses may actually be *increased* by the federal flood control program due to the failure to limit new development in areas protected against ordinary flooding but still vulnerable to catastrophic events.

Flood control projects may inadvertently worsen flood hazards by altering natural processes. A flood control dam blocks sediments from reaching the lower river. This creates a siltation problem behind the dam, reducing storage capacity, and may cause increased bank erosion downstream at times of high flow as the stream seeks to dissipate its excess energy. Where silt does reach the lower river from tributaries, the elimination of lower-magnitude floods due to flood control projects allows the stream channel to silt up during periods of low flow without being periodically scoured. A storm of extraordinary magnitude or one that is centered downstream from the flood control reservoirs may overtax the stream channel's reduced capacity. Overbank flooding may then exceed what would have occurred under natural conditions.

Haphazard local protection measures also may exacerbate flooding. A property owner or municipality that tries to fend off erosion by stabilizing the outer bank of a meander curve may simply transfer the erosion to a downstream neighbor. If the stream cannot dissipate its energy where it chooses, it adjusts its flow to scour more vulnerable areas with even greater force. A channelized stream flowing into a nonchannelized downstream segment bears great erosive force. Similarly, coastal jetties and groins may cause downdrift shorelines to erode due to lack of sand supply.

Federal response to the flood problem has been episodic (Burby and French, 1985, p. 6). Consistent with the model in Chapter 2, new laws, policies, and initiatives have tended to follow closely major floods or series of floods, as perceived by policy makers with the help of expert advice (Table 12–4). The most significant turning point in federal flood policy occurred in the mid-1960s following a series of disasters that were driving federal disaster relief costs to over $1 billion annually. After Hurricane Betsy in 1965, Congress enacted the Southeast Hurricane Disaster Relief Act of 1965, which directed the Secretary of the new Department of Housing and Urban Development to examine the feasibility of a federal flood insurance program. At the same time, the Bureau of the Budget formed the Presidential Task Force on Federal Flood Control. These two studies (U.S. Congress, 1966a and 1966b) were directed respectively by Marion Clawson and Gilbert White.

The White task force questioned the wisdom of exclusive reliance on structural flood control in view of the limitations cited in the foregoing. Instead, it proposed a multi-means approach to flood hazards including:

- *flood control works,* where economically justified and where benefited interests share the full costs of construction;
- *floodproofing* of existing structures or new structures that must be built in flood plains;

- improvement of *forecast and warning* systems;
- *land-use management,* including floodplain zoning and land acquisition;
- *flood insurance* provided at affordable rates by the federal government but subject to community floodplain management to control further development in hazard areas;
- *relief and rehabilitation* following a flood disaster.
  (U.S. Congress, 1966b; White, 1975, Table II-1)

It further urged caution in the establishment of a national flood insurance program:

A flood insurance program is a tool that should be used expertly or not at all. Correctly applied, it could promote wise use of flood plains. Incorrectly applied, it could exacerbate the whole problem of flood losses. For the Federal Government to subsidize low premium disaster insurance . . . would be to invite economic waste of great magnitude. Further, insurance coverage is necessarily restricted to tangible property; no matter how great a subsidy might be made, it could never be sufficient to offset the tragic personal consequences which would follow enticement of the population into hazard areas (U.S. Congress, 1966b, p. 17).

The task force report was forwarded to Congress by President Lyndon Johnson together with Executive Order 11296 directing federal agencies to evaluate flood hazards before funding new construction or the purchase or disposal of land. Congress then adopted the National Flood Insurance Act of 1968 (P.L. 90-448, Title XIII).

The NFIP: A New Approach

The National Flood Insurance Program (NFIP) has been the mainstay of federal response to floods since the early 1970s. It represents a dramatic departure from the long-standing reliance on structural flood control projects. Instead, the NFIP offers an array of nonstructural forms of response to flood hazards including (1) floodplain mapping; (2) floodplain management; (3) flood insurance; and, to a modest extent (4) floodplain land acquisition.

The NFIP is really two programs rolled into one. On the one hand, it seeks to reallocate a portion of the costs of floods from the victims themselves and federal taxpayers (who fund disaster relief) to all property owners in hazard areas through the mechanism of insurance. Flood insurance has been unavailable from private insurers because flood losses, when they occur, are catastrophic and require billions of dollars of reserves. The NFIP, operating through private insurance companies, provides insurance against flood damage to buildings and their contents. Routine claims and administrative costs are paid out of insurance premiums (which are subsidized for older structures and actuarial for new ones, as explained in the following). If a multi-billion-dollar hurricane occurs, the NFIP is backed up by the federal Treasury.

Table 12–4 Chronology of major floods and public response

Year	Major Flood Disasters	Significant Events in National Response
1925	Lower Mississippi—1927 New England—1927	Lower Mississippi Flood Control Act of 1928
1930		Tennessee Valley Authority Act of 1933 Report of Water Resources Committee of National Resources Board, 1934
1935	Kansas River—1935 Upper Susquehanna—1935 Eastern United States—1936 Ohio/middle Mississippi—1937 New England—1938	Flood Control Act of 1936 Flood Control Act of 1938
1940		Flood Control Act of 1944
1945		
1950	Kansas and Missouri rivers—1951 New England—1954	President's Commission on Water Resources Policy—1950 Watershed Protection and Flood Prevention Act of 1954
1955	New England—1955	Flood Insurance Act—1956
1960	Gulf Coast—1960 Southwest and Midwest—1961 Atlantic coast—1962 Louisiana—1964	Floodplain Information Program, Corps of Engineers—1961
1965	Mississippi-Louisiana—1965 Upper Mississippi—1965 Upper Mississippi—1969 Mississippi-Louisiana—1969	Southeastern Hurricane Disaster Relief Act of 1965 Water Resources Planning Act of 1965 HUD Study on Flood Insurance—1966 Report of Task Force on Federal Flood Control Policy—1966 Executive Order 11296—1966 National Flood Insurance Act of 1968
1970	Rapid City, South Dakota—1972 Hurricane Agnes—1972 Upper Mississippi—1973	Flood Disaster Protection Act of 1973 Water Resources Development Act of 1974 Federal Disaster Assistance Act of 1974
1975	Mid-Atlantic—1975 Massachusetts coasts—1978 Southern California—1978 Pearl River, Mississippi—1979 Red River—1979 Texas Gulf—1979	Executive Orders 11988, 11990 (1977) Creation of Federal Emergency Management Agency—1979
1980	Hurricane Frederic	OMB Directive on Post-Flood Mitigation Assessments—1980 Coastal Barrier Resources Act of 1982

On the other hand, the NFIP promotes floodplain management to limit flood risk to new construction along the nation's rivers and shorelines. As noted in the White report, unless new encroachments are controlled, the NFIP could inadvertently subsidize new floodplain development. But the idea of federal control over land use in floodplains has always been politically unviable. Therefore, a clever back-door approach is employed by the NFIP. It does not impose mandatory national land-use standards, equivalent to EPA's air and water quality standards. Instead it establishes minimum standards for local adoption as a condition to the availability of flood insurance within a given community. Communities are free to ignore the floodplain management guidelines and deny flood insurance to owners of property at risk within their jurisdictions.

At first, the NFIP began with a whimper: Only four communities entered the program and 20 policies were sold during its first year. Two obstacles impeded its progress. First, most communities lacked detailed maps of their flood hazard areas on which land-use regulations could be based. Second, few property owners were interested in buying flood insurance; thus there was little incentive for communities to adopt politically unpopular floodplain regulations in order to enter the program.

To address the first problem, the NFIP has spent $1 billion over two decades mapping the nation's flood plains at a scale suitable for local land-use regulations (1 to 400). Unlike the Coastal Zone Management Program, which has funded states to prepare their own maps, the Federal Emergency Management Agency (FEMA), which administers the NFIP, has directly prepared flood hazard maps under contract with other federal agencies and private engineering firms. The maps are based on standard engineering models of stream flow and coastal flooding (HEC-2 and SLOSH are the usual models). Using available hydrologic or oceanographic data, the models estimate the elevation of the 100-year flood at specific points along a stream or coast. Elevation data are then converted to horizontal estimates of the geographic extent of the floodplain, using topographic data. These calculations are of course subject to several sources of error (Dingman and Platt, 1977). The typical NFIP map depicts the 100-year flood plain ("base flood" in NFIP jargon) and also in some cases a "floodway," a "coastal high hazard area" (on open ocean coasts), and a 500-year floodplain. They also provide data on the level of risk from which actuarial insurance rates may be calculated (Figure 12-2).

NFIP maps are prepared for individual political units: municipalities or unincorporated portions of counties. Prior to completion of a map study, a community may qualify for provisional participation in the NFIP. Once the map is complete and opportunity for appeal of errors has been provided, the map is formally published. From the date of publication, the community has one year to adopt full-scale floodplain management measures or lose all NFIP benefits.

The other obstacle, buyer resistance, was addressed in the Flood Disaster Protection Act of 1973 (PL 93-234), which closely followed Tropical Storm Agnes. This Act required that anyone borrowing money from a federal or federally related source for purchase or improvement of a structure in a floodplain identified by the NFIP must purchase a flood insurance policy. Since many lending institutions are insured or regulated by federal agencies, this covered a large proportion of mortgage loans involving flood-prone structures.

With these two mid-course corrections, the NFIP began to grow rapidly. By 1975, 9,600 communities had joined and a half-million policies were in effect.

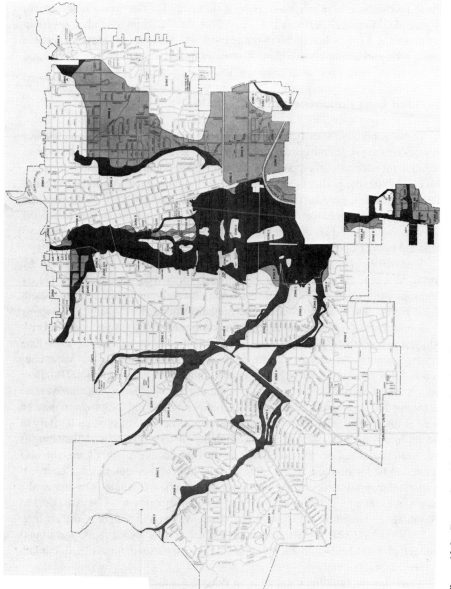

Figure 12.2 Excerpt from Flood Insurance Rate Map for Boulder, Colorado. (Dark area is regulatory 100-year floodplain; light gray is 500-year floodplain). *Source:* Federal Emergency Management Agency.

These numbers virtually doubled by 1979. By 1988, of nearly 22,000 communities in which flood hazards had been identified, 18,200 had joined the NFIP and 16,470 of those had satisfied the minimum federal standards for floodplain management (Congressional Research Service, 1987, p. 5). A total of 2.1 million policies were in effect, covering $156 billion in flood-prone property. Also the program had become self-supporting in average loss years. It actually turned a small profit in 1987 and 1988 due to increased premium rates and low flood activity.

Flood Loss Reduction under the NFIP

These impressive statistics on the expansion of the NFIP provide no evidence as to whether the program has affected the national level of flood losses for better or worse. As mentioned earlier, average annual losses continued to rise during the 1970s and 1980s, but perhaps less than if the NFIP did not exist. Also, the impact of NFIP would be felt gradually as new structures are built to NFIP standards, and older ones are destroyed or upgraded. It is therefore not simple to evaluate the effectiveness of the NFIP in terms of flood loss reduction. But we can at least understand how it operates.

The NFIP seeks to reduce flood losses to structures in two ways: (1) by charging actuarial rates for new or substantially improved structures and (2) by setting minimum standards for local regulation of development and redevelopment in flood plains. These are intended to be mutually supportive.

Actuarial rates are insurance premiums calibrated to the assumed level of risk to which a structure is exposed. Thus, a greater likelihood of loss would necessitate a higher premium. Actuarial rates are charged for flood insurance on "new construction," for example, that begun after the community entered the NFIP. Older buildings may be insured to a specified amount at a nonactuarial, subsidized rate. This means that structures existing when a community entered the program may be insured at low rates unrelated to risk, whereas new structures must be built safely to avoid prohibitively high rates (and unavailability of federally related financing if flood insurance is not purchased). There is a strong incentive, therefore, for new structures to be located and designed so as to minimize the actuarial rate charged. Elevation of coastal structures is the most conspicuous example: the higher the ground-floor elevation, the lower the flood insurance premium. The Atlantic and Gulf shorelines are lined with post-NFIP structures standing high in the air atop substantial pilings. This in part reflects the influence of actuarial rates, and in part the effect of local minimum elevation requirements pursuant to NFIP floodplain management standards.

Land-use and building regulations in floodplains were rare before 1970. In the mid-1950s, Hoyt and Langbein (1955, p. 95) noted: "Floodplain zoning, like almost all that is virtuous, has great verbal support, but almost nothing has been done about it." Murphy (1958) surveyed national experience with local floodplain management and found only about 50 communities with a local program. The negative outcome in the *Morris County* case (discussed in Chapter 9) reflected the lack of experience and judicial support for floodplain and wetland regulations in the early 1960s. The judicial outlook brightened considerably however with the *Turnpike Realty* and *Just* decisions in 1972.

The NFIP entered the turbulent waters of floodplain management cautiously. Section 1361 of the Act mandates the development of federal criteria for state and local measures to:

- "Guide the development of proposed construction away from locations which are threatened by flood hazards,"
- "Assist in reducing damage caused by floods," and
- "Otherwise improve the long-range land management and use of floodprone areas."

Such criteria were finally published in 1976 and have been relatively unchanged ever since (44 CFR Secs. 60.3 et seq.). They specify progressively more stringent levels of floodplain management to be satisfied by local communities as the NFIP provides more detailed flood hazard data. The fundamental requirement is that when elevations of the local 100-year flood are determined, the community must require the lowest floor of new residential construction to be at or above that elevation. New commercial structures must be elevated or designed to withstand flood damage ("floodproofed") to that level. This dovetails with the actuarial rate incentive mentioned in the foregoing. (Other technical requirements are not discussed here.)

Tighter control is required in areas delineated by the NFIP as "floodways" or "coastal high hazard areas." A *floodway* is the inner portion of a riverine floodplain adjoining the channel, which experiences frequent and sometimes fast-moving inundation. (See Figure 9–4 supra.) No structures are to be permitted in such areas which "would result in any increase in flood levels within the community . . ." during the 100-year flood. Some states and communities have simply banned all new development in floodways, which the NFIP does not require.

Coastal high hazard areas ("V zones") along many open ocean coastlines are defined by the estimated reach of a three-foot breaking wave during a 100-year storm. Such areas are indeed hazardous and also are often subject to increasing risk due to coastal erosion. Nevertheless, the NFIP does not prohibit new construction in V zones. New structures must be landward of mean high tide and must be elevated above the estimated reach of waves during a 100-year storm surge. Also, "manmade alteration" of dunes or mangrove stands is prohibited in V zones. Subject to these minimal requirements, new buildings may be constructed in V zones and insured by the NFIP. Like the flood control program, if the design limits of NFIP floodplain management criteria are exceeded, the resulting loss to structures allowed in the hazard area may be greater than if insurance were not available and investors stayed away from the water's edge.

Another issue is consistency of effort and level of effectiveness from one community to another. In 1979 and 1983, researchers at the University of North

Table 12–5 Action instruments used by communities of more than 5000 inhabitants to manage flood hazard areas (1979)

Action Instruments	Percent Using
Police Power Regulations	
Elevation requirements	77%
Subdivision regulations	75
Zoning	75
Floodproofing requirements	59
Special floodway regulations	52
Septic tank permits	42
Sedimentation and erosion control regulations	29
Wetlands protection regulations	25
Critical areas designation	18
Density exchange/cluster development regulations	6
Sand dune regulations	6
Capital Improvement Policies	
Location of public facilities outside of flood hazard areas	22
Land Acquisition	
Land acquisition for open space, parks and other public uses	34
Relocation of existing hazard area development	3
Incentives	
Preferential taxation	9
Information and Advice	
Public information about hazard	37

SOURCE: Burby and French, 1985, Table 3–2

Carolina surveyed the floodplain management practices of local governments in the NFIP (involving 1,415 and 956 respondent communities respectively) (Burby and French, 1985). The results, displayed in Table 12–5 indicate a wide diversity in the means used to meet or exceed NFIP floodplain management standards. Table 12–5 also indicates that smaller communities (reflected in the 1983 data) are less likely to employ even the basic tools of minimum elevation and zoning. Some discretion is allowed for local communities to select management techniques that are compatible with their physical, fiscal, political, and land-use characteristics. But undue flexibility may defeat the flood-loss-reduction goal of the NFIP by making it impossible to assess and monitor the effectiveness of each community's program.

The Need for Orchestration

Ironically, the nation's response to floods has shifted from being excessively narrow as to approach and agency mission, to its present state in which it is perhaps overly diffuse. Responsibilities and authority are fragmented among the federal, state, regional, and local levels of government. Federal flood-related efforts alone are diffused among 27 agencies and nine program purposes (Federal Emergency Management Agency, 1986, p. VII-2). The NFIP is the cornerstone of federal response, but many functions, such as flood prediction and warning, mapping, and emergency planning, cross program and agency boundaries. Since first articulated (U.S. Congress, 1966b), a "unified national program" for managing flood losses has been an elusive goal. The U.S. Water Resources Council (1976) urged intergovernmental coordination in *A Unified National Program for Flood Plain Management,* which was revised in 1979 and 1986. (The latter version was published by the Federal Emergency Management Agency). Nevertheless, a National Review Committee on Floodplain Management (1989) (also chaired by Gilbert White) concluded that "there is no truly unified national program . . ." and further:

> There is no central direction for the Unified National Program. No agency has the charter or capability to carry it out in its entirety, and no agency has authority for assuring coordination of the numerous programs targeted on its objectives. There are serious overlaps, gaps, and conflicts among programs aimed at solving the same problem.

The Coastal Barrier Resources Act

In the case of coastal barriers, Congress was persuaded that various federal programs were hopelessly at loggerheads and decided to cut through the bureaucratic muddle. Coastal barriers are elongated spits or islands, composed mainly of sand, which fringe much of the Atlantic and Gulf coastlines (Figure 12-3). In a natural state, they provide important habitat for marine life and birds, and dramatic expanses of beach, dune, and salt marsh. When developed, they are exposed to the destructive force of hurricanes and winter storms that sometimes overwash an entire barrier. They also tend to erode readily when beach material is obstructed and in response to sea level rise.

The Coastal Barrier Resources Act of 1982 (PL 97-348) declared that:

> Coastal barriers contain resources of extraordinary scenic, scientific, recreational, natural, historic, archeological, cultural, and economic importance, which are being irretrievably damaged and lost due to development on, among, and adjacent to such barriers (Sec. 2(A) (2)).

The Act further stated that federal assistance was contributing to "the loss of barrier resources, threats to human life, health, and property, and the expenditure of millions of tax dollars each year . . ."; and therefore:

> A program of coordinated action by Federal, State, and local governments is critical to the more appropriate use and conservation of coastal barriers (Sec. 2(A) (5)).

To effectuate such a program, the Act established a Coastal Barrier Resources System (CBRS) within which certain federal benefits would be withheld. The

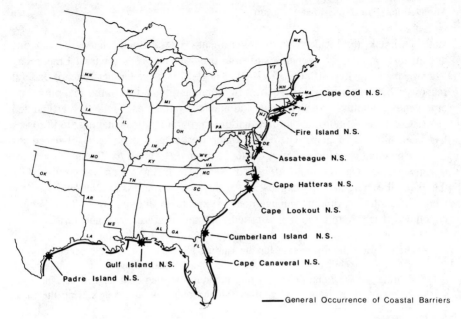

Figure 12.3 Location of coastal barriers and national seashores on Atlantic and Gulf of Mexico shorelines.

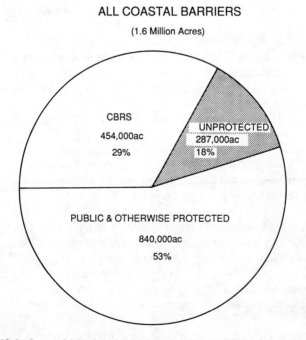

Figure 12.4 Status of Atlantic and Gulf Coast barriers, 1987. CBRS = Coastal Barrier Resources System.

CBRS was based on an inventory of nonpublic, nonprotected, undeveloped coastal barriers prepared by the Department of the Interior (Figure 12-4). These included 186 geographic units extending along 656 miles of oceanfront shoreline (about 24 percent of the total U.S. barrier coastline along the Atlantic and Gulf of Mexico) (Platt, et al., eds. 1987). Within the CBRS, the Act prohibits most federal incentives to growth including flood insurance, road and bridge funds, sewer and water construction, and beach protection. Areas within the CBRS may still be developed without federal involvement if local law allows. But the denial of flood insurance and other federal benefits is an innovative approach to slowing growth in sensitive areas (Kuehn, 1984). It is also a bold, if fairly crude, method of "orchestrating" federal policies for specified geographical circumstances.

FEDERAL WETLANDS PROGRAMS

Many floodplains are also wetlands and vice versa, although the two are not synonomous. Despite extensive areas of geographical overlap, federal programs concerning wetlands have evolved independently from those addressing floods. Each involves a different set of statutes, agencies, goals, procedures, and terminology. During the 1980s, the need for coordination of federal floodplain and wetland management efforts with each other and with the Coastal Zone Management Program was frequently expressed.

As discussed in Chapters 1 and 9, wetlands comprise an important subset of the total land and water resources of the United States. The term encompasses a variety of ecological and hydrological regimes generally characterized by: (1) the presence of water; (2) predominance of saturated hydric soils; and (3) prevalence of vegetation adapted to wet conditions (hydrophytes) (Mitsch and Gosselink, 1986, pp. 15–16). Many kinds of physical features share these broad characteristics, for example, red maple swamps and black spruce swamps in the northern states (associated with glaciation); estuarine salt marshes behind coastal barriers; bottomland hardwood forests in the lower Mississippi Valley; prairie potholes in the Great Plains; playa and riparian wetlands in the West, and wet tundra in Alaska (Tiner, 1984, p. vii). Depending on their physical type, size, and location, wetlands provide many natural values including habitat for flora and fauna; natural flood detention (inland wetlands) or shoreline buffering (coastal wetlands); aquifer recharge and pollution filtration; scenic beauty; and open space (Kusler, 1983; Mitsch and Gosselink, 1986).

Since 1937, Congress has supported the acquisition, restoration, and maintenance of wildlife areas including wetlands under the Pittman-Robertson Act (16 USCA Sec. 66a). This Act authorizes grants of up to 75 percent of the cost of projects from excise taxes on firearms and ammunition. The 1950 Dingell-Johnson Act (16 USCA Sec. 777) similarly has supported wetland acquisition for fish habitat restoration. The 1958 Fish and Wildlife Coordination Act mentioned earlier required consideration of impacts on wetlands and other wildlife habitat before a federal water development project could be approved.

The National Wetlands Inventory

A long-standing federal activity has been to identify the types, extent, and distribution of wetlands and to estimate their rates of change. The U.S. Department of Agriculture (USDA) conducted crude surveys of wetlands as early as 1906 and

Figure 12.5 Excerpt from a National Wetlands Inventory map. Abbreviations refer to various types of wetlands. *Source:* U. S. Fish and Wildlife Service.

1922 to see how much cropland could be created by draining wetlands. The first conservation-oriented wetland survey was conducted during the mid-1950s by the U.S. Fish and Wildlife Service (FWS) (1956).

In 1974, FWS launched the National Wetlands Inventory (NWI), a massive project to classify and map virtually all the nation's wetlands. The NWI uses color-infrared aerial photography at scales ranging from 1:60,000 to 1:130,000. Wetland data thereby obtained is superimposed on conventional topographic base maps of 1:24,000 and smaller scales (Mitsch and Gosselink, 1986, p. 471) (Figure 12-5). As of September, 1986, the NWI had mapped wetlands for about 10,000 standard U.S. Geological Survey quadrangles. This covered all of the nation's coastal wetlands (except portions of Alaska). Mapping was also complete for several entire states including Massachusetts, Connecticut, Rhode Island, Vermont, Arizona, Connecticut, New Jersey, Delaware, and Hawaii. Some area maps have been digitized for use in geographic information systems.

Measuring rates of wetland loss is hampered by differing definitions from one survey to another as well as by technical inaccuracy of measurement techniques. The NWI has estimated total wetlands in the early 1980s to comprise about 99 million acres in the lower 48 states, of which 93.7 million were inland freshwater wetlands, and the rest coastal (Frayer et al., 1983). This represented a loss since the mid-1950s of approximately 14.8 million acres of freshwater wetlands and 482,000 acres of saltwater wetlands (Figure 12-6) (Conservation Foundation, 1988).

A 1981 USDA study estimated nonfederal wetlands to amount to 70.5 million acres, with an apparent loss of 14 percent between 1956 and 1980. Annual loss of wetlands nationally due to dredging, filling, drainage, and conversion to agricultural or urban purposes is roughly estimated to be 300,000 acres (U.S. Congress, 1985, p. 1).

Nationally, approximately one-fourth of wetlands are fully protected under public ownership. Extensive areas of salt marsh are contained within National Seashores and National Wildlife Refuges. Other federal, state, and local land holdings include diverse types of wetlands. Certain wetlands are owned by private conservation organizations such as the Nature Conservancy, the Massachusetts Audubon Society, and the National Audubon Society. The remaining wetlands are privately owned, and in many areas are ripe for drainage or filling for development. There is little technical difficulty in filling or dredging wetlands and alteration is practically irreversible. Governmental intervention in the private land market through wetland regulatory programs has sought to moderate this process.

The Section 404 Program

Section 404 of the Federal Clean Water Act (33 USCA Sec. 1344) established a rather odd approach to the management of wetlands. Although it is generally known as the "Federal Wetlands Program," the term "wetland" does not actually appear in the law. It regulates dredge and fill in "waters of the United States" but the latter have been construed to include bottom and hardwood forests, mangrove

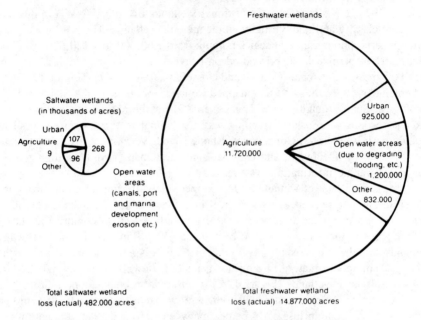

Figure 12.6 Wetlands conversions: mid-1950s to mid-1970s. *Source:* Protecting America's Wetlands: An Action Agenda, Final Report of the National Wetlands Policy Forum, 1988. Reprinted by permission of The Conservation Foundation.

swamps and prairie potholes that are dry most of the year. The program is administered jointly by an uneasy alliance between the U.S. Army Corps of Engineers (COE) and the U.S. Environmental Protection Agency (EPA). Also, the U.S. Fish and Wildlife Service of the Department of the Interior and the National Marine and Fisheries Service of the Department of Commerce are notified of 404 applications and given opportunity to offer their own recommendations regarding their respective agency interests in a particular wetland.

Section 404 was enacted in 1972 as part of extensive amendments to the Federal Water Pollution Control Act, later renamed the Clean Water Act. It was intended to supplement the COE's existing authority under Section 10 of the 1899 Rivers and Harbors Act (33 USCA Sec. 403) to prohibit "unauthorized obstruction or alteration of any navigable water of the United States" in the absence of a permit from the COE. In 1968, the COE amended its Section 10 regulations to require permit reviews to consider: " . . . the effect of the proposed work on navigation, fish and wildlife, conservation, pollution, aesthetics, ecology and the general public interest." The denial of a permit to fill eleven acres of mangrove swamp under these criteria was upheld by a Federal Court of Appeals in *Zabel* v. *Tabb* 430 F.2d 199 (5th Cir., 1970).

But when Section 404 was adopted in 1972, the Corps interpreted its authority as extending only to traditional "navigable waters." This narrow interpretation of "waters of the United States" in Section 404 was invalidated in *Natural Resources Defense Council v. Callahan* 392 F. Supp. 685 (D.C., 1975) which held that the phrase was intended by Congress to apply broadly to wetlands.

The geographic reach of Section 404 was eventually defined by COE administratively (33 CFR 323.2) to include a wide range of land-water phenomena, for example:

- "Navigable waters" (including waters which once were navigable or that could be made navigable, and including all tidal waters)
- All interstate waters including interstate wetlands
- "All other waters . . . the use, degradation or destruction of which could affect interstate or foreign commerce"
- Impoundments of waters described above
- Tributaries of waters described above
- The territorial sea
- Wetlands adjacent to the waters described above

The term "wetlands" was defined by the Corps to include:

Those areas that are inundated or saturated by surface or groundwater at a frequency and duration sufficient to support, and that under normal circumstances do support, a prevalence of vegetation typically adapted for life in saturative soil conditions. Wetlands generally include swamps, marshes, bogs, and similar areas (33 CFR Subpart E).

One might expect that the jurisdiction of Section 404 (that is, an area where a permit is required to dredge or fill) would be determined according to the maps of the National Wetlands Inventory. No such direct connection between the NWI and Section 404 has been established however. The Corps' permit jurisdiction under 404 must be established by site visit in most cases. NWI maps are indicative of areas that probably fall under 404 jurisdiction. But unlike the flood hazard maps prepared by the National Flood Insurance Program, they have no legal weight. (In some cases, the NWI maps are difficult to use for regulatory purposes due to problems of scale, coding of wetland data, and inaccuracy.)

Parties subject to 404 are defined broadly to include "any individual, commercial enterprise, organization, or governmental agency" that intends to construct or fill in areas covered by Section 404. Even federal agencies other than the COE itself must obtain a permit (33 CFR 323.3(b)). The Corps' 404 review process is illustrated in Figure 12-7.

As stated above, the Section 404 Program is jointly administered by the Army Corps of Engineers and the U.S. Environmental Protection Agency. While the former receives permit applications and performs a "public interest review," its

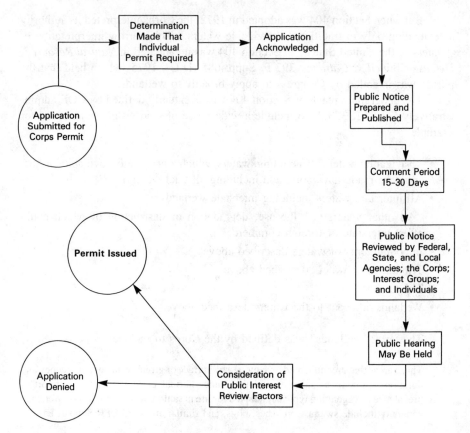

Figure 12.7 Approval process for Section 404 Federal wetlands permit.
Source: U. S. Army Corps of Engineers.

decision in each case must conform to "environmental guidelines" issued by EPA under Section 404(b)(1) (found at 40 CFR Part 230). These guidelines specify additional policies and procedures to be followed by the Corps. It is unusual for one federal agency to have statutory authority over the administrative actions of another. Furthermore, EPA has the right to review any case submitted to COE, and if it wishes, to veto a permit which has already been issued by the latter. Furthermore, EPA may utilize "advance identification" to prohibit COE permits in specified wetlands (Platt, 1987).

The EPA Guidelines articulate two key management concepts: "water dependency" and "mitigation." Water dependency refers to activities that require access or proximity to water to fulfill their purpose, for example, marinas or fishing docks. For activities that are not water dependent, it is presumed that an alternative nonwetland site is available unless the applicant proves otherwise.

Mitigation refers to actions that minimize adverse effects of fill or discharge into regulated areas. EPA's guidelines specify a series of remedial measures to reduce the effects of the proposed fill. These address:

1. the location of the discharge;
2. the type of material to be discharged
3. control of the material after discharge
4. dispersion of material;
5. adapting discharge technology to the needs of the site;
6. restoration of alternative degraded sites (40 CFR, Subpart H).

The last of these options has introduced a number of experiments with the rehabilitation of degraded wetlands and in a few cases, attempts to create entirely new wetlands to offset the loss of areas to be dredged or filled.

Both of these concepts were applied in a landmark case involving a proposal to fill part of "Sweedens Swamp" in southeastern Massachusetts for a regional shopping mall in the mid-1980s. After considerable deliberation, the Army Corps of Engineers concluded that the only alternative site was owned by a competing shopping center developer. It issued a Section 404 permit for the project contingent on the creation of substitute wetland habitat on a tract two miles from the project site. The EPA then vetoed the permit under its 404(c) authority, finding that the alternative site could have been purchased by the developer and that the proposed mitigation by creating a replacement wetland was unreliable. On appeal, the Federal District Court in Boston upheld the EPA as acting reasonably on the basis of its administrative expertise (*Bersani* v. *EPA*, 674 F.Supp. 405, 1987).

Assessment of the 404 Wetlands Program, like the Coastal Zone Management and National Flood Insurance Programs, is difficult at a national scale. Most 404 permit applications are approved: Of 11,000 applications received annually, COE denies only about 3 percent. However, of those approved, about one-third are modified from their original form and 44 percent of the 11,000 applications are withdrawn without action (Baldwin, 1987). The federal Office of Technology Assessment (1984) has credited the 404 Program with preventing 50,000 acres of wetland loss each year, primarily by requiring modifications in projects. But such estimates, while comforting, must be taken with salt: They rely on evidence supplied by the COE itself which may be self-serving.

Some wetland conversions slip through the 404 net, either legally or illegally. Agricultural drainage of wetlands and several other activities are exempted by statute or "general permit" from 404 review. In states that have their own wetland laws, the state review may serve as a substitute for COE review in minor cases. In other cases, state and federal reviews may both be required. And in really important cases, an Environmental Impact Statement may be required prior to decision on a federal 404 permit. (See discussion of Sears Island, Maine on pages 306–7.)

Unquestionably, Section 404 compliance is a cumbersome and often time-consuming process. Sometimes it duplicates state or local reviews, and can be used as a lever for tying up unpopular projects in court. It is not an effective substitute for preserving key wetlands through acquisition. Nor is it a surrogate for advance comprehensive planning at a state or regional level (Platt, 1987). It is however a "Rube Goldbergian" attempt to impose a *de facto* national wetlands policy through the medium of water pollution legislation where federal jurisdiction is stronger than on dry land.

AGRICULTURAL LAND

Agriculture is a long-standing national concern dating back to the debate between Jefferson and Hamilton regarding the disposition of the incipient public domain. The most visible and controversial federal agricultural programs of the twentieth century are those concerned with price supports and subsidies for selected crops. While those programs indirectly influence land usage and cropping practices in various regions, they are beyond the scope of this discussion. But Congress has addressed the conservation of agriculture land more directly in a series of measures beginning with the creation of the Soil Conservation Service of the U.S. Department of Agriculture in 1935. Another federal role has been to sponsor large-scale irrigation projects in arid western states beginning with the establishment of the Bureau of Reclamation in 1902.

In the 1970s and 1980s, several new initiatives were taken in response to mounting concern regarding stewardship of the nation's land resources for the production of food and fiber. The two principal issues have been the conversion of productive farmland to urban development and other irreversible and non-agricultural uses, and the loss of topsoil due to water and wind erosion. Federal response has taken the form chiefly of a series of inventories documenting the extent and condition of agricultural lands and, through the Food Security Act of 1985, a bold initiative to remove large areas of highly erodible lands from production.

The Cropland Base and Productivity

As discussed in Chapter 1, cropland is the most sensitive and valuable of the nation's rural land resources. Estimates of the extent of cropland and its variation over time are based on periodic inventories of agricultural land conducted by the U.S. Department of Agriculture (USDA). Land is classified according to its actual usage at a given time (for example, harvested crops, pasture, forest, range, etc.) and according to its physical suitability for various uses, particularly crops. The physical advantages and limitations of specific land are assessed by means of the Land Capability Classification (LCC) of the USDA Soil Conservation Service. The LCC, which has been in use since the mid-1960s, rates soils in terms of adequacy of moisture, capacity to hold moisture without waterlogging, susceptibility to erosion, freedom from excessive alkilinity, acidity, and salinity, stoniness, and climatic variables (Furuseth and Pierce, 1982, p. 20). The LCC classes are listed in Table 12–6.

Agricultural Land

Table 12–6 U.S. soil conservation service land capability classification

	Capability Classes
Class I	Soils with few limitations that restrict their uses.
Class II	Soils which have moderate limitations that reduce the choice of plants or that require moderate conservation practices.
Class III	Soils that have severe limitations that reduce the choice of plants, require special conservation practices, or both.
Class IV	Soils having very severe limitations that reduce the choice of plants, require very careful management, or both.
Class V	Soils that are not likely to erode but have other limitations, impractical to remove that limit their use largely to pasture, range, woodland, or wildlife.
Class VI	Soils having severe limitations that make them generally unsuited to cultivation and limit their use largely to pasture or range, woodland, or wildlife.
Class VII	Soils which have very severe limitations that make them unsuited to cultivation and that restrict their use largely to pasture or range, woodland, or wildlife.
Class VIII	Soils and landforms with limitations that preclude their use for commercial plants and restrict their use to recreation, wildlife, water supply, and esthetic purposes.

SOURCE: Furuseth and Pierce, 1982, Table 2 (Derived from U.S. Soil Conservation Service).

Cropland is predominant on the Class I–III soils but some crop cultivation is attempted on lower level soils as well. Also, some of the higher level soils are actually in forest, range, or pasture condition rather than crops (Figure 12-8).

The 1977 National Resources Inventory (NRI) estimated the total extent of land that was or could be used for the production of crops—the "cropland base"—to be 540 million acres. This included 413 million acres of readily available "cropland" and 127 million acres of "potential cropland" which would require substantial investment to convert it from other uses (Figure 12-9) (National Agricultural Lands Study, 1981).

The USDA also identified 345 million acres as "prime farmland" on which crops can be produced for the least cost and with the least damage to the resource base (USDA Economic Research Service, 1984). Of that total, 230 million acres were in cropland and the rest in other rural status (Figure 12-10) (USDA Soil Conservation Service, 1981, p. 5).

American farmers have steadily raised the productivity of the nation's land. In 1957, it was estimated that the doubling of American agricultural output that characterized the first half of the twentieth century would be repeated in the second half (Harris, 1957). The U.S. Farm Output Index in fact rose by 46 percent in 34 years from 73 in 1950 to 111 in 1984 (1977 = 100) (U.S. Bureau of the Census, 1986, Table 1165). This increase occurred without any net increase in harvested acreage, which actually declined from 377 million acres in 1950 to 337 million in 1984. While much of the rise in productivity before 1950 resulted from mechanization, the post-1950 rise has resulted largely from the wider use of irrigation, artificial fertil-

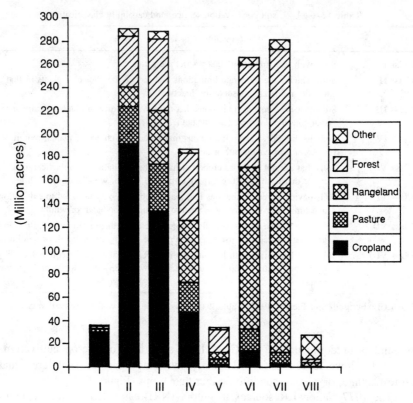

Figure 12.8 Use of nonfederal rural land by soil capability class. *Source:* U. S. Department of Agriculture, 1987.

izer and pesticides, and genetic improvements. Meanwhile, inputs of labor per unit of production have steadily declined, thus dramatically raising the productivity of farm labor (Conservation Foundation, 1985, p. 154–7). (Figure 12-11.)

Three qualifications to the foregoing statements are necessary. First, estimates of total cropland are fairly inexact and difficult to compare over time. Even in the 1980s, cropland data were not obtained from remote sensing sources (aerial or satellite imagery), but rather from estimates of individual agricultural extension agents in most of the nation's 3,041 counties. More scientific surveys, such as the National Resources Inventory discussed in the following, employ a more precise sampling methodology. But national-level estimates of cropland and other land-use categories are inherently approximate.

Second, national estimates of total cropland mask regional variations in the quality of land under cultivation. As prime land is lost in the Northeast and north central states, acreage of lower quality requiring drainage or irrigation and heavy use of pesticides has been brought into production in the South and Southwest. While the number of acres may appear to be unchanged, the average quality of land in production is thus diminished (Batie and Healy, 1983).

Agricultural Land

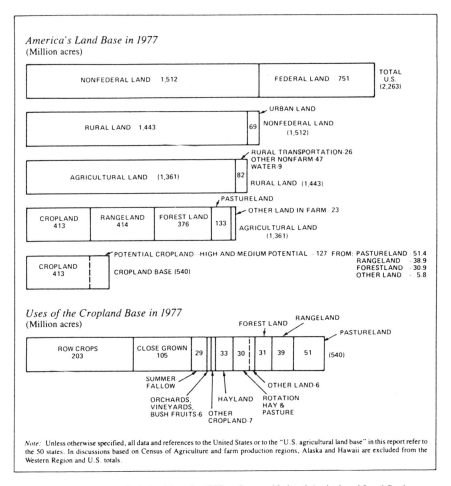

Figure 12.9 America's land base in 1977. *Source:* National Agricultural Land Study, 1981.

Third, within each region much land shifts into and out of cropland usage over time. When prices are high, as in the late 1970s, marginal land is planted in crops despite severe limitations, such as steep slopes, erodibility, aridity, wetness, or poor soil quality. The total "harvested cropland" was thus temporarily boosted (e.g., from 330 million acres in 1975 to 391 million acres in 1981 due to high export demand) but the physical limitations of the newly added "cropland" are not indicated. With government land retirement incentives or lower commodity prices, such marginal lands may be withdrawn from usage—unless a permanent loss of better lands and/or public subsidies (e.g., for irrigation) makes retention of such poor lands economically justified.

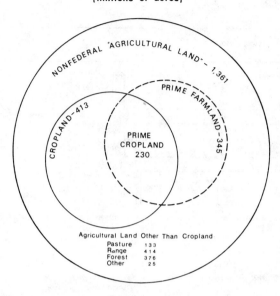

Figure 12.10 Utilization of nonfederal agricultural land, 1977. *Source:* Platt, 1985, Fig. 2. Reproduced by permission of Association of American Geographers.

Urbanization of Agricultural Land

Concern about farmland conversion is largely a post-1960 phenomenon. The primary farmland issue of the 1930s was overpopulation of land, which was desiccated in some regions by drought, inundated elsewhere by floods, and widely eroding. President Roosevelt's National Resources Board (1934) urged that 75 million acres of marginal land be removed from agricultural usage with the impoverished inhabitants relocated elsewhere. Ironically, the Board urged certain public measures that today are proposed to *retain* land in farming, for example, land-use zoning, purchase of easements, and tax incentives (Platt, 1985, p. 434).

The past three decades, however, have witnessed a wordy, sometimes acrimonious, and often confusing debate over the impact of urbanization on agricultural land. There is no question that metropolitan growth has paved, flooded, or built over much farmland—approximately 900,000 acres per year, of which perhaps one-third is prime farmland (U.S. Department of Agriculture, 1987, p. 3–9). But land-use scholars have differed on the significance of this trend and whether federal, state, or local intervention is justified.

Much of the problem has been a confusion of objectives. Crosson and Haas (1982) identify two dimensions of the farmland adequacy issue: (1) the role of the land in producing food and fiber, which they term the *capacity issue* and (2) the role

Agricultural Land

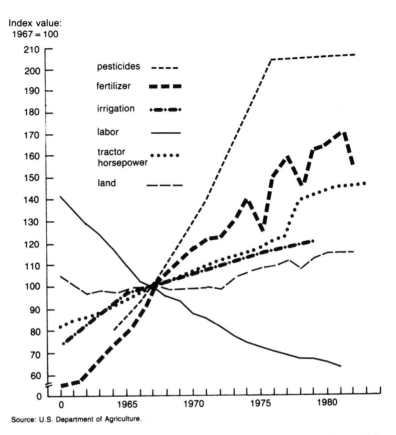

Figure 12.11 Resources Used in Agricultural Production: 1964–1983. *Source:* U.S. Department of Agriculture. Reprinted from *State of the Environment: An Assessment at Mid-Decade*, by permission of the Conservation Foundation.

of the land in producing intangible values, which they label the *amenity issue*. The latter may preferably be termed the *land-planning issue*, since more than amenity is involved. In any event, these two issues have often been argued at cross purposes.

The distinction between food production and land planning was recognized by Clawson (1963, pp. 16 and 61), who estimated that the nation could feed 10 to 20 times its existing population, but that urban growth was "withdrawing" twice as much land from rural use as was actually needed for urban purposes, generating harmful externalities such as visual blight, water pollution, and adverse fiscal impacts. Anderson, Gustafson, and Boxley (1975) likewise dismissed concerns about food production but emphasized land-planning concerns. Alluding to the parable of the blind men and the elephant, they urged that a "range of perspectives" exists regarding the need to conserve farmland.

The U.S. Department of Agriculture, Soil Conservation Service (1977) stirred up a professional uproar with its *Potential Cropland Study* (PCS), which reported a net conversion of 24 million acres to urban and water uses from various agricultural categories between 1967 and 1975. Of this total, 8 million acres was "prime farmland" and 5.5 million acres came from crop production—an average annual conversion of about 675,000 acres of cropland. Unfortunately, the findings were widely misunderstood, often recited as a loss of 3 million acres of cropland or prime land per year. Brewer and Boxley (1981, p. 888) challenged researchers to "sharpen the semantics of agricultural land policy" and noted that in any event the period 1967–1975 was atypical in its level of land conversion.

The PCS was followed by a swarm of additional federal studies including the National Agricultural Lands Study (NALS), the National Resource Inventories of 1977 and 1982, and studies conducted under the Soil and Water Resources Conservation Act of 1977 (P.L. 95-102). The NALS, a joint venture of the USDA and the Council on Environmental Quality, was a highly publicized and polemical exploration of "the economic, environmental, and social consequences, of agricultural land conversion and methods used to attempt to restrain and retard conversion" (National Agricultural Lands Study, 1981). The report, issued in the waning hours of the Carter Administration, was weakened by fiscal, methodological, and ideological difficulties (Brewer and Boxley, 1981; Fischel, 1982).

NALS, however, did accomplish several tasks: (1) It documented regional differences in farmland conversion; (2) it inventoried farmland protection measures by states and local governments; and (3) it identified 90 federal programs that contribute to farmland conversion (Platt, 1985). The latter findings supported adoption of the Farmland Protection Act of 1981 (PL 97-98, Subtitle 1), which required all federal agencies to:

> . . . develop criteria for identifying the effects of Federal programs on the conversion of farmland to nonagricultural uses.
>
> . . . consider alternative actions, as appropriate, that could lessen such adverse effects. . . .

This law was little noticed. During the Reagan Administration, the impact of urbanization on farmland ceased to be a federal concern and was left to the states and local governments.

Two recent studies cast new light on the viability of agriculture in metropolitan areas. Lawrence (1988) examined the land area and production value of selected commodity types in a sample of 15 metropolitan counties, located in both the center and periphery of 7 metropolitan areas between 1949 and 1982. He found a general decline in dairy and poultry production, a rise in both area and value of horticultural specialties (mostly nonedible), and uneven changes in field crops. Generally the latter declined in value and area in inner counties (not surprisingly) while rising in value and sometimes area in the outer counties. Heimlich (1989), using USDA national-level data, found metro agriculture characterized by smaller farm units, more intensive production, a focus on high value commodities, and prevalent off-farm employment. Both studies indicate that agriculture in the vicinity

Agricultural Land

of urban development may prosper economically, albeit producing a number of nonedible items such as chrysanthemums, Indian corn, and turf. Landscape amenity is thereby preserved through farmland conservation policies, but not necessarily agricultural production.

Soil Erosion

Congress in the 1930s and again in the 1980s has been aroused by the specter of widespread degradation of rural land through soil erosion. Soil erosion is the rapid weathering of the soil mantle through the action of water, wind, or both. Water-caused erosion occurs in several forms: 1) overland sheet flow caused by rainfall or rapid snowmelt; 2) rill and gully formation whereby overland flow becomes concentrated into distinct pathways or channels of diverse size; and 3) streambank caving at times of high water.

Water-caused erosion is most serious in hilly or rolling agricultural regions in the midwest, south central, and southeastern states. Soil erosion by water not only degrades the source land surface but also causes deterioration of water quality and sedimentation of reservoirs, river channels, and harbors downstream (Clark, Haverkamp, and Chapman, 1985).

Wind erosion, which scours exposed topsoil to form dust clouds, is characteristic of the west central states, especially the High Plains. The infamous "Dust Bowl" of that region prompted Congress in a unanimous vote to establish the Soil Conservation Service (SCS) in 1935. The formation of the Tennessee Valley Authority in 1933 was also motivated in part by perception of excessive soil erosion on that region's hilly farmlands.

Congress in 1935 granted SCS broad powers to combat soil erosion, finding:

> . . . that the wastage of soil and moisture resources of farm, grazing, and forest lands of the Nation resulting from soil erosion, is a menace to the national welfare. . . (16 USCA sec. 590a).

It empowered SCS:

> (1) To conduct surveys, investigations, and research relating to the character of soil erosion and the preventive measures needed. . . ;
>
> (2) To carry out preventive measures, including but not limited to, engineering operations, methods of cultivation. . . and changes in use of land;
>
> (3) To cooperate or enter into agreements with, or to furnish financial or other aid to any agency . . . or any person . . . for the purposes of this Act; and
>
> (4) To acquire lands, or rights or interests therein . . . for the purposes of this Act. (16 USCA sec. 590a).

Despite the portentous phrase "to carry out . . . changes in land use," SCS has never engaged in federal land-use regulation for erosion-prone areas. Instead, its modus operandi has been to fund and otherwise assist "soil and water conservation districts," which were ultimately established in nearly all the nation's counties. These districts are corporate entities established under state, not federal, law. SCS assists the districts, technically and financially, to prepare and carry out plans for the mitigation of erosion and protection of farmland within their respective counties. In practice, SCS, state, and county officials all work cooperatively with local landowners to promote better land-use practices such as contour plowing, terracing, and conservation tillage.

Nevertheless, the National Resources Inventories conducted by SCS in 1977 and 1982 revealed a shocking level of soil erosion still afflicting U.S. agricultural land. The 1982 survey disclosed that 25 percent of the nation's cropland, or 105 million acres, was subject to sheet and rill erosion at rates exceeding natural soil replenishment. Wind erosion was affecting 14 percent of cropland, or 67 million acres, (some of which was also subject to water-induced erosion). For all nonfederal land, water erosion affected nearly 200 million acres; and wind erosion, 90 million acres (U.S. Department of Agriculture, 1987, Figure 5). Over one-half of the water-induced erosion occurs in the Corn Belt and northern plains, the nation's primary rainfall-based grain region (Clark, Haverkamp, and Chapman, 1985 p. 5).

What does this signify in terms of loss of productivity? One estimate is that sheet and rill erosion alone, based on the rate in 1977, is removing the *equivalent* of about 1 million acres per year from the cropland base, with another quarter-million acre-equivalents removed by wind. This would add up to about 62 million acres of lost productivity over 50 years, assuming constant rates of loss (Sampson, 1981, p. 131). This is considerably higher than the estimated rate of loss of cropland to urbanization. Of course, erosion losses are actually much more widely distributed and more gradual in any given location than this formulation would indicate. It is merely a device for describing the effects of incremental losses spread over more than 125 million acres.

In the Food Security Act of 1985 (PL 99-198), Congress mounted a new assault on soil erosion. First, it required SCS to identify "highly erodible" land and to develop conservation plans for such areas in cooperation with state and local interests by 1990. Landowners who fail to conform with such conservation plans are ineligible for USDA crop and other subsidies. Total land involved in such plans is approximately 118 million acres.

A second feature of the Act is the Conservation Reserve Program, under which USDA is authorized to lease highly erodible land from private owners to remove it from crop usage and to convert it back to grassland or woodland. About 45 million acres are to be covered by such leases by 1990, of which 23 million acres were signed up by mid-1988 (Brown, 1989, p. 48).

The Act also denies federal agricultural benefits to landowners who grow crops on highly erodible land that is plowed for the first time after the date of the Act (known as the "sodbuster law"). A parallel provision denies benefits to owners who grow crops on newly drained wetlands ("swampbuster").

Conclusion

Recent federal agricultural land-management efforts thus have focussed on urbanization in the 1970s and soil erosion in the 1980s. The former chiefly involved studies of trends in land usage with encouragement of nonfederal programs to retard urban encroachment on prime farmland. The soil erosion problem has been addressed chiefly through economic incentives and disincentives. In neither area has Congress grasped the nettle of federal regulation of agricultural land.

SOLID AND HAZARDOUS WASTE MANAGEMENT

Earlier sections of this chapter have examined the federal role in diverse land management contexts: the coastal zone, floodplains, wetlands, and agricultural land. Although differing in terms of objective and form of federal involvement, each of those topics was geographically selective. The federal policy or program was directed to a particular land type, distinguished by its physical and human-usage characteristics. To the extent that public authorities have intervened in the private land market in those areas, such actions have been based on a perception of their specific geographic attributes, for example, ecological characteristics, scenic amenity, hazardousness, resource productivity, and so forth.

Public response to the problem of solid and hazardous waste management is of a different nature. Here the focus of attention is not on particular types of land (although some sites are clearly better or worse than others for waste disposal), but rather on the activity—production and disposition of waste—that generates the problem. Furthermore, the nature and degree of governmental response also varies greatly with the type of waste considered, ranging from mundane garbage and sludge to exotic and lethal chemical, biological, and radiological wastes. The selection of management option must also consider the need to avoid transferring the problem to a different medium—air to water—which would subvert national pollution abatement goals for those resources. The practice of ocean dumping, for example, a long-time practice of coastal cities, has been substantially limited by the Marine Protection, Research, and Sanctuaries Act of 1972 and may be totally banned in the United States by the end of the century. This throws the solid waste disposal problem back "on shore." And lurking in ambush for any site-specific proposal is the growing phenomenon of "NIMBY"—not in my backyard.

The management of solid and hazardous wastes thus presents complex policy issues that impinge both on the conduct of economic activities that generate the wastes and on geographic resources that receive them. The following discussion considers first the problem of ordinary solid waste as a land-use issue, followed by a brief review of federal efforts to prevent or remedy public health hazards due to unsafe disposal of toxic and hazardous wastes.

Conventional Solid Wastes

The United States probably leads the world as a throw-away economy, although Western Europe and Pacific Rim countries have expanded their capacity to generate (and sometimes reuse) wastes. Currently, the United States is estimated to produce about 160 million tons of conventional trash and garbage, or about 3.5 pounds per capita per day. The U.S. Environmental Protection Agency (1988, p. 81) reports a daily total of 4.0 pounds per capita for New York City, as compared with 3.0 pounds in Tokyo, 2.4 pounds in Paris, 1.9 in Hong Kong, and 1.3 in Rome. Cities in less developed nations produce much less per capita waste, for example, 1.1 pounds in Calcutta and Manila and 1.0 pounds in Kano, Nigeria. (These lower figures, however, probably mask a larger percentage of uncollected waste in poorer societies.)

The 160 million tons of household waste in the United States includes food and yard wastes, packaging, newsprint and catalogues, polystyrene containers, and miscellaneous other unwanted household items. Paper products account for 41 percent of the total; and plastics, 30 percent. A small but dangerous proportion of this total consists of hazardous household products such as paint, cleaning agents, pesticides, medicines, petroleum products and organic compounds. According to *Newsweek* (November 27, 1989, p. 67), one year's total of such household wastes, if accumulated in one place, would reach:

> . . . 30 stories high over 1,000 football fields, enough to fill a bumper-to-bumper convoy of garbage trucks halfway to the moon.

This gargantuan pile of refuse in fact is only the tip of the solid waste iceberg. Collected municipal waste plus sewerage sludge comprise only about three percent of total U.S. solid waste production. The EPA estimates a total annual waste burden of 6 billion tons, half generated by agriculture, 39 percent in mining operations, 6.4 percent by industrial production, and 1.2 percent by utilities (U.S. Environmental Protection Agency, 1988, p. 79; Conservation Foundation, 1987, pp. 107–115). Of course, this may overstate the issue since much of the agricultural "wastes" may beneficially be applied to land as manure, and mining overburden may be returned to excavations to restore the natural land surface (as now widely required under the federal Surface Mining Control and Reclamation Act and its state counterparts). But even excluding these two sources from the total leaves about 642 million tons of municipal, industrial, and utility wastes annually—or about 5,350 pounds annually per capita (or 14.6 pounds daily per capita) which would easily reach *beyond* the moon!

But for new laws, programs, and policies at each level of government over the past two decades, the total quantity of waste would be much greater and in fact it was greater in 1970 than at present. In that year, the first Annual Report of the new U.S. Council on Environmental Quality (CEQ) estimated a total volume of 190 million tons of collected residential waste, representing a daily average of 5 pounds per capita as compared with 3.5 pounds currently. Although intertemporal comparisons are approximate due to differences in definition and estimation method, it appears that collected residential waste has been reduced both absolutely and per capita since 1970 through a variety of measures such as state "bottle bills" and newspaper recycling efforts. (Some of the apparent gain, however, may be attributable to the proliferation of garbage disposals, which convey food wastes to the wastewater stream and thus are not "collected solid waste" except for the portion that becomes sewage sludge.)

Aside from the total quantity of conventional solid waste, which remains immense, the most pronounced difference since 1970 has occurred with respect to land disposal practices. In that year, over three-fourths of all collected solid wastes (146 million tons) was deposited in open dumps, which numbered about 14,000 nationally (U.S. Council on Environmental Quality, 1970, pp. 110–11). Such dumps were notorious blights and health hazards on the outskirts of American cities and small towns. Most were operated privately or by municipal governments, with no federal or state management standards. The average dump was a clearing in a field or woods or perhaps a filled wetland where trash and garbage was simply piled. Combustible material was burned to reduce its volume, thus emitting smoke and other air pollutants. Precipitation on dumps conveyed biological and chemical pollutants into nearby surface waters or into ground water in the form of "leachate." Surrounding areas were affected by numerous unpleasant externalities such as windborne trash, odors, rats and other vermin, visual blight, and vehicular traffic.

The first Congressional recognition of the growing problem of solid waste and the hazards of open dumps was expressed in the Solid Waste Disposal Act of 1965 (PL 89-872, Title II), an amendment to the still-primitive federal air pollution law. This act noted a "rising tide of scrap, discarded and waste materials" and the adverse effects of "inefficient and improper methods of disposal" (Sec. 202(a)). In response to this perception, Congress declared solid waste disposal to be a federal concern for the first time:

> (6) that while the collection and disposal of solid wastes should continue to be primarily the function of state, regional and local agencies, the problems of waste disposal . . . have become a matter national in scope and in concern . . . (Sec. 202(a)).

With reference to our model in Chapter 2, a new level of authority—the federal government—was thus added to the "Legal" circle, but revisions to the resource-management vector were initially very slight. In the 1965 Act, Congress merely offered financial and technical assistance, demonstration projects, and matching grants for state and interstate solid waste planning. Five years later, a national survey by the federal Bureau of Solid Waste Management (1970) found that 90 percent of all urban wastes were being deposited on land disposal sites and that 94 percent of these sites were unsatisfactory. About 88 percent were estimated to be open and burning, and most had poor drainage (McLean, 1971, p. 25).

These findings helped persuade Congress to expand the federal role considerably in the Resource Recovery Act of 1970 (PL 91-152), which strengthened the 1965 law. Among other changes, the 1970 law mandated the development of "guidelines for solid waste collection, transport, separation, recovery, and disposal systems" (Sec. 101). The term *guidelines* marked the beginning of a shift from merely encouraging nonfederal authorities to upgrade their programs to imposing direct federal regulation of land disposal practices. Such direct regulation emerged during the 1970s under the Clean Air Act, which required states to ban burning of open dumps, and the Clean Water Act, addressing the problem of leachate pollution of surface and ground water. The new Environmental Protection Agency assumed responsibility for federal solid waste efforts after 1970.

By 1980, the smoldering open dump was rapidly disappearing from the American landscape, except in rural areas. In its place, federal and state policy favored the *sanitary landfill*. A properly designed and managed sanitary landfill, according to the American Society of Civil Engineers, is:

> a method of disposing of refuse on land without creating nuisances or hazards to public health or safety, by utilizing the principles of engineering to confine the refuse to the smallest practicable area, to reduce it to the smallest practical volume, and to cover it with a layer of earth at the conclusion of each day's operation. . . . (quoted in McLean, 1971, p. 28).

Later design requirements for sanitary landfills have included impermeable liners to prevent leachate from entering ground water. Modern landfills also provide for the venting or capture of methane, which is sometimes used as an energy source for heating or electrical generation. The proportion of collected municipal wastes deposited in sanitary landfills rose from 13 percent in 1970 (U.S. Council on Environmental Quality, 1970, p. 111) to about 85 percent in 1987 (Conservation Foundation, 1987, p. 111).

But the era of the sanitary landfill itself is soon to end. The number of such facilities declined from 16,000 in 1984 to only about 6,000 in 1989 (*Newsweek*, November 27, 1989, p. 70). Existing landfills in many areas are rapidly being filled to capacity, leaving rounded grassy hills known in some places as "Mount Trashmore." The safe closing of sanitary landfills with proper provision for methane venting and monitoring of ground water is a major concern of EPA. But even more perplexing is what to do with the trash as available landfill capacity shrinks. Modern sanitary landfills typically have been regional facilities of considerable size, generating much truck traffic and depressing property values in their localities.

Neighboring property owners and entire communities object to receiving trash from other municipalities: Sanitary landfills are classic examples of "locally unwanted land uses" (LULUs), which are stoutly resisted by NIMBYites everywhere. Connecticut has had no new landfills established since 1978. Facilities in New Jersey declined from 310 in 1977 to 128 in 1984 (Conservation Foundation, 1987, p. 111). A notable example is the closing of the landfills in the Hacksensack

Meadowlands in northern New Jersey across from New York City. As of late 1986, three huge landfills in this area received 11,000 tons of solid waste daily, one-third of New Jersey's total collected wastes. By 1990, these landfills had been capped off and a proposed regional incinerator was yet to be completed. Much of New Jersey's waste was being trucked out of state at costs of hundreds of dollar per ton. New York, Pennsylvania, and New Jersey export 8 million tons of solid waste annually; Long Island townships each spend an average of $23 million a year shipping garbage out of state (*Newsweek,* November 27, 1989, p. 67). The saga of the infamous garbage barge from Islip, New York, which wandered the oceans in 1987 looking in vain for a place to unload before returning to Islip—still loaded— symbolized the growing solid waste crisis.

Beyond landfills, two options for disposal of certain wastes are incineration and land treatment. Primitive incinerators of course are outlawed, since they are major sources of air pollution. Modern state-of-the-art resource recovery incinerators utilize extremely high temperatures (1,800°F) for efficient combustion and employ sophisticated air pollution emission controls. Heat produced by the combustion is used to generate either steam or electricity. In 1987, some 70 such facilities were in operation nationally, as compared with 20 in 1977, and these handled 4 to 5 percent of total municipal wastes including sludge (Conservation Foundation, 1987, p. 113). As of early 1989, Connecticut had five such plants in operation, handling 60 percent of the state's refuse. Another five plants were operative in New Jersey, Long Island, and Westchester County, New York, with another 21 plants projected to open in that region by 1993 (*The New York Times,* February 12, 1989, p. 24). Nevertheless, even these facilities leave one-quarter of the total weight of trash in the form of highly toxic ash, which must be landfilled or processed for recycling. (Disposal at sea is increasingly restricted by EPA.)

Land treatment is the application of organic wastes to farmland or other rural land in order to recycle the nutrients into some form of biotic productivity. This is a modern version of the ancient practice of spreading raw human wastes on agricultural fields surrounding urban settlements, which is still practiced in China and other less developed nations today (Wong and Whitton, eds., 1987). In numerous forms, such techniques are said to "close the nutrient cycle."

Direct application of partially treated wastewater from a municipal sewage system to farmland has been practiced in several locations in the United States. The primary purpose of such systems is to provide a high degree of "renovation" or pollution abatement of the wastewater before it returns to streams or coastal waters. A secondary objective is to enhance the productivity of farmland and reduce dependence on commercial fertilizers.

A prototype system serving several cities and towns in Muskegon County, Michigan was constructed in the early 1970s. It was designed for a 1992 population of 170,000, generating an average flow of 43.3 million gallons per day (mgd) of wastewater including an industrial flow of 24.3 mgd. The system involves four physical elements: (1) intercepting sewerage facilities; (2) lagoon treatment and storage facilities; (3) irrigation facilities; and (4) drainage facilities. After collection, aeration, and settling out of solids (equivalent to "primary treatment"), wastes are stored in lagoons until seasonal conditions permit irrigation. The latter function is performed by conventional center-pivot devices. The wastewater is chlorinated before application to fields. The total land area occupied by storage lagoons and irrigated fields is approximately 10,800 acres. The land to which the wastewater is applied was formerly marginal (and therefore cheaply acquired by the county for this purpose). Nutrients in the wastewater are absorbed by the soil mantle and taken up by corn planted during the growing season. Upon percolation below the soil level, the now purified irrigation water is collected by drains and discharged to local streams to avoid excess soil waterlogging. The quality of such effluent is monitored: The system was designed to provide the equivalent of tertiary treatment, including decomposition of viruses. (Bauer Engineering, Inc., 1974)

The Muskegon system has produced bumper crops of corn for livestock feed, thereby defraying part of its construction and operating costs. The development of similar systems elsewhere has been impeded, however, by concerns as to the public health effects of industrial pollutants and viruses that may reach the agricultural crop. It has become more common to apply wastewater to forests, golf courses, wetlands, and other lands not yielding products entering the human food chain.

Nutrients may be recaptured from sewage treatment through the reprocessing and land application of sludge, a semi-solid by-product of the nation's 15,000 sewage treatment plants. Sludge is often handled as a solid waste and either deposited in landfills or incinerated after dewatering. Increasingly, however, sludge is being converted into various forms of usable fertilizer to be applied to land in place of standard commercial fertilizers. In 1984, the EPA estimated that 19,000 sites were employed for this purpose and that about one-quarter of municipal sewage sludge was being applied to land (Conservation Foundation, 1987, p. 114). Chicago's Metropolitan Sanitary District, for example, converts sludge produced in its treatment plants into a liquid fertilizer, which it applies to several thousand acres of marginal cropland in southern Illinois.

A comprehensive approach to solid waste management must employ a broad variety of strategies including, but not limited to, landfills, incineration, and land application. Four categories of techniques applicable to the problem include: (1) reduction in the amount of waste generated; (2) waste treatment; (3) waste containment; and (4) waste dispersal (Conservation Foundation, 1987, p. 433). The first option is receiving increasing emphasis in many states that either mandate or facilitate recycling of certain materials to prevent them from entering the waste stream. Bottle bills requiring deposits on beverage containers are in this category. Composting of yard and food wastes is becoming common in urban-rural fringe areas. Newspaper recycling is now so widely utilized that the market for used newsprint is saturated. Waste engine oil is sometimes recycled or converted into other petroleum-based products. Even ash from incinerators can be converted into paving material. As landfill space dwindles and costs of trucking solid wastes to remote locations escalate, mandatory separation and recycling of wastes is increasingly prevalent, especially in the Northeast and in the upper Midwest.

Congress has generally let the states choose their own strategies in dealing with conventional wastes, subject to minimum standards for landfills, air quality, and water quality. Hazardous and toxic wastes, however, are another story.

Hazardous and Toxic Substances

The United States, like other high-technology societies, was shockingly slow to recognize the risks to life, health, and property associated with the careless use and disposal of chemical and other hazardous substances. Not until the mid-1970s, when immense actual and potential damage had already been caused, was remedial legislation adopted by Congress. And it was well into the 1980s before this legislation, as subsequently amended, began to exert some impact on the management and disposal of hazardous substances. As with so many other legal reforms discussed in this book, the necessary prerequisite was a heightened understanding, perception, and indeed fear of the magnitude of the threat to the physical environment and human health.

For one class of hazardous chemicals, namely pesticides, Rachel Carson's *Silent Spring* (1962) sounded a clear warning. Her disclosures regarding DDT and other agricultural poisons helped to ignite the environmental movement generally, and reform of federal pesticide law in particular. As early as 1947, Congress had adopted the Federal Insecticide, Fungicide, and Rodenticide Act (7 USCA Secs. 135 et seq.), which merely required labeling of pesticides sold in interstate commerce. This Act was considerably strengthened by the Federal Environmental Pesticide Control Act of 1972 (PL 92-516), which regulates the sale and application of agricultural chemicals.

The U.S. Council on Environmental Quality's (1970) first Annual Report devoted 10 pages to pesticides, with a direct reference to *Silent Spring*, but mentioned no other types of hazardous substances. Where was the Rachel Carson for asbestos, mercury, PCBs, lead, vinyl chlorides, radioactive wastes, hospital wastes, and dozens of other hazardous substances? In recognition of this void in public awareness, the CEQ itself prepared a brief report on the problem entitled *Toxic Substances* (U.S. Council on Environmental Quality, 1971). It concluded that regulatory mechanisms to protect the public health from a wide variety of chemical hazards were inadequate. While not as charismatic as Carson's book, this report was a major impetus to the adoption in 1976 of the Toxic Substances Control Act (TSCA) (PL 94-469). Citing the CEQ report, the committee report on that law noted that:

> It is estimated that there are presently 2 million recognized chemical compounds in existence with nearly 250,000 new compounds produced each year. . . . approximately 1,000 new chemicals will find their way into the marketplace and subsequently into the environment through use or disposal. As the chemical industry has grown, we have become literally surrounded by a man-made chemical environment. (*U.S. Code, Congressional and Administrative News*, "Legislative History of PL 94-469", p. 4493.)

TSCA directed the EPA to test potentially hazardous substances, particularly those thought to be carcinogens. The Act also requires labeling of such substances to disclose risks to the user or the public.

The year 1976 also marked the adoption of the Resource Conservation and Recovery Act (RCRA) (PL 94-580). This represented a major expansion of the earlier federal solid waste legislation and the first explicit Congressional action on hazardous wastes. According to Rodgers (1988, p. 510), RCRA:

> . . . prescrib[ed] in 45 pages a regime for the management of solid and hazardous wastes that included many features (grants, planning, compliance orders, citizen suits, imminent hazards) borrowed from the Clean Air and Clean Water Acts. For the first time in 1976, serious regulatory measures appeared in the federal law—a qualified prohibition on the open dumping of solid or hazardous wastes, and the initiation (in a mere 7 pages in Subtitle C) of the famed "cradle to grave" regime for the control of hazardous wastes.

The "cradle to grave" provision instituted a system of manifests or documentation to accompany designated hazardous materials through each stage of their "lifetime": manufacture, transport, use, and disposition. This allows federal and state authorities to track the quantities and location of hazardous substances and to monitor their safe disposal. It also requires generators of hazardous materials to certify that they are minimizing the amount and toxicity of their waste, and that the method of treatment, storage, or disposal they have chosen will minimize the risk to human health and the environment. The implementation of this system, however, was to be delayed and frustrated for years by problems of definition, accountability, and the immense number of generators involved. Inevitably, considerable quantities of wastes continued to be dumped illegally down drains or wells, in fields, or on highways.

The magnitude of the problem in terms of volume of hazardous materials to be managed was vastly underestimated at first. Estimates of total amounts of such substances—defined by the U.S. Environmental Protection Agency (1986, p. 5) to include ignitable, corrosive, reactive, or toxic materials—have risen from 9 million metric tons in 1973, to between 27.5 and 41.25 million metric tons in the 1976 RCRA hearings, to 150 million metric tons in 1984, to 260 million metric tons in subsequent studies (Rodgers, 1988, p. 519). This expansion reflects in part a broadening of the definition of "hazardous substances" to include additional materials, and in part the discovery that some materials are more abundant than originally recognized. In 1988, EPA estimated that there were some 3,000 facilities handling a total of 275 million metric tons of RCRA hazardous wastes (EPA, 1988, p. 85).

RCRA was significantly amended in 1980 and 1984. The former comprised a retrenchment with the exemption of several classes of wastes from the terms of the Act. But the 1984 Hazardous and Solid Waste Amendments (HSWA) (PL 98-616) was an important mid-course correction that greatly strengthened RCRA. This included a major revision and tightening of the "cradle to grave" regulations for tracking hazardous materials. It extended RCRA's coverage to "small generators" of hazardous wastes, adding some 100,000 firms producing between 220 and 2,200 pounds of hazardous waste monthly to the 15,000 large generators (U.S. Environmental Protection Agency, 1988, p. 89). It initiated regulations for leaking underground storage tanks ("LUSTs"), which were estimated to number about 400,000 out of a total of 5–6 million underground tanks for storing petroleum products or hazardous chemicals (U.S. Environmental Protection Agency, 1988, p. 102). And land disposal of certain hazardous wastes was totally prohibited. Rodgers (1988, p. 510) describes the 1984 HSWA as "passionate, confident, and demanding."

States play a key role in the administration of RCRA. By late 1986 EPA had certified 40 state hazardous waste programs to assume day-to-day responsibility for implementing RCRA.

Superfund

RCRA applies prospectively to hazardous wastes from the time of their generation until their final disposition. A major gap in its coverage was the problem of wastes already discarded improperly—in pits, lagoons, injection wells, or leaking containers strewn on the ground behind chemical plants (Figure 12-12).

National perception was finally directed to this issue by the tragic disclosures of the Love Canal disaster in Niagara Falls, New York. As recounted in *The Atlantic* by Michael H. Brown (1979), Love Canal was an artificial ditch dating back to the

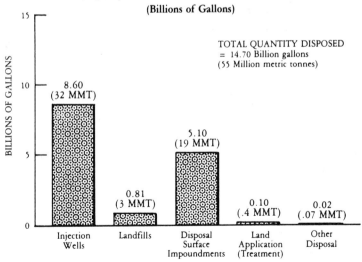

Figure 12.12 Methods of disposing hazardous wastes, 1981. *Source:* U. S. Council on Environmental Quality, 1984, Fig. 3-1.

late nineteenth century that was used for decades as a waste dump by the Hooker Chemical Co., which owned it. In 1977, signs of desperate problems began to appear in the residential neighborhood that bordered the canal. Children were born with birth defects, high rates of cancer were prevalent, and foul-smelling chemical wastes oozed through basement walls and formed puddles in yards and playgrounds. After much denial by Hooker, the City of Niagara Falls, and the County Health Commissioner, the crisis was finally addressed by state and federal authorities. A preliminary investigation led to the declaration of a "major disaster" by President Jimmy Carter in August, 1978, the first such declaration for a technological disaster. The state, with the help of federal funds, then acquired the homes and assisted the relocation of 237 families from the vicinity. The canal was sealed off and a remedial drainage project initiated. Subsequently, it became apparent that the chemicals had spread much further through the ground than originally estimated. And the brew of toxic chemicals was found to contain, in addition to benzene and a dozen other carcinogens, a measurable amount of the most deadly chemical ever synthesized: dioxin. The Love Canal Homeowners Association campaigned for further public assistance. Eventually, more than 600 homes were acquired.

Media attention soon disclosed many other chemical waste horrors lurking in America's industrial backyards. Woburn, Massachusetts was characterized as "a tangle of dumps and disease" (*Boston Globe,* 1980). Times Beach, Missouri was found to be contaminated by dioxin contained in oil used to resurface the streets. (The entire town was bought by the Federal Emergency Management Agency and its residents relocated.) Even the staid journal *Science* (Maugh, 1979) described Love Canal and its counterparts as "an environmental time bomb gone off."

Seldom can a legislative enactment be attributed so directly to a traumatic jolt in public environmental perception. In 1980, within months after the second round of Love Canal findings, Congress adopted the Comprehensive Environmental Response, Compensation, and Liability Act (CERCLA) (PL 96-510), popularly known as *Superfund.* The Act's purpose was:

> To provide for liability, compensation, cleanup, and emergency response for hazardous substances released into the environment and the cleanup of inactive hazardous waste sites (Sec. 2).

The Act authorized $1.6 billion over five years for a comprehensive program to clean up the worst abandoned or inactive waste dumps in the nation (U.S. Environmental Protection Agency, 1988, p. 80). The funds were to be derived from an excise tax on the sale or use of petroleum and 42 chemicals used commercially to produce hazardous substances. States must provide at least 10 percent of cleanup costs for a site within their jurisdictions. Wherever a "responsible party" could be identified they would either be required to perform remedial actions themselves, or the government would do so and sue them for the costs plus penalties. In 1986, Superfund was reauthorized by PL 99-499 and expanded to a potential funding level of $8.5 billion. (The Superfund law is codified at 42 USCA Secs. 9601 et seq.)

"Throwing money at the problem" was a necessary but not sufficient response to the threat of hazardous waste contamination. Since the problem of abandoned waste dumps had only recently been recognized, no accurate inventory of the location, size, and contents of such sites had been compiled. By definition, such sites tend to be concealed and often below ground where wastes have been buried or injected into ground water. Furthermore, it was often very difficult and dangerous to ascertain which of several hundred hazardous substances were present at any given site. A massive task for the EPA during the 1980s was to identify sites and to allocate cleanup resources. Some 19,300 sites were identified as potentially eligible for Superfund by the end of 1984: By late 1989, this list had grown to over 31,500 sites nationally. EPA has conducted preliminary assessments on most of these. Based on evaluation of a complex array of risk factors, the "worst" sites are placed on a National Priorities List (NPL), which qualifies them for Superfund cleanup, eventually. By late 1989, over 1,200 sites were listed on the NPL. The Office of Technology Assessment (1989, p. 6) has estimated that as many as 9,000 additional sites are eligible for the NPL, and the potential cost to all parties of cleaning them up could amount to $500 billion over several decades.

Meanwhile, actual progress in implementing the goals of the 1980 CERCLA and its 1986 amendments has been slow. The startup of Superfund was impeded by controversy surrounding the actions or nonactions of President Reagan's first EPA Administrator Ann Burford and her Superfund deputy, Rita Lavelle. The latter was convicted of perjury regarding EPA's dealings with Stringfellow Acid Pits, a notorious waste dump in California. In 1983, Burford resigned after being charged with contempt of Congress. By 1985, EPA had cleaned up only six sites, with questionable success (*Time,* 1985).

By 1988, EPA reported that it had initiated over 1,000 short-term remedial actions (such as fencing), but that only 200 sites had been investigated in detail and long-term cleanup initiated on about 140 sites (EPA, 1988, pp. 83 and 95). Only a few dozen cleanups were completed by 1990.

A permanent cleanup of a hazardous waste site is immensely difficult and costly. Obviously, risks to the cleanup workers and to the surrounding area must be minimized. Removal of wastes to another location may simply transfer the problem elsewhere unless the receiving site is properly located and designed. Few approved hazardous waste sites were available. Other than removal to land disposal sites, management options are complex and expensive. They include destruction through thermal, biological, or chemical treatment; stabilization; engineering controls, institutional controls, or natural treatment (OTA, 1989, p. 45). A management approach must be formulated individually for each site with the participation of all levels of government and affected private interests. Also involved in the decision process for each site are a variety of professional consultants, for example, chemists, soil scientists, biologists, hydrologists, and, of course, lawyers.

The OTA (Office of Technology Assessment) (1989) has identified a "Superfund syndrome" that impedes progress in making the program work effectively. This is a state of "constant confrontation" among the interested parties, produced by the high economic costs, the scientific uncertainties, the emotionalism, and what OTA considers to be "excessive flexibility" in Superfund implementation:

> Unless everyone breaks out of the Superfund syndrome, most cleanups will seem to do too little or too much. Billions will be spent. Hardly anyone will be satisfied. Hardly anyone will feel treated fairly. Hardly anyone will seem in control. (p. 3.)

Cleaning up Superfund sites and the location of new storage, treatment, and disposal facilities inherently pose geographical as well as legal and public health issues. At the heart of the problem is the pervasive issue of externalities: Each site poses potential off-site risks to the surrounding area, as in the possibility of ground water pollution. Even where actual risks are minimized through appropriate treatment and facility design, a zone of perceived risk may extend widely beyond the area of actual risk. Such spatial distribution of actual or perceptual risks apply, not only to the sites of generation, use, and disposal, but also in linear form along the routes of transport (e.g., highway, rail, barge) of hazardous substances (Zeigler, Johnson, and Brunn, 1983).

Recognition of the spatial patterns of opposition (NIMBYism), whether or not limited to actual zones of risk, is essential. Ultimately the resolution of the Superfund syndrome will depend on (1) effectively reducing the actual risks of hazardous wastes through proper management practices and (2) allaying the fears of those who perceive themselves and their property values to be at risk, through public involvement, education, and, in appropriate cases, compensation.

SURFACE MINE RECLAMATION

This chapter concludes with a brief summary of federal involvement in a regionally specific land-use activity: surface mining of coal. Coal strip mining, as distinct from deep tunnel mining, accounts for about 60 percent of U.S. coal production. This technique involves stripping away topsoil and rock overburden to reach coal deposits within a few hundred feet of the surface. As compared with tunnel mining, strip mining reduces labor costs through large-scale mechanization and is less hazardous to those working in the mine: black lung disease, cave-ins, and explosions are rare in surface mining.

On the other hand, surface mining without reclamation leaves the land devastated with alternating excavations and piles of infertile overburden. All agricultural use of the site is extinguished and its capability to support biotic communities is greatly modified. The spoil heaps are inherently unstable formations and occasionally may collapse into debris flows. The failure of a spoil dam was responsible for heavy loss of life downstream in the 1973 Buffalo Creek, West Virginia disaster.

Another problem is sulfuric acid pollution of streams draining out of strip-mined lands, which destroys fish and plant life and poses a public health hazard.

In the early 1970s, the federal Bureau of Mines estimated that the U.S. had some 2 million acres of unreclaimed and unproductive strip-mined lands (Hibbard, 1978, p. 296). In 1977, surface mining of coal was disrupting some 4,000 acres of land every week (U.S. Congress, 1977, p. 3).

Environmental and energy policies clashed head on in the mid-1970s. On the one hand, coal strip mining was degrading the land and water resources of extensive territories in Appalachia, the midwest and western plains states. On the other hand, the Middle East oil price rises of 1974 compelled the nation to expand reliance on its domestic energy sources, including coal. The need, therefore, was to facilitate expansion of strip mining operations while regulating the mining industry to compel responsible management and reclamation of strip-mined lands.

Like most land-use issues, the regulation of strip mining before 1976 was left to the states. In fact, 48 states had adopted laws to regulate strip mining by the mid-1970s (Hibbard, 1978, p. 297). But the stringency of requirements and the level of enforcement effort varied widely among states. Furthermore, mining companies could play one state off against another by threatening to close their mines in a state that proposed to "get tough." Against the customary objections to federal intervention in a land-use problem, this inconsistency among state programs was a major argument for a national approach. Federal jurisdiction was also invoked on the basis of the pollution of surface waters from mine drainage. In the west, much of the coal accessible to strip mining lay beneath federal lands, which offered a further justification for a federal surface mining law. (Platt, Macinko, and Hammond, 1983, p. 144).

An epic legislative battle in Congress culminated in the Surface Mining Control and Reclamation Act of 1977 (PL 95-87). According to the Act's chief sponsor, Representative Morris K. Udall:

> It took six years of tenacity and bitter debate to pass Public Law 95-87. The history of the Act would serve as a textbook for any national legislator desiring to thwart the clear will of the majority of the Congress. . . . a legislative endeavor involving 183 days of hearings and legislative consideration, eighteen days of House action, three House-Senate Conferences and Reports, two Presidential vetoes, approximately fifty-two recorded votes in the House and Senate, and the machinations (and statesman-like conduct) of three Presidents is an activity ripe for scholarly analysis, let alone the stuff for a pretty good novel (Udall, 1979, p. 554, as quoted in Platt, Macinko, and Hammond, 1983, p. 145).

The result was an 88-page law that established a national program for regulating surface mining and the surface effects of underground mining. It addressed three phases of mining activities: (1) preliminary planning; (2) mining operations; and (3) restoration of the disturbed land. The fundamental purpose of the Act was to require that topsoil and other overburden be stockpiled for replacement on excavated land that was to be returned to its original contours and level of productivity. The statute and its extensive regulations articulate mandatory national performance standards for strip mining.

But like RCRA, the federal surface mining law relies on states to adopt the national standards into their own laws and thereafter enforce them, subject to federal surveillance. Thus, the facade of state authority is maintained and the federal role is exercised indirectly, but forcefully.

REFERENCES

ABBEY, E. 1968. *Desert Solitaire*. New York: Simon and Schuster.

ANDERSON, W. D., G. C. GUSTAFSON, AND R. F. BOXLEY. 1975. "Perspectives on Agricultural Land Policy." *Journal of Soil and Water Conservation* 30(1): 36–43.

BALDWIN, M. F. 1970. "The Santa Barbara Oil Spill," in *Law and the Environment* (M. F. Baldwin, Jr. and J. K. Page, Jr., eds.). New York: Walker and Co.

———. 1987. "Wetlands: Fortifying Federal and Regional Cooperation." *Environment* 29(7): 16–20; 39–43.

BATIE, S. S. AND R. G. HEALY. 1983. "The Future of American Agriculture." *Scientific American* 248(2): 45–53.

BAUER ENGINEERING, INC. 1974. "Update on Muskegon County, Michigan Land Treatment System." Paper presented at ASCE and National Environmental Engineering Convention, Kansas City, MO, October 21–25, 1974.

BEAR, D. 1989. "NEPA at 19: A Primer on an 'Old' Law with Solutions to New Problems," *Environmental Law Reporter*, 19: 10060–70.

BESTON, H. 1949. *The Outermost House*. New York: Rinehart.

Boston Globe. 1980. "Woburn: A Tangle of Dumps and Disease." (June 28), 1 and 18.

BREWER, M. F. AND R. F. BOXLEY. 1981. "Agricultural Land: Adequacy of Acres, Concepts and Information." *American Journal of Agricultural Economics* 63: 879–888.

BROWN, L. R. 1989. "Reexamining the World Food Prospect." In *State of the World—1989* (L. R. Brown, ed.). Washington, D.C.: Worldwatch Institute.

BROWN, M. H. 1979. "Love Canal and the Poisoning of America." *The Atlantic*. Vol. 244 (December): 33–47.

BURBY, R. AND S. FRENCH. 1985. *Flood Plain Land Use Management: A National Assessment*. Boulder: Westview Press.

BUREAU OF SOLID WASTE MANAGEMENT. 1970. *Comprehensive Studies of Solid Waste Management: First and Second Annual Reports*. Washington, D.C.: U.S. Government Printing Office.

CALDWELL, L. K. 1989. "A Constitutional Amendment for the Environment." *Environment* 31(10): 6–11; 25–28.

CARSON, R. 1962. *Silent Spring*. Boston: Houghton Mifflin.

CLARK, E. H. II, J. A. HAVERKAMP, AND W. CHAPMAN. 1985. *Eroding Soils: The Off-Farm Impacts*. Washington, D.C.: The Conservation Foundation.

CLAWSON, M. 1963. *Land for Americans*. Chicago: Rand McNally.

References

COMMISSION ON MARINE SCIENCE, ENGINEERING, AND RESOURCES. 1969. *Our Nation and the Sea*. Washington, D.C.: U.S. Government Printing Office.

COMMONER, B. 1971. *The Closing Circle*. New York: Knopf.

CONGRESSIONAL RESEARCH SERVICE. 1987. *Managing Coastal Development*. Report no. 88-354. Washington, D.C.: Library of Congress, Congressional Research Service.

CONSERVATION FOUNDATION. 1977. "The Coasts Are Awash with Disputes." *Conservation Foundation Letter* (March).

———. 1985. *State of the Environment: An Assessment at Mid-Decade*. Washington, D.C.: The Conservation Foundation.

———. 1987. *State of the Environment: A View Toward the Nineties*. Washington, D.C.: The Conservation Foundation.

———. 1988. *Report of the National Wetland Forum*. Washington, D.C.: The Conservation Foundation.

COOLEY, R. A. AND G. WANDESFORDE-SMITH (EDS.). 1970. *Congress and the Environment*. Seattle: University of Washington Press.

CROSSON, P. R. AND R. B. HAAS. 1982. "Agricultural Land." In *Current Issues in Natural Resource Policy* (P. R. Portney and R. B. Haas, eds.). Washington, D.C.: Resources for the Future.

CULHANE, P. J., H. P. FRIESEMA, AND J. A. BEECHER. 1987. *Forecasts and Environmental Decision-Making*. Boulder: Westview Press.

DARLING, F. F. (ed.). 1965. *Future Environments of North America*. Garden City: The Natural History Press.

DINGMAN, M. L. AND R. H. PLATT. 1977. "Floodplain Zoning: Implications of Hydrologic and Legal Uncertainty," *Water Resources Research* (June): 519–23.

DOLGIN, E. L. AND T. G. P. GUILBERT. 1974. *Federal Environmental Law*. St. Paul: West Publishing Co.

ENGEL, R. 1985. *Sacred Sands: The Struggle for Community in the Indiana Dunes*. Middetown, CT: Wesleyan University Press.

ENVIRONMENTAL LAW INSTITUTE. 1989. "News and Analysis," *Environmental Law Reporter* 19: 10060.

FEDERAL EMERGENCY MANAGEMENT AGENCY. 1985. *FEMA Floodplain Community Report. State Ranking, 1985*. Washington, D.C.: Federal Emergency Management Agency.

———. 1986. *A Unified National Program for Managing Flood Losses*. Washington, D.C.: Federal Emergency Management Agency.

FISCHEL, W. A. 1982. "The Urbanization of Farmland: A Review of the National Agricultural Lands Study." *Land Economics*, 58: 236–259.

FRAYER, W. E., T. J. MONAHAN, D. C. BOWDEN, AND F. A. GRAYBILL. 1983. *Status and Trends of Wetlands and Deepwater Habitats in the Conterminous United States: 1950s to 1970s*. Fort Collins: Colorado State University, Department of Forest and Wood Sciences.

FURUSETH, O. J. AND J. T. PIERCE. 1982. *Agricultural Land in an Urban Society*. Washington, D.C.: Association of American Geographers.

GRUNTFEST, E. C. 1987. *What We Have Learned Since the Big Thompson Disaster*. Special Pub. no. 16. Boulder: University of Colorado, Natural Hazards Research Center.

HARDIN, G. 1968. "Tragedy of the Commons." *Science* 162: 1243–1248.

HARRIS, C. D. 1957. "The Pressure of Residential-Industrial Land Use." In *Man's Role in Changing the Face of the Earth* (W. L. Thomas, Jr., ed.). Chicago: University of Chicago Press.

HEALY, R. G. AND J. ZINN. 1985. "Environment and Development Conflicts in Coastal Zone Management." *Journal of the American Planning Association* 51(3): 299–311.

HEIMLICH, R. E. 1989. "Metropolitan Agriculture: Farming in the City's Shadow." *Journal of the American Planning Association* 55(4): 457–466.

HIBBARD, W. R., JR. 1978. "Environmental Impact of Mining." In *Sourcebook on the Environment* (K. A. Hammond, G. Macinko, and W. B. Fairchild, eds.). Chicago: University of Chicago Press.

HOYT, W. G. AND W. B. LANGBEIN. 1955. *Floods*. Princeton: Princeton University Press.

KUEHN, R. R. 1984. "The Coastal Barrier Resources Act and the Expenditures Limitation Approach to Natural Resources Conservation: Wave of the Future or Island Unto Itself?" *Ecology Law Quarterly* 11(3): 583–670.

KUSLER, J. A. 1983. *Our National Wetland Heritage*. Washington, D.C.: Environmental Law Institute.

LAWRENCE, H. W., 1988. "Changes in Agricultural Production in Metropolitan Areas." *The Professional Geographer* 40(2): 159–175.

LEOPOLD, A. 1949/1966. *A Sand County Almanac*. New York: Oxford University Press.

MARSH, G. P. 1864/1975. *Man and Nature: Physical Geography as Modified by Human Action*. (D. Lowenthal, ed.). Cambridge: Belknap Press of Harvard University Press.

MATUSZESKI, W. 1985. "Managing the Federal Coastal Programs: The Planning Years." *Journal of the American Planning Association* 51(3): 266–274.

MAUGH, T. H. 1979. "Toxic Waste Disposal a Growing Problem." *Science* 204 (May 25): 819–824.

MCHARG, I. 1968. *Design with Nature*. Garden City, N.Y.: Natural History Press.

MCLEAN, M. 1971. *Planning for Solid Waste Management*. Planning Advisory Service Report no. 275. Chicago: American Society of Planning Officials.

MCPHEE, J. 1971/1976. "Encounters with the Archdruid." In *The John McPhee Reader* (W. L. Howarth, ed.). New York: Random House Vintage Books.

MITSCH, W. J. AND J. G. GOSSELINK. 1986. *Wetlands*. New York: Van Nostrand Reinhold.

MURPHY, F. C. 1958. *Regulating Flood-Plain Development*. Monograph no. 56. Chicago: University of Chicago Department of Geography.

NATIONAL ACADEMY OF SCIENCES. 1990. *Managing Coastal Erosion*. Washington, D.C.: National Academy Press.

NATIONAL AGRICULTURAL LANDS STUDY. 1981. *Final Report*. Washington, D.C.: U.S. Government Printing Office.

NATIONAL OCEANOGRAPHIC AND ATMOSPHERIC ADMINISTRATION. 1975. "The Coastline of the United States." Washington, D.C.: U.S. Government Printing Office.

NATIONAL RESOURCES BOARD. 1934. *Final Report*. Washington, D.C.: U.S. Government Printing Office.

NATIONAL REVIEW COMMITTEE ON FLOODPLAIN MANAGEMENT. 1989. "Action Agenda For Managing the Nation's Floodplains." (mimeo: available from Natural Hazards Center, University of Colorado, Boulder).

Newsweek. 1989. "Buried Alive." (November 27): 67–75.

New York Times. 1989. "Why the Garbage is Never Gone." (February 12) (Sec. 7, p. 24).

OFFICE OF OCEAN AND COASTAL RESOURCE MANAGEMENT. 1988. *Coastal Management: Solutions to Our Nation's Coastal Problems.* Technical Assistance Bulletin no. 101. Washington, D.C.: Office of Ocean and Coastal Resource Management.

OFFICE OF TECHNOLOGY ASSESSMENT. 1984. *Wetlands: Their Use and Regulation.* Washington, D.C.: U.S. Government Printing Office.

———. 1989. *Coming Clean: Superfund Problems Can Be Solved.* Washington, D.C.: U.S. Government Printing Office.

PLATT, R. H. 1972. *The Open Space Decision Process.* Research Paper no. 142. Chicago: University of Chicago Department of Geography.

———. 1978. "Coastal Hazards and National Policy: A Jury-Rig Approach." *Journal of the American Institute of Planners* 44(2): 170–180.

———. 1982. "The Jackson Flood of 1979: A Public Policy Disaster." *Journal of the American Planning Association* 48(2): 219–231.

———. 1987. "Coastal Wetland Management: The Advance Designation Approach." *Environment* 29(9): 16–20; 38–43.

———. 1985. "The Farmland Conversion Debate: NALS and Beyond." *The Professional Geographer* 37(4): 433–442.

PLATT, R. H., S. G. PELCZARSKI, AND B. K. R. BURBANK, EDS. 1987. "Cities on the Beach: An Overview." In *Cities on the Beach: Management Issues of Developed Coastal Barriers.* Research Paper no. 224. Chicago: University of Chicago Department of Geography.

PLATT, R. H. ET AL. 1980. Intergovernmental Management of Floodplains. Monograph no. 30. Boulder: University of Colorado Institute of Behavioral Science.

PLATT, R. H., G. MACINKO, AND K. HAMMOND. 1983. "Federal Environmental Management: Some Land Use Legacies of the 1970s." In *United States Public Policy: A Geographical View* (J. W. House, ed.). Oxford: Clarendon Press.

PLATT, R., SR. 1942/1988. *This Green World.* New York: Dodd Mead.

———. 1947. *Our Flowering World.* New York: Dodd Mead.

———. 1966. *The Great American Forest.* Englewood Cliffs, N.J.: Prentice-Hall.

REISNER, M. 1986. *Cadillac Desert.* New York: Penguin.

RODGERS, W. H., JR. 1988. *Environmental Law: Vol 3—Pesticides and Toxic Substances.* St. Paul: West Publishing Co.

SAMPSON, R. N. 1981. *Farmland or Wasteland: A Time to Choose.* Emmaus, PA: Rodale Press.

Scientific American. 1965. "*Cities.*"(Special Issue).

TEAL, J. AND M. TEAL. 1969. *Life and Death of the Salt Marsh.* New York: Ballantine Books.

Time. 1985. "A Problem That Cannot Be Buried." (October 14): 76–84.

TINER, R. W., JR. 1984. *Wetlands of the United States: Current Status and Recent Trends.* Washington, D.C.: U.S. Government Printing Office.

UDALL, M. K. 1979. "The Enactment of the Strip Mining Control and Reclamation Act of 1977 in Retrospect." *West Virginia Law Review* 81(4): 553–7.

U.S. BUREAU OF THE CENSUS. 1986. *Statistical Abstract of the United States.* Washington, D.C.: U.S. Government Printing Office.

U.S. CONGRESS. 1966a. *Insurance and Other Programs for Financial Assistance to Flood Victims.* Committee Print no. 43., (89th Congress, 2d Session.) Washington, D.C.: U.S. Government Printing Office.

———. 1966b. *A Unified National Program for Managing Flood Losses.* House Document 465. (89th Congress, 2d Session.) Washington, D.C.: U.S. Government Printing Office.

———. 1973. *National Land Use Policy Legislation, 93rd Congress: An Analysis of Legislative Proposals and State Laws.* (Prepared by the Library of Congress Congressional Research Service for the Senate Committee on Interior and Insular Affairs.) (93rd Congress, 1st Session.) Washington, D.C.: U.S. Government Printing Office.

———. 1975. *Oversight on Federal Flood Insurance Programs.* (Senate Committee on Banking, Housing, and Urban Affairs). (94th Congress, 1st Session). Washington, D.C.: U.S. Government Printing Office.

———. 1977. *Hearings on the Surface Mining Control and Reclamation Act of 1977* before the House Committee on Interior and Insular Affairs. (95th Congress, 1st Session.) Washington, D.C.: U.S. Government Printing Office.

———. 1985. *Oversight Hearings of Section 404 of the Clean Water Act.* Senate Committee on Environment and Public Works, Subcommittee on Environmental Pollution. (99th Congress, 1st Session.) Washington, D.C.: U.S. Government Printing Office.

U.S. COUNCIL ON ENVIRONMENTAL QUALITY. 1970. *Environmental Quality.* Washington, D.C.: U.S. Government Printing Office.

———. 1971. *Toxic Substances.* Washington: U.S. Government Printing Office.

———. 1984. *Environmental Quality.* Washington: U.S. Government Printing Office.

U.S. DEPARTMENT OF AGRICULTURE. 1987. *The Second RCA Appraisal: Soil, Water, and Related Resources on Nonfederal Land in the United States.* Washington, D.C.: U.S. Department of Agriculture.

U.S. DEPARTMENT OF AGRICULTURE, ECONOMIC RESEARCH SERVICE. 1984. *U.S. Cropland, Urbanization, and Land Ownership Patterns.* Washington, D.C.: U.S. Government Printing Office.

U.S. DEPARTMENT OF AGRICULTURE, SOIL CONSERVATION SERVICE. 1977. *Potential Cropland Study.* Statistical Bulletin no. 578. Washington, D.C.: U.S. Department of Agriculture, Soil Conservation Service.

———. 1981. *America's Soil and Water: Conditions and Trends.* Washington, D.C.: U.S. Department of Agriculture.

U.S. ENVIRONMENTAL PROTECTION AGENCY. 1986. *Solving the Hazardous Waste Problem.* Washington, D.C.: Environmental Protection Agency.

———. 1988. *Environmental Progress and Challenges: EPA's Update.* Washington, D.C.: U.S. Government Printing Office.

U.S. FISH AND WILDLIFE SERVICE. 1956. *Wetlands of the United States.* Circular 39. Washington, D.C.: U.S. Government Printing Office.

U.S. WATER RESOURCES COUNCIL. 1976. *A Unified National Program for Floodplain Management.* Washington, D.C.: Water Resources Council.

———. 1977. *Estimated Flood Losses:* Nationwide Analysis Report. (mimeo).

References

WHITE, G. F. ET AL. 1958. *Changes in Urban Occupance of Flood Plains in the United States.* Research Paper no. 57. Chicago: University of Chicago Department of Geography.

———. 1964. *Choice of Adjustment to Floods.* Research Paper no. 93. Chicago: University of Chicago Department of Geography.

WHITE, G. F. 1975. *Flood Hazard in the United States.* Monograph no. NSF RA-E-75-006. Boulder: University of Colorado Institute of Behavioral Science.

WONG, M. H. AND B. A. WHITTON (EDS.). 1987. Special Issue, "Recycling of Organic Wastes for Fertilizer, Food, Feed and Fuel." *Resources and Conservation* 13(2–4).

ZIEGLER, D. J., J. H. JOHNSON, JR., AND S. D. BRUNN. 1983. *Technology Hazards.* Washington, D.C.: Association of American Geographers.

13

STATE LAND USE PROGRAMS

Last but by no means least, we address the roles of state governments in the management of the nation's land resources. States of course are the sovereign units of the United States. The federal government gained its existence and authority through the delegation of certain powers by the original thirteen states in the Constitution of 1787. Also, municipalities and special districts are creatures of the states, exercising powers expressly or implicitly delegated to them by state law. Thus, having divested some of their powers upward to the federal level and others downward to municipalities, what functions do states retain to exercise on their own?

It is difficult to generalize about state land-use programs because they differ widely from one state to another. Although basically similar in their legal status and potential capabilities, states have evolved differently in their land-use policies and efforts, as articulated in their respective constitutions, legislation, administrative regulations, executive orders, and court decisions. It is a strength of the federalist system of the United States that states are free, within the bounds of the U.S. Constitution, to develop their individual approaches to land-use problems according to their respective physical and fiscal resources, politics, and perceived needs. For example, certain "high amenity" states such as Hawaii, Oregon, Vermont, Maine, and Florida during the 1960s and 1970s launched ambitious state land-use management programs, while more "resource-dominated" states such as Texas, Ohio, West Virginia, and Colorado did not (DeGrove, 1984).

However, certain fundamental roles of state government may be identified which are performed with differing levels of zeal and effectiveness in all or most of the 50 states. These are parallel with equivalent functions of the federal government:

- land ownership and management;
- funding of state and local public infrastructure, for example, highways, water and sewer systems;
- regulation and permitting of specified land and water uses; for example, power plant siting, waterway encroachments, landfill siting and design;
- planning and technical assistance; and
- taxation.

In many cases, states operate as agents for or partners with the federal government. Many of the post-1960 federal programs discussed earlier, such as the Land and Water Conservation Fund and the Coastal Zone Management Program, rely on state initiative to promulgate national policies and goals and to utilize federal funds. But states may also act autonomously to influence the use of land and patterns of metropolitan development within their jurisdictions. Or they may sit back, and let the federal and local levels of government exercise their respective powers with little state input.

STATE PARKS

Before 1970, practically the only land-management function of most states was the operation of state parks and recreation areas. Perhaps the earliest, and certainly one of the largest, state facilities was the Adirondack State Forest Preserve established in New York State in 1885. (Nash, 1982, p. 119). During the 1930s, many state parks were added, expanded, and improved with federal assistance. Much of the infrastructure of older state parks—including roads, trails, restrooms, recreation fields—dates back to the Civilian Conservation Corps and other public works programs of the 1930s.

Assisted by the federal Land and Water Conservation Fund, and state-level bond issues, many states have substantially enlarged and improved their park systems since the 1960s. In 1984, the 50 state park systems collectively comprised about 10 million acres. Excluding Alaska from both federal and state totals, state parks have about one-third the acreage of the National Park System but more than twice the level of visitation (in "recreation user-days") (Conservation Foundation, 1987, p. 287). State parks tend to be smaller, more widely distributed, and more accessible to metropolitan populations than national parks. Day trip usage is therefore much higher.

Many states have incorporated new kinds of facilities and administrative concepts into their open space preservation programs. These include historic sites, scenic and wild rivers, trail corridors, and greenways along metropolitan streams.

Massachusetts has pioneered the concept of the "heritage park" to combine the functions of urban renewal, cultural preservation, and recreation in several old mill cities such as Lowell, Holyoke, and North Adams. New York in 1971 established an Adirondack State Park Agency to plan and control land use for the 6-million-acre Adirondack Park region surrounding the actual state-owned forest reserve (Figure 13-1). The use of less-than-fee interests, for example, conservation easements, have been used to protect existing parks from visual encroachment. Also, a number of states have asserted ownership rights over the beds of rivers, lakes, and coastal waters under the doctrine of public trust.

Sometimes, states have served as the cornerstone of cooperative intergovernmental approaches to land preservation. Thorn Creek Woods in the south suburbs of Chicago was saved from development in the early 1970s through public purchase by the State of Illinois, the Will County Forest Preserve District, two municipalities (one of which was Park Forest—the home of William H. Whyte's celebrated "Organization Man"), and a new state university. None of these public entities could

Figure 13.1 Location of the Adirondack State Park (areas in black are owned by the state). *Source:* Commission on the Adirondacks, 1990.

geographically or fiscally preserve the entire woods on their own. The entire joint acquisition was facilitated by federal grants under the HUD Open Space and New Communities programs amounting to 70 percent of the total costs. Thorn Creek Woods State Park is today managed by the state under cooperative agreements with the other landholding entities.

"THE QUIET REVOLUTION"

Between 1965 and 1975, many states dramatically expanded their land planning and management roles in tandem with the reforms at the federal level. This surge of state activism became known as the "Quiet Revolution" after the title of a report prepared for the new Council on Environmental Quality (Bosselman and Callies, 1971). The introduction to this report stated the situation as follows:

> This country is in the midst of a revolution in the way we regulate the use of our land. It is a peaceful revolution, conducted entirely within the law. It is a quiet revolution, and its supporters include both conservatives and liberals. It is a disorganized revolution, with no central cadre of leaders, but it is a revolution nonetheless.
>
> The *ancien regime* being overthrown is the feudal system under which the entire pattern of land development has been controlled by thousands of individual local governments, each seeking to maximize its tax base and minimize its social problems, and caring less what happens to all the others.
>
> The tools of the revolution are new laws taking a wide variety of forms but each sharing a common theme—the need to provide some degree of state or regional participation in the major decisions that affect the use of our increasingly limited supply of land. (Bosselman and Callies, 1971, p. 1.)

In fact, there was a "cadre of leaders," albeit informal and open-ended in its membership. Parallel with the city planning movement in the early twentieth century, the Quiet Revolution was motivated by a group of lawyers and planners in several states with a center of gravity in Chicago, the location of the Ross, Hardies law firm that included Richard F. Babcock, Marlin Smith, Fred Bosselman, David Callies, and John Banta among its land-use specialists. Closely allied with this group was William K. Reilly, editor of *The Use of Land* (1973) and a senior staff member of the federal CEQ which commissioned the Quiet Revolution report. (Reilly later headed the influential Conservation Foundation and then was appointed administrator of the U.S. Environmental Protection Agency by President Bush.)

Stimulus was provided to the Quiet Revolution through the drafting of a Model Land Development Code by the American Law Institute (1975). The Code was developed as a model for new state legislation to update the old Planning and Zoning Enabling Acts of the 1920s. Its most notable innovation was Article 7, which proposed state review and possible override of local zoning decisions concerning (1) areas of particular concern, (2) large-scale developments, and (3) developments of regional benefit. In the first two categories, the role of the state was likely to be more protective (less permissive) than local government. The third category raised the possibility of state veto of local zoning that prohibited a locally unwanted but regionally needed facility (e.g., a power plant, prison, incinerator or wastewater treatment facility). Early drafts of the Code, beginning in 1968, generated widespread praise from "quiet revolutionaries" and alarm among some local governments.

The Code was not literally adopted anywhere. A few states (e.g., Vermont, Maine, Oregon, and Florida), however, adopted state land planning laws based in spirit on Article 7. It served also as the inspiration for the state management requirements of the federal Coastal Zone Management Act of 1972.

The momentum of the Quiet Revolution diminished in the late 1970s as development pressures abated due to high interest rates and energy issues dominated public attention. But the state programs that it motivated remained in place and in several instances were expanded, even during the conservative 1980s (Popper, 1988). Some of these programs are statewide in their application, for example, Vermont's Act 250 and Act 200, Oregon's Act 100. Florida's Land and Water Management Act and its 1984 State and Regional Planning Act, and the progenitor of state planning laws, the Hawaii Land-Use Act of 1961.

The relative scarcity of comprehensive state planning laws has fostered the view among proponents of centralization that the Quiet Revolution fizzled out prematurely. The Reagan Administration, however, felt that land use was overregulated and advocated regulation-free "enterprise zones" to assist private investment. Popper (1988) disputes both the "liberal" view that we have too little centralized land management and the "conservative" position that we have too much. He argues that since 1970 the nation has achieved a harmonious new balance of local-versus-centralized planning control. In the process many, if not most, states have gained a crucial role in land-use control, not usually on a statewide basis, but selectively. The handful of comprehensive state planning programs are augmented by a multitude of state laws focused on either specific geographic areas or specific land-use issues.

SPECIAL AREA PROGRAMS: THE PINELANDS CASE

The original Quiet Revolution report studied several state programs focused on specific geographic areas, for example:

- San Francisco Bay Conservation and Development Commission (CA),
- Adirondack State Park Agency (NY),
- Twin Cities Metropolitan Council (MN),
- Tahoe Regional Planning Agency (CA and NV), and
- Hackensack Meadowland Development Commission (NJ).

Each of the foregoing was alive and apparently well in the late 1980s. Except for the Twin Cities Metro Council, which does not regulate land use, each survived political and legal challenges to its land-management functions. Yet they all influenced to some degree the pattern of development that would otherwise have occurred in their regions of concern.

The New Jersey Pinelands Commission, which appeared in the late 1970s, provides a more recent example of a state program, initiated with some federal nudging, that is achieving a remarkable degree of control over the use of sensitive

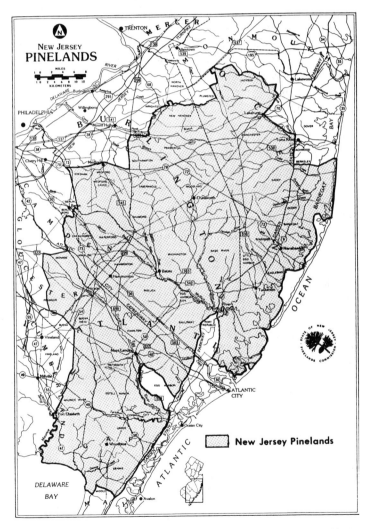

Figure 13.2 Map of the New Jersey Pinelands Region. *Source:* New Jersey Pinelands Commission.

land with an apparently high level of public acceptance. The Pinelands or "Pine Barrens" is a region of about 1 million acres of sandy, wooded terrain in southern New Jersey, of which about two-thirds is privately owned (Figure 13-2). The region is underlain by a vast aquifer estimated to hold some 17 trillion gallons of fresh water. Cranberry and blueberry cultivation are prominent agricultural activities in portions of the region. The Pinelands are sparsely settled. Its special historic and cultural identity was portrayed in John McPhee's *The Pine Barrens* (1968).

Lying between Philadelphia and the coast, the region during the 1970s attracted growing interest from real estate developers. With jurisdiction over the Pinelands divided among seven counties and 55 towns and villages, efforts to protect its natural resources at the local level would have been fruitless. Instead, preservationists, armed with McPhee's book and great persistence, achieved a unique three-fold designation of the Pinelands Region as: (1) a National Reserve; (2) an International Biosphere Reserve; and, most important, (3) a special state planning region administered by the New Jersey Pinelands Commission.

This process began with the designation of the Pinelands as the first National Reserve in Section 502 of the National Parks and Recreation Act of 1978. Within the 1.1 million acres so designated, Congress authorized $26 million for land acquisition and planning by the state and localities, provided the state established a special planning commission for the area. In response, New Jersey's Governor Brendon Byrne, by executive order, established the state-level Pinelands Commission and banned new development in the region until a comprehensive plan and land-use regulations were prepared. These powers were confirmed in the New Jersey Pinelands Act of 1979 despite vocal opposition (Babcock and Siemon, 1985, Ch. 8).

The Pinelands Comprehensive Management Plan as finally approved by the commission, the governor, and the secretary of the interior in 1980–1981 was a landmark in state land management. The 1979 law had designated a state Pinelands Area of 933,000 acres (slightly smaller than the Pinelands National Reserve) and divided that into a central "preservation area" of 368,000 acres and a surrounding "protection area" of 565,000 acres. The commission's plan further divided the Pinelands into nine land-use zones based on the physical and human characteristics of each area. The preservation area, also known as the "heart of the pines," is described by the commission as follows:

> Here are the unique forests of pygmy pines and oaks known as the East and West Plains. The ruins of long-deserted towns and factories poke through the underbrush amidst a maze of twisting, barely passable sand roads. The solemn gloom of cedar swamps gives way to the flowery brilliance of inland marshes and bogs (Pinelands Commission, 1989, p. 4).

Within this core area, no new residential development is permitted except that "Pineys"—long-term inhabitants of the area—may construct a new home under certain conditions.

The remainder of the Pinelands, the "protection area", is divided into: (1) Forest Areas; (2) Agricultural Protection Areas; (3) Regional Growth Areas; and (4) Rural Development Areas. Additional areas are designated as Pinelands Towns and Villages and for military purposes.

A key feature of the plan is the transfer of development rights (TDR) from the inner restricted areas to peripheral locations where growth is encouraged. This plan allows owners of restricted land to sell "Pineland Development Credits" (PDCs) to developers in designated Regional Growth Areas at a specified ratio. For example, one dwelling unit credit may be sold for each 39 acres of upland owned in the Preservation Area, and two credits for an equivalent area in the Agricultural Production Areas. For wetlands, credits are only one-fifth of the upland levels. It is estimated that as many as 24,000 units of housing could be built in the Growth Areas through transfers of credits from restricted areas of the Pinelands, in addition to 86,000 units otherwise permitted to be built in the Growth Areas. Another 74,000 new units are permitted among the designated Pinelands Towns and Villages.

To facilitate the sale of PDCs, two clearinghouses have been established: the State Pinelands Development Credit Bank and the Burlington County Pinelands Development Credit Exchange. PDCs may also be sold privately between property owners in the appropriate zones. As of September, 1989, 21 development projects using PDCs were approved or constructed in Pinelands Growth Areas, with another 29 developments proposed ("The Pinelander," 1989).

One of the drafters of the Pinelands Plan, Richard F. Babcock, reports that despite continuing opposition in certain parts of the region, "the CMP has been by all accounts a great success." As evidence of this success he notes that 96 percent of all development subsequent to the Plan's adoption has been located in areas designated for growth and as of 1985 no litigation against the Plan had been filed (Babcock and Siemon, 1985, p. 156). Also by that time, 36 municipalities and four counties had amended their local zoning and were certified as conforming with the Plan. Babcock credits the professionalism and responsiveness of the Commission staff for the relatively smooth transition to the new era of land planning and control in the Pinelands.

SPECIAL-ISSUE STATE PROGRAMS

The progress of state-level land-use controls noted by Popper (1988) and DeGrove and Stroud (1987) has been reflected far more pervasively in state laws focused on specific management issues than in those focused on specific geographic regions. For every Pinelands-type regional management program, there are dozens of state laws that address generic issues or clusters of issues. Principal examples are laws concerning farmland preservation, coastal zone management, wetland and floodplain management, housing, siting of large-scale facilities, and the disposal of solid and hazardous wastes. These of course have a geographic dimension to the extent that all land-use laws affect land somewhere. But rather than circumscribing the state program within a specially defined subarea of the state, to the exclusion of other areas with substantially similar problems, issue-focused laws apply wherever such issues arise.

States differ as to the comprehensiveness and administration of their land-use laws. As mentioned earlier, a few states—notably Vermont, Oregon, Hawaii, and Florida—have land-use statutes that are relatively comprehensive, covering many issues within the review process mandated by a single law and administrative process. Thus Vermont's Act 250 requires review of developments falling within its purview in terms of water supply, soil erosion, highway congestion, schools, governmental services, aesthetics, and historic preservation, among other matters (Healy and Rosenberg, 1979, p. 45). Florida's State and Regional Planning Act of 1984 established a statewide planning framework keyed to a legislatively adopted state comprehensive plan. The plan is "a relatively concise statement of goals and policies for 25 major state issues ranging from problems of the elderly to protection of property rights to transportation issues" (DeGrove and Stroud, 1987, p. 7). Regional and local plans are required to conform to the state plan, as are state functional plans including the State Water Use Plan and the State Land Development Plan.

Far more numerous are states that have addressed land-use issues selectively in independent legislation. All 50 states have statutes of this nature, which are most readily found and compared in the Bureau of National Affairs *Environmental Law Reporter*. Many such laws have been adopted in response to or in imitation of federal legislation, as, for instance, the several states with environmental impact laws modeled on NEPA. Others have been passed in response to a severe state problem and/or at the insistence of a strong governor. Both factors applied to the adoption, for instance, of New Jersey's Fair Housing and State Planning acts in 1985.

As noted in the last chapter, the issue of protecting agricultural land from urbanization has largely been left to the states to address as they wish. The most widespread response has been to alleviate the burden of state and local taxes that might otherwise force a landowner to sell farmland prematurely. According to Furuseth and Pierce (1982, Table 8), 42 states (and five Canadian Provinces) have adopted some form of preferential assessment or deferred payment of property taxes on productive farmland in locations subject to development pressures. Twenty-eight states have provided some form of relief from state inheritance taxes to avoid forcing farmland to be sold at the death of its owner.

Other state-level approaches to farmland preservation have been more direct. Certain states or counties have established programs for the purchase of farmland development rights—notably Massachusetts; Connecticut; New Jersey; Maryland; Suffolk County, NY; and King County, WA. These programs involve the appropriation of funds to acquire development rights at a negotiated price from interested farm owners. The latter continue to own the land and may, but are not compelled, to continue farming it. It may be sold subject to the agricultural restriction, as with other forms of transfer of development rights. Since the amount of land offered for purchase of development rights usually exceeds what can be purchased with available funds, selection criteria are imposed by the state in terms of land quality, location, and productivity.

Agricultural districts and zoning are additional techniques used in a few states. New York since 1969 has authorized counties to establish, upon petition of farmland owners, "agricultural districts" within which public projects that facilitate development are discouraged and a variety of other farmland protection measures apply. Districts have included as much as 1 million acres of New York farmland. Maryland also has utilized the district approach on a farmer-initiated basis. Building and zoning regulations are relaxed on farm-related activities within districts in exchange for a commitment by owners to maintain their land in agriculture for at least five years (Furuseth and Pierce, 1985, p. 64). California has employed both agricultural districts and agricultural zoning. Exclusive agricultural zoning has also been extensively used under Oregon's state land-use planning program.

Floodplain, wetland, and coastal zone management have been widely addressed through state legislation. As discussed in the preceding chapter, these three policy areas share considerable geographic overlap, yet are administratively fragmented at both the federal and state levels. Furthermore, states differ greatly in their particular approaches to each of the three areas. Generalization and comparison of state programs is hazardous.

A recent survey of state regulatory standards for riverine floodplains (L. R. Johnston Associates, 1989, Table 11–3) found remarkable diversity among approaches. While a state-level program is not required for eligibility for benefits under the National Flood Insurance Program (unlike the federal CZM program), 26 states reported that they restrict new development in designated floodways, either directly or through local conformity to state standards. Nine states define "floodways" more restrictively than the federal NFIP standards. Twenty states regulate building in the "floodway fringe" (balance of the 100-year floodplain) and 18 impose state-level building construction standards in floodplains. Sixteen states however reported having none of the foregoing measures at the state level, although most provide technical assistance to local floodplain management activities. The same survey disclosed comparable diversity among states with respect to coastal and lakeshore flooding.

State wetland laws date back to the early 1960s and may be very complex within even a single state. Massachusetts, for instance, has a wetland *protection* law regulating both inland and coastal wetlands through review of development proposals by local conservation commissions. Greater control is provided through separate coastal and inland wetland *restriction* laws under which Massachusetts may limit all development of designated areas through a recorded restriction. The latter approach has been primarily applied to certain coastal salt marshes and very rarely for inland wetlands.

According to Mitsch and Gosselink (1986), at least 17 states have some form of coastal wetland program, mostly requiring permits to be obtained from the state or local authorities for disturbance of such areas (Table 16–5). Some 25 states are reported to have laws that directly or indirectly regulate the alteration of inland wetlands (Table 16-6). The relationship between state wetland laws and the federal Section 404 program is not easy to define. In a few states, formal agreements exist between the state wetlands program and the Army Corps of Engineers for joint application and review procedures. Under certain circumstances, the Corps may defer to the state determination in wetland regulation. But often, developers and landowners must obtain both federal and state permits, necessitating both delay and expense.

The federal Coastal Zone Management Program has approved the coastal plans of 29 states and territories, qualifying them for ongoing program support. (The status of all coastal jurisdictions under the CZM program in 1985 is displayed in Table 12-2). To gain plan approval, each jurisdiction had to assert state-level control over significant coastal development decisions, as discussed in the previous chapter. Some states such as California and North Carolina accomplish this through comprehensive state coastal laws. Others piece together a network of specialized statutes, regulations, and executive orders. Some of these involve the flood hazard and wetland programs already mentioned. Other objects of state coastal planning include recreation and public access, pollution abatement, economic development, protection of endangered species and other wildlife habitat, commercial and sport fisheries, and navigability.

To accomplish these objectives, states employ several types of development controls including permit requirements, detailed review of large projects, specialized regulations for critical areas, mitigation of wetland loss, land acquisition, and the promotion of particular types of development. Techniques used by each approved state coastal programs were indicated in Table 12-3.

Coastal erosion afflicts most of the nation's tidal and Great Lakes shorelines to varying extents. Federal policies have sought to control the effects of erosion *in situ* through seawalls, groin fields, jetties, breakwaters, and beach nourishment. The National Flood Insurance Program, while insuring losses due to coastal erosion, has not required setback or other precautions to protect new structures from undermining or rising flood risk due to erosion. In this situation, at least 11 states have seized the initiative by adopting their own erosion setback regulations (National Research Council, 1990). North Carolina, a leader among coastal state programs, requires new homes to be set back a distance at least equal to 30 times the average annual erosion rate (AAER) and larger structures 60 times that rate (Owens, 1985). The 30-year AAER standard has also been applied by Florida, Michigan, Rhode Island, and New York. South Carolina has adopted a 40-year AAER standard, which posed a substantial problem for owners of shorefront homes destroyed by Hurricane Hugo in 1989. (It was subsequently weakened by amendment.)

CONCLUSION

States, long viewed as retarded and useless participants in the process of public land-use control, have thus assumed a variety of roles, powers, and functions since the 1960s. They have rightly served as the lead level of government on agricultural land protection. They have responded creatively to federal incentives in coastal zone management. They have performed tolerably, and in some cases very effectively, as managers of wetlands and floodplains. "Nation-states" such as California, Florida, and New York operate a galaxy of land-related programs. Amenity-oriented states such as Vermont, Oregon, and Maine have carved out enduring state land-use control functions, albeit less ambitious and effective than early "quiet revolutionaries" had hoped.

New Jersey has demonstrated that even states that are not bastions of environmentalism may engage in state land planning as in the Pinelands, the Hackensack Meadowlands, and in its fair housing and coastal laws. Other states also have served as laboratories for innovation in land-use control, as in the case of coastal erosion setbacks. While often accused of being handmaidens to the federal government and neglectful parents of local governments, it is clear that many states are land-use managers to be reckoned with.

REFERENCES

AMERICAN LAW INSTITUTE. 1975. *A Model Land Development Code*. Philadelphia: American Law Institute.

BABCOCK, R. F. AND C. L. SIEMON. 1985. *The Zoning Game Revisited*. Boston: Oelgeschlager, Gunn, and Hain.

BOSSELMAN, F. AND D. CALLIES. 1971. *The Quiet Revolution in Land Use Control*. Washington, D.C.: U.S. Government Printing Office.

CONSERVATION FOUNDATION. 1987. *State of the Environment: A View Toward the Nineties*. Washington, D.C.: Conservation Foundation.

DEGROVE, J. M. 1984. *Land Growth and Politics*. Chicago: American Planning Association Planners Press.

DEGROVE, J. M. AND N. E. STROUD. 1987. "State Land Planning and Regulation: Innovative Roles in the 1980s and Beyond." *Land Use Law* (March): 3–8.

FURUSETH, O. J. AND J. T. PIERCE. 1982. *Agricultural Land in an Urban Society*. Washington, D.C.: Association of American Geographers.

HEALY, R. G. AND J. S. ROSENBERG. 1979. *Land Use and the States* (2nd ed.). Baltimore: Johns Hopkins Press for Resources for the Future.

HEALY, R. G. AND J. A. ZINN. 1985. "Environment and Development Conflicts in Coastal Zone Management." *Journal of the American Planning Association* 51(3): 299–311.

L. R. JOHNSTON ASSOCIATES. 1989. *A Status Report on the Nation's Floodplain Management Activity*. (Mimeo).

MCPHEE, J. 1968. *The Pine Barrens*. New York: Farrar, Straus & Giroux.

MITSCH, W. J. AND J. G. GOSSELINK. 1986. *Wetlands*. New York: Van Nostrand Reinhold.

NASH, R. N. 1982. *Wilderness and the American Mind* (3rd ed.). New Haven: Yale University Press.

NATIONAL RESEARCH COUNCIL. 1990. *Managing Coastal Erosion*. Washington: National Academy Press.

OWENS, D. W. 1985. "Coastal Management in North Carolina: Building a Regional Consensus." *Journal of the American Planning Association.* 51(3): 322–329.

Pinelander, The. 1989. "PDC Sales Good News for Landowners." 9(2): 2.

PINELANDS COMMISSION. 1989. "A Brief History of the New Jersey Pinelands and the Pinelands Comprehensive Development Plan." (Mimeo).

POPPER, F. J. 1988. "Understanding American Land Use Regulation Since 1970." *Journal of the American Planning Association* 54(3): 291–301.

REILLY, W. (ED.). 1973. *The Use of Land: A Citizen's Guide to Urban Growth.* New York: Thomas Y. Crowell.

CONCLUSION: SUCCESSES, FAILURES, AND UNFINISHED BUSINESS

This book has reviewed the evolution of public controls over land use in the United States from their medieval origins in England to the latest land use initiatives of municipalities, states, and the federal government in the 1980s. A persistent theme of the book has been the tendency of a capitalist society, through its various levels and instruments of public authority, to intervene in the private land economy only in response to perception of socioeconomic or environmental dysfunction. This typically takes the form of adverse externalities inflicted by the actions of certain land use managers on others who may be immediate neighbors or a broader geographic constituency defined by physical or socioeconomic criteria, for example downstream occupants of a floodplain, downwind residents in an airshed, or households seeking new residential opportunities in a regional housing market. The nature and geographic scale of a "dysfunction" may vary from local traffic congestion and fiscal imbalance to a regional soil erosion problem, to a nationwide proliferation of abandoned toxic waste dumps, or to global warming. Corresponding avenues of public response to these particular concerns have included, respectively, local "growth management" policies, soil erosion control programs, the Superfund Act of 1980, and proposals (as yet uneffectuated) for limits on deforestation and carbon dioxide emissions.

Clearly this process of societal response to perceived threat is highly uneven in its application and effectiveness, and often leads to conflict among policies and measures instituted by individual units or different levels of public authority. Adverse externalities arise from the inconsistent perceptions and actions of municipalities, states, and nations as they do from the actions of individual property owners. These include, for example, the distorting of housing markets and the economic and environmental costs of long-distance commuting, due in part to municipal zoning and slow growth policies. Furthermore, attempts to solve social problems through manipulation of the spatial arrangement of land use may only address the visible tip of the iceberg. For example, programs to mandate or facilitate "affordable housing," while commendable in purpose, do little to alleviate underlying conditions of poverty. (And even "affordable" housing may be well out of reach of the poor: "Affordable" in the New York region means a house costing less than $100,000, and in Toronto, a rental below $800 a month.)

It has been three centuries since the "first modern building law"—the Act for Rebuilding London of 1667—and 150 years since the first general sanitary laws appeared in England and the United States. Eight decades have elapsed since the First National Conference on City Planning and Congestion in 1909, six decades since the *Euclid* decision, and two decades since NEPA. What has all this legacy of societal response to inefficiency in the private land market accomplished, and where do we stand today?

A statistical response to this question is difficult since the results of the 1990 Census are not available at this writing, and federal data on housing and urban development is skimpy for the 1980s. But some generalizations may be drawn from accounts by writes and journalists, from recent research findings (some of which are cited in Chapters 8 through 13), and from anecdotal evidence.

Like the proverbial blind men describing the elephant, characterization of the present state of metropolitan and rural land use must depend on what is viewed and by whom. The contemporary U.S. landscape is a mosaic of public policy successes and failures, some of which change character depending on the function, scale, and location considered. Let us optimistically begin with some successes, as viewed by some beholders at least, before turning to the failures.

- *Central business districts* (CBDs) of many cities have new vitality, architectural distinction, cultural diversity, and economic viability due to a combination of public and private innovations in the design and use of urban space. The public plaza (indoor or outdoor) has become a standard fixture of new commercial construction since the 1960s due to zoning incentives. If suitably designed, such spaces promote street life, civic pride, and urbanity (Whyte, 1988). Skywalks, underground shopping concourses, lobby exhibit areas, and public art may help to enliven downtowns. Even medium-sized cities such as Hartford, Connecticut; Portland, Oregon; Charlotte, North Carolina; or San Antonio, Texas offer a cosmopolitan ambience to those who work, visit, or increasingly reside in their CBDs.
- Meanwhile, new *outlying regional centers* are challenging the dominance of the older CBD's in many metropolitan areas. Houston's Galleria, Atlanta's Perimeter Mall and Cumberland/Galleria Mall, and South Coast Plaza in Orange County, California are prototypes of twenty-first century urbanization.

These complexes, situated at strategic crossings of metropolitan highways, contain opulent retail shops, millions of square feet of office space, major hotels, restaurants, and convention facilities, all interconnected by climate controlled pedestrian concourses. Evolving in tandem with these super-regional malls has been the continued growth of "mushroom suburbs" or "urban villages" such as Tyson's Corner, Virginia or Irvine, California. According to Leinberger and Lockwood (1986), South Coast Plaza in 1986 had the largest sales volume of any shopping center in the nation, and was close to surpassing downtown San Francisco in total sales. Office space available in the Irvine region in Orange County that year totalled 21.1 million square feet, mostly built since 1980, which nearly equalled the entire San Francisco CBD. Orange County in 1989, however, had an office vacancy rate of 25.1 percent (*The New York Times,* March 12, 1989, p. 38).

- American cities have become increasingly accessible to *handicapped* individuals due to federal and state building requirements applicable to new construction as well as renovation of older structures. Commercial, governmental, transportation, educational, cultural, and religious facilities are now widely usable by mobility-impaired persons.

- *Historic preservation* has accomplished much over the past three decades to protect and enhance much of the nation's cultural heritage. National and state registers of historic landmarks list thousands of structures and sites of special historic or architectural importance. Not all of these have been saved from destruction or alteration, but many have been. Beginning with the Battery District of Charleston, South Carolina, and New Orleans' French Quarter, historic districts have been widely established under federal, state, or local laws which regulate in minute detail the exterior appearance of structures. Many historic properties have been acquired by public agencies or private nonprofit foundations for permanent preservation. In older communities that have experienced economic transition, many commercial structures such as mills, railroad depots, wharfs, and warehouses have been "adaptively reused" as retail outlets, restaurants, offices, condominiums, and new kinds of manufacturing space. This process has been fostered in part by public tax credits and other legal and financial incentives.

- Many of the nation's *older bedroom suburbs* are aging gracefully, to the delight of their fortunate inhabitants. These paragons of the Garden City influence and Euclidean zoning still offer large older homes, shady quiet streets, mature landscaping, good schools, ample water and sewerage service, parks, hospitals, commuter trains, and other amenities. While most such communities remain largely white, some have accepted a modest degree of racial integration by middle-class nonwhite families. In a few suburbs such as Oak Park, Illinois this process has been voluntary and proactive; elsewhere it has resulted from federal or state fair housing legislation or has not occurred at all.

By and large, these communities continue to embody the American Utopia, as subsidized by the federal tax code through deductions for mortgage interest and local taxes (Jackson, 1985).

- Since 1960, many *natural sites and open spaces* have been protected through public measures. Since 1965 the federal Land and Water Conservation Fund (LWCF) has allocated over $3 billion for matching grants to states and local governments for the purchase and improvement of open land (although this program has been greatly reduced by the Reagan and Bush Administrations). The acreage in the National Park System in the lower 48 states has increased by 40 percent since 1960, reflecting the establishment of 14 new national seashores and lakeshores, several national recreation areas, and many other new facilities. The National Forests and National Wildlife Refuge Systems have been augmented by new land purchases funded by the LWCF. More than 50 million acres of federal land have been classified as "wilderness" areas under the 1964 Wilderness Act.

  Private nonprofit organizations, from The Nature Conservancy at the national level to local land trusts, have bought or accepted gifts of thousands of acres of natural land. Land philanthropy by private owners has been encouraged by federal income tax deductions for the value of such gifts.

- As contemporary sequels to Olmsted's Emerald Necklace Plan for Boston, stream corridors and coastlines are receiving increasing attention as multifunction resource systems. *"Greenways"* have been established along many metropolitan streams, for example, in the Bay Area of California, Portland, Seattle, Atlanta, San Antonio, Denver, and Washington, D.C. The greenway approach to resource management combines several public objectives: water quality improvement, protection of riparian and fish habitat, public recreation, flood hazard reduction, and visual separation of urban districts. Means used to effectuate greenways include public acquisition, regulation, and financial incentives. The $30 billion federal *sewage treatment* program established in 1972 has significantly contributed to the improvement of surface water quality in many urban areas.

- Federal and state *coastal zone and wetlands* programs have sought to improve public access and protect natural resources in coastal areas, although extremely high property values impede public management of the nation's shorelines. In addition, many wetlands, particularly coastal saltmarshes, have been protected under federal, state, and local regulatory programs.

- Smoldering *garbage dumps,* commonplace in the 1960s, have largely been abolished by solid waste management laws, although their toxic contents may continue to pose a threat to groundwater. Sanitary landfills have replaced dumps as the approved method of land disposal of solid waste. These, however, are themselves being replaced by energy recovery facilities, materials recycling programs, composting, and other approaches to waste management.

Conclusion: Successes, Failures, and Unfinished Business

The foregoing represent accomplishments of U.S. land use laws and programs over the past several decades. While no arena of land use policy can be deemed an unqualified success, these have been areas of significant progress or at least absence of deterioration. There is, however, a dark side to the U.S. landscape in the 1990s which represents a legacy of public policy failures or chronic neglect:

- The same central cities whose downtowns and wealthier suburbs flourished in the 1980s have experienced continued deterioration of their inner-city, *low-income neighborhoods*. The urban renewal programs of the 1950s and 1960s left a patchwork of now-dilapidated "housing projects," unredeveloped vacant lots, and surviving tenement dwellings, many of which are now abandoned except by crack dealers. While these areas retain basic public services—water, sewerage, streets, parks—which are being replicated in outlying areas at great public cost, the condition of inner-city public facilities, particularly schools, is generally abysmal.
- As discussed in Chapter 11, 50 years of *public housing* programs by 1987 had yielded only 950,000 family units and 350,000 elderly units nationally—in comparison with 32.5 million people officially counted as living below the federal poverty level that year. And that omits millions of illegal immigrants and other uncounted individuals who further crowd the dwindling lower-cost housing supply.
- Federal subsidies for new construction and rehabilitation of *affordable housing*, which flourished briefly during President Johnson's "War on Poverty," have since languished. Similarly, state programs such as the New York Urban Development Commission which offered much promise two decades ago have fallen far short of meeting the demand for lower-cost units, a situation that will surely worsen as the economies of urban states such as New York, Massachusetts, and New Jersey weaken in the 1990s. In 1988, New Jersey required 145,000 new affordable units to comply with *Mount Laurel*-related mandates for communities, and a housing panel estimated that Connecticut needed 169,000 such units. But only a few thousand units priced below $100,000 were then under construction in those states. The median price of a home in the New York metropolitan region was then $192,000, affordable by only the upper 20 percent of the region's households (*The New York Times*, October 2, 1988).
- Inevitably, the growing scarcity of affordable housing has precipitated a rapid increase in the *homeless* population, which has been estimated to amount to some 3 million people nationally (Rowe and Wolch, 1990). Paradoxically, since the homeless take up "residence" in the public spaces of mainstream urban America, they have attracted more public attention than the more numerous but less visible occupants of ghettos and slum districts. (For example, a homeless person was recently encamped beside the elite Skull and Bones

secret society at Yale University, of which President Bush was a member.) To date, however, public awareness has not translated into public action to alleviate homelessness. According to Rowe and Wolch (1990, p. 202): "The current lack of homeless political power leads policy makers away from meeting their needs for greater assistance, such as additional low-cost housing, social services, job-training and employment." The phenomenon has perhaps been regarded more as a pathetic but benign spectacle rather than a threat to the larger society. Lacking evidence of significant externalities, little public response has been forthcoming.

- Public *infrastructure* of many kinds is displaying widespread evidence of inadequate capacity, lack of maintenance, and obsolescence. For instance, the Massachusetts Special Commission on Infrastructure Finance (undated) estimated that repair and replacement of public infrastructure (for transportation, environmental management, state buildings and grounds, public housing, etc.) in that state would require expenditures totalling $37.4 billion during the 1990s. A broad scholarly review of infrastructural needs in the United States (National Academy of Engineering, 1988, p. 4) found that:

> . . . we do not know well the condition of our infrastructure. Simple, partial inventories are available for some components of the infrastructure. For many key systems, and for many cities, we are unable to specify much at all. Our diagnosis of the vital systems of cities and the nation remains primitive. . . .

- A closely related concern is the reliability of vital *"lifeline"* systems, such as transportation facilities, electrical power, water and sewage service, and other utilities in the face of natural or technological disaster. The major source of prolonged disruption resulting from the "World Series Earthquake" in California in October 1989 was not the collapse of individual structures which were built for the most part to seismic code standards, but the failure of the Oakland Bay Bridge and the Nimitz Expressway, both aging but vital links in the Bay Area highway system.

- On the subject of *disasters,* American cities have achieved a measure of safety against some types of hazards while allowing the risks of others to be magnified. Urban conflagration, the scourge of wood-built cities from the colonial period until the San Francisco Earthquake of 1906, has ceased to be the primary threat today, thanks to planning and building regulations, reliable water supply systems, and modern fire-fighting technology. Urban flood disasters, however, have increased markedly during the twentieth century due to the expansion of cities into their riverine and coastal floodplains. Smaller metropolitan streams flood more often and more dangerously as paving and development upstream contribute more runoff to natural drainage ways during storms. As discussed in Chapters 9 and 12, local floodplain management prompted by the National Flood Insurance Program since the mid-1970s has modified building practices for new or rebuilt structures in many floodplains.

But millions of older buildings and their inhabitants remain at risk along rivers, lakes, and ocean shorelines. And in the wake of mass devastation, as along the South Carolina coast following Hurricane Hugo in 1989, governmental restrictions on rebuilding may be loosely applied out of sympathy for the victims.

- The proliferation of new commercial malls, office parks, and condos that comprise the "mushroom suburb" or "urban village" discussed earlier is scarcely an unmitigated blessing. While internally planned for the comfort and convenience of their occupants (not always effectively), such developments inflict diverse adverse externalities on surrounding areas. These include proverbial *traffic congestion* (few are served by nonhighway mass transit), *air pollution, energy inefficiency, visual discord,* and *loss of natural habitat or agricultural land.* Economically, such developments create large numbers of low-wage service jobs but do not provide affordable housing, schools, parks, medical clinics, and the other necessities of working class urban communities. In short, the people who qualify for jobs in such areas usually have to commute long distances, often by very inadequate public transportation from the inner city or lower cost suburbs (Leinberger and Lockwood, 1986). As documented by William H. Whyte (1988), a number of major New York corporations have moved their headquarters from Manhattan to affluent hinterland locations, which happened to be close to the estates of their chief executive officers but far from affordable communities for their employees. In the process, the central city loses not only jobs but corporate taxes as well. Despite all the lip-service given to regional planning since the 1920s, metropolitan growth in the 1980s continued to reflect the dominance of private capital and its handmaiden, local municipal autonomy, with regard to location, planning, and taxation of new development.

- Even worse, the private real estate development process was substantially distorted during the 1980s by federal *deregulation of mortgage lending,* particularly by savings and loan institutions (S & L's). By 1990, several dozen S & L's primarily in the South and Southwest had failed, with many more likely to follow. Estimates of the potential liability of federal taxpayers to bail out the depositors of failed S & L's range as high as $500 billion over several decades at this writing. Meanwhile the physical results of improvident speculation litter the American landscape. These include condominium projects, hotels and resorts, and office developments, often uncompleted or vacant. Arguably, many of these project would have not been built if the investor had been prudent in terms of locational analysis—applying basic economic and marketing geography principles. The Reagan Administration deemed that federal restrictions on the soundness of S & L lending practices were obstructing economic growth, but S & L depositors continued to be insured by the federal government against loss. The availability of huge sums of investment funds untied to normal standards of economic soundness thus fueled the infamous S & L building binge (*Newsweek,* May 21, 1990, pp. 20–32). Far from restraining improper development through land use planning and regulation, the public sector actually managed to short-circuit even the normal caution of the private market with respect to financial risk-taking, with disastrous results.

- *Solid and hazardous wastes,* as discussed in Chapter 12, pose a galaxy of unsolved land-related problems. Conventional solid waste management in 1990 is at a threshold: Land disposal sites are filling up and few new facilities are being established due to NIMBY objections among other factors. Several new approaches are coming into wider usage, including source reduction, waste stream separation, recycling, incineration with energy recovery, composting, and other forms of land treatment. The political acceptability and economic feasibility of many of these approaches, however, depend on the isolation and safe disposal of toxic and hazardous substances which are commonly intermingled with conventional wastes. The number of substances deemed to be toxic or hazardous and estimates of the total volume of such wastes have increased exponentially over the past two decades. Meanwhile, since the passage of the Resource Conservation and Recovery Act of 1976, " . . . not one state has succeeded in siting a hazardous waste disposal facility and few commercial facilities have obtained operating permits" due to intense local political opposition (Lake and Johns, 1990). Meanwhile, the U.S. Environmental Protection Agency has identified over 30,000 unsafe toxic and hazardous waste disposal sites, of which over 1,800 have been placed on the National Priority List for eventual attention by the Superfund Program. The latter, however, as discussed in Chapter 12, has been mired in bureaucratic, scientific, and political difficulties and to date has only cleaned up a handful of sites.

The foregoing chronicle of successes and failures comprises a crude "report card" on the status of several areas of land use policy in the United States. It is admittedly subjective and selective; other issues not mentioned here were discussed elsewhere in the book or have been omitted altogether. But this summary is meant to indicate areas of unfinished business, or even unstarted business, for professionals in the field of land use control over the next two or three decades.

Among the many techniques and approaches discussed in the latter half of this book, the most ubiquitous and least promising is, of course, Euclidean zoning. In the tradition of Reps and Delogu, I would agree that zoning has outlived its usefulness and probably causes more problems than it resolves. It legitimates and anoints the actions of the most geographically irrational unit of American land use management—the local government. However, like the premature reports of Winston Churchill's death, proclamations of zoning's imminent demise have been greatly exaggerated. Zoning thrives, tenacious as a bittersweet vine, and about as useful. The attempt to reform or overthrow zoning that occurred in the 1970s (e.g., their allies recognized the evils and dangers associated with urban conditions of their times. Each in diverse ways helped to initiate or promote new forms of public response to those conditions as recounted in Chapter 4.

As "sanitary reform" was the catalyzing issue in the 1840s, "congestion" in the first decade of this century, "open space" in the 1960s, and "environmental quality" in the 1970s, "urban sustainability" could be the focal issue of the 1990s. Cities and their metropolitan hinterlands are increasingly dysfunctional in light of social, economic, and ecological limitations. New perceptions, policies, doctrines, and institutions are required to address the interrelated issues of housing, transportation, economic growth, waste management, energy usage, and vulnerability to natural and technological disaster. A national commitment to the reexamination of these issues and the formulation of new expressions of public response requires a Presidential mandate and Congressional support comparable to the New Deal, the Manhattan Project, the Moon Program, or the Savings and Loan bailout.

While national objectives and fiscal resources are needed to mount remedial efforts, the national scale is too large for effective action, just as the municipal scale is too small. Sweeping reforms at the state or regional levels are required, with the help of federal funding and broad policy guidance. One precedent at the state level is the revolution in housing doctrine initiated in New Jersey through the *Mount Laurel* litigation. Given federal fiscal backing, this undertaking could yield significant levels of new affordable housing, widely distributed, as Justice Hall originally envisioned.

An example of "sweeping reform" at the regional scale is the recent plan to abate air pollution in metropolitan Los Angeles proposed by the California South Coast Air Quality Management district (Weisman, 1989). This plan, published in 1989 in a 45 volume, 3-foot high stack of reports, apparently took literally Daniel Burnham's adage: "Make no little plans, for they have no magic to stir men's blood." If Angelenos are stirred to limit their use of cars and backyard barbecues as this plan urges, anything is possible.

REFERENCES

LAKE, R. W. AND R. A. JOHNS. 1990. "Legitimation Conflicts: The Politics of Hazardous Waste Siting Law". Unpublished paper presented at the Annual Meeting of the Association of American Geographers, Toronto, Ontario, April, 1990. Available from the authors at the Center for Urban Policy Research, Rutgers University, New Brunswick, NJ 08903.

LEINBERGER, C. B. AND C. LOCKWOOD. 1986. "How Business is Reshaping America." *The Atlantic* 258(4): 43–52.

MASSACHUSETTS SPECIAL COMMISSION ON INFRASTRUCTURE. UNDATED. *A Survey of Massachusetts Infrastructure Needs for the 1990s*. Boston: The Commission. Mimeo.

NATIONAL ACADEMY OF ENGINEERING. 1988. *Cities and Their Vital Systems: Infrastructure, Past, Present, and Future*. (J. H. Ausubel and R. Herman, eds.). Washington: National Academy Press.

Conclusion: Successes, Failures, and Unfinished Business

through the Model Land Development Code) succeeded chiefly in bringing states into the land use control process in certain locations and on certain issues. Meanwhile many local governments, despairing of zoning but unable to rid themselves of it, have resorted to a number of supplementary strategies under the broad rubric of growth management. As considered in Chapter 9, devices such as transfer of development rights and impact fees attempt to "privatize" the provision of public needs though incentives or exactions directed at developers. The jury is still out as to how widely and effectively the "beyond zoning" techniques pioneered in the 1970s and 1980s will prove to be utilized, and what their external effects beyond the implementing municipality will be.

The primary need for American land use planning is to get back to basics: protecting the public health, safety, and welfare, in light of the new urban and global realities of the 1990s. In the nineteenth century, light, air, water, sanitation, and structural soundness were the goals of urban reformers who produced the modern city planning movement. Today, building and sanitary codes are virtually ubiquitous, although enforced unevenly for new construction and sometimes not at all for older buildings. Meanwhile, professionals in the field of land use control have long since left matters of health and safety to code enforcement officials and have concerned themselves increasingly with "quality of life" issues, particularly relating to more affluent sectors of the population who support planning. Amenities, not merely survival, have dominated much of the substance of public land use control since World War II.

There is, of course, nothing wrong with amenities, and this writer strongly maintains that aesthetics, historic preservation, billboard controls, and ecological protection are sufficiently related to the "public health, safety and (especially) welfare" to justify the use of public regulations with appropriate standards to achieve them. These efforts must continue.

But as small victories are gained in the areas listed earlier as "successes," social and economic disasters loom in the areas denoted as "failures." Since the end of Lyndon Johnson's "War on Poverty" in 1969, this country has ideologically abandoned its central cities. The federal government allocates $300 billion annually to defense while its Fiscal Year 1990 budget for the Department of Housing and Urban Development was $16.7 billion, much of which is earmarked for programs unrelated to central cities. The proliferation of homeless people, the deteriorating conditions in lower-income urban neighborhoods, and the costly, time-wasting climate-altering scourge of long-distance highway commuting all relate to the absence of a national commitment to rational and equitable urban development. T ultimate irony is that we are now planning to spend hundreds of billions of doll to bail out investors for building totally unneeded and useless structures wh millions of people have little or no shelter because the government does not cho to support housing construction.

In a sense, we stand where Chadwick stood in the 1830s, confrontir cesspool of public neglect which threatens to engulf the national society. Chadw Olmsted, Haussmann, Pullman, Howard, Riis, Marsh, Burnham, Bettmann,

ROWE, S. AND J. WOLCH. 1990. "Social Networks in Time and Space: Homeless Women in Skid Row Los Angeles." *Annals of the Association of American Geographers* 80(2): 184–204.

WEISMAN, A. 1989. "L.A. Fights for Breath." *The New York Times Magazine.* (July 30): pp. 15ff.

WHYTE, W. H. 1988. *City: Rediscovering the Center.* New York: Doubleday.

INDEX AND LIST OF CASES CITED

Acquisition, of public land, 161, 242, 246–53, 327
Acronyms, list of, 294
Act for Rebuilding London, 35, 37, 56–58, 376
Ad valorem tax, 119, 159
Adirondack State Park, 263, 363–64, 366
Adverse possession, 125
Aesthetic considerations, in zoning, 212–17, 224, 383
Affordable housing. *See* Housing, Low-income
Agriculture, 4–7, 334–43, 371 (*See also* Cropland)
Air rights, 114, 116
American Law Institute, 309, 365–66
A-95 Clearinghouse Process, 289–90
Annexation, 87, 143–49, 154, 155–57
Apartments. *See* Multi-family dwellings
Appalachian Redevelopment Act, 290
Appalachian Regional Commission, 290–93
Architectural regulations, 216–17
Army Corps of Engineers, 9, 241, 263, 305, 316, 330–32, 372

Babcock, Richard F., 204, 205, 212, 220, 365, 369
Balancing test, 196–99
Balancing the equities, 53, 117
Barrier beaches, 308–9, 325–27
Bennett, Edward, 168–70
Bettman, Alfred, 197, 202, 205–6, 213
Billboards, 181, 193, 213–16
Bond issues, 159–60
Boston
  growth, 28, 30–33, 67–68, 75, 143–45
  parks, 96–98, 378
  water supply, 88–91
Boundaries
  of jurisdiction, 19–20, 152–57
  legal, 126–32
  zone, 183, 188
Brandeis, Louis, 197
Building regulations, 119, 172–74, 177, 179, 181, 183–85, 197
  early history, 32, 54–58, 73, 84
Building permits, 189, 230
"Bundle of sticks," 114–19
Bureau of Land Management, 8, 9, 265–66, 312
Bureau of Outdoor Recreation, 283

Bureau of Reclamation, 9, 263, 264, 316, 334
Bureau of the Census, 10–12
Burnham, Daniel H., 86, 167–70, 175, 202, 251, 288, 384
Bush, George, 365, 378, 380

California South Coast Air Quality Management District, 384
Carson, Rachel, 300, 349
Central city, 227, 278, 285–88, 301, 379, 383–84
Central Park (New York), 92–96, 248
Central place theory, 20
Chadwick, Edwin, 39, 61, 72–75, 175
Charles II, 56
Charters, municipal, 48–49, 138
Chicago
  1909 Plan, 86, 168–70, 251, 288
  1969 regional plan, 251–52
  Sanitary District, 145–46, 151–53, 348
Chicago World's Fair, 86, 98–99, 105, 167
Citations, legal, 27–28
Cities. *See* Central city, Urban areas
Cities, medieval, 37, 45–52
City Beautiful movement, 86, 166–69, 176
City planning movement, 175–77, 365, 383
Civil Rights movement, 279, 285–87
Clean Air Act, 346
Clean Water Act, 241, 312, 329–34, 346
Clearinghouse, regional, 289–90
Clinton, DeWitt, Jr., 89–90
Cluster zoning, 188, 189–91
Coastal areas, 269–70, 307–9, 323, 371–373, 378
Coastal Barrier Resources Act, 308, 325–27
Coastal Zone Management Act, 279, 299, 301, 309–14, 327, 366, 372
Columbian Exposition of 1893, 86, 98–99, 105, 167
Commercial Club of Chicago, 168, 251, 288
Committee on Congestion of Population, 175–77
Common law, 39–44, 51–53
Commons, 39–45, 51, 54, 55–56, 265
Compensation, 56–57, 195–99, 247–48, 250, 286
Comprehensive plan, 178, 201–4, 229–30, 289–90
Condemnation. *See* Eminent domain
Conditional use permit, 179, 189
Condominiums, 116, 221, 224

Index

Conference on City Planning and Congestion, 175–77, 201
Conservation easement, 122, 249, 364
Conservation movement, 279–81
Conservation Reserve Program, 342
"Conservation," as resource management, 264–65
Constitution, U.S., 25–26
  commerce clause, 299–300
  Fifth Amendment, 26, 196–99, 214, 246, 248
  First Amendment, 215
  Fourteenth Amendment, 199–201, 285
  Tenth Amendment, 275
Conversion, land-use, 15, 125–26
  of cropland, 5–6, 334, 338–340, 343, 370–71
  of wetlands, 6, 9, 242–43, 328–34, 371–72
Council on Affordable Housing (NJ), 211
Council on Environmental Quality, 302, 304, 340, 344–45, 349, 365
Councils of governments, 289–90
Counties, 136–37, 140–43
Covenant, restrictive, 59–60, 122–23, 217, 223–24
"Cradle to grave" provision, 350
Craig, James, 29–30, 59
Cropland, 4–7, 237, 334–43, 370–71 (See also Agriculture)
Cumulative zoning, 179–80, 193

Decision agent, 113, 126, 133
Deed restrictions, 32, 60, 122–23, 217, 223–24
Density bonus incentives, 190–91, 235
Density regulations, 179, 181
Development rights, 122, 216, 234–37, 369, 370–71, 383
Dillon's Rule, 158
Dimensional regulations, 172–74, 177, 179, 181, 183, 185–87, 189–91
Dingell-Johnson Act, 327
Douglas, William O., 214, 248
Due process, 199–201
Dumps, 345–47, 378

Easements, 121–22, 125, 223–24, 364
Economic rent, 126
Edinburgh, Scotland, 28–29, 59
Elevation requirements, 322–23, 324
Elizabeth I, 54, 58
Emerald Necklace (Boston), 96–98, 378
Eminent domain
  history, 32, 84, 91
  power of, 124–25, 160, 196, 214, 247–49
Enclosure laws, 37, 41, 44, 54
Environmental impact statement, 266, 304–7, 333
Environmental movement, 279, 300–301
Environmental Protection Agency, 241, 298–99, 301, 307, 330–34, 346, 350–53
Erosion, coastal, 317, 325, 373
Erosion, soil, 6–7, 275–77, 334, 341–43
Estuarine sanctuaries, 308
Euclid decision. *See* List of cases
Exactions, 224–27, 228
Excess condemnation, 84, 248
Exclusionary zoning, 204, 207–12, 287–88
Externalities, 17, 22–23, 25, 34, 53, 117–19, 375–76

Fair Housing Act (NJ), 211–12, 370, 373
Fair Housing Act (U.S.), 282, 287
Farmland. *See* Cropland
Farmland Protection Act, 340
Farr, William, 71–72
Federal Emergency Management Agency, 241, 320–21, 352
Federal Housing Administration, 278
Federal Land Survey, 30, 129–32, 142, 257
Federal Tort Claims Act, 251
Federalism, 25, 300, 309
Fee simple absolute, 59, 114, 119–20
Feudal commons, 39–44, 51
Fifth Avenue Association, 177, 180, 196
Fiscal zoning, 204, 226–27
Fish and Wildlife Coordination Act, 303, 327
Fish and Wildlife Service, 265, 303, 329, 330
Flood Control Act of 1936, 315
Flood control projects, 238, 275–77, 292–93, 316–18
Flood Disaster Protection Act, 320
Flood insurance, 241, 308, 315–27, 371, 373, 380
Floodplains, 154, 236, 237–46, 314–27, 371, 380
Floor area ratio (FAR), 181
Foreclosure, 124–25
Forests, 7–8, 263, 265–67
Freehold ownership, 51–52

Garden Cities movement, 106–10, 132, 166, 169–71, 277
Greenbelt, 54, 102, 106–10, 277
Greenways, 378
Grid street pattern, 28–30, 131–32
Growth management, 139–40, 227–34, 376, 383

Hall, Justice, 209–10, 243
Hardin, Garrett, 40–41, 55, 300
Hardship, 188–89, 192
Haussmann, Georges-Eugene, 81–88, 248
Hazardous waste, 349–54, 378
Height limitations, 173–74, 177
Hetch Hetchy Valley, 262, 264, 269
Historic preservation, 215–17, 237, 377
Holmes, Oliver Wendell, 196, 197–98, 242
Home Owners Loan Corporation, 277–78
Home rule, 25, 135–36, 158, 178
Homestead Act, 260–61
Hoover, Herbert, 178, 276
Housing
  federal policies, 277–78, 285–88
  low-income, 191, 204, 208–12, 220, 376, 379–80, 383
  regional markets, 208–12, 232
Housing Acts, federal, 278–79, 282, 286–89
Housing of the Working Classes Act, 75
Howard, Ebenezer, 39, 72, 98, 100, 101, 102, 105–10, 169, 277
Hundred-year floodplain, 239, 293, 314, 320, 323, 371
Hurricanes, 309, 314, 316–20, 373, 381

Impact fees, 225–27, 383
Improvement commissions, 60–62

Inner city. *See* Central city
Interstate Highway System, 14, 278, 295, 302–3
Inverse condemnation, 199

Jefferson, Thomas, 21, 140–41, 259, 334
Johnson, Lyndon Baines, 286–87, 290, 301, 318, 379, 383
Jurisdiction, 19, 23–24, 50, 153–62

Land and Water Conservation Fund, 253, 272, 281, 283–84, 363, 378
Land Capability Classification, 334–35
Land Ordinance of 1785, 30, 129, 257–59
Land trusts, 249, 253, 378
Legal citations, key, 27–28
Legal description, 114, 126–32, 154
Letchworth (England), 101, 108–9, 169
License, 121, 122
London
  Georgian period, 58–59
  Great Fire, 54–58
  Nash improvements, 76–81, 248
  sanitary reform, 67–76
Los Angeles, 145, 146–49, 171
Lot size, 181, 189–91, 208, 209, 221
Love Canal, 352
Low-income housing. *See* Housing, low-income

McMillan Commission, 166–67
Marine Protection, Research, and Sanctuaries Act, 343
Marsh, Benjamin C., 175–77, 201
Marsh, George Perkins, 261–63, 279–80, 300
Master plan, 201–4, 229–30
Megalopolis, 280–81
Metes and bounds, 127–29
Metropolitan Health Act of 1866 (NY), 75, 166
Metropolitan planning, 87, 288–90
Metropolitan Statistical Area, 1–2, 10–12, 145, 279
Mineral rights, 113–15, 120, 121, 196
Mobile homes, 204, 209
Model Cities program, 282, 286–87
Model Land Development Code, 309, 365–66, 382–83
Mount Laurel decisions, 209–12, 379, 384
Muir, John, 262–63, 264
Multi-family dwellings, 14, 179, 207–8
Municipal corporations, 45–51, 60, 135–37, 153–62, 165–66
"Mushroom suburbs," 151, 377, 381

Nash, John, 59, 78–81, 248
National Agricultural Lands Study, 340
National Conference on City Planning, 197, 376
National Environmental Policy Act, 266, 268, 295, 298, 299, 301, 302–307, 370
National Flood Insurance Program, 241, 293, 318–25, 371, 373, 380
National forests, 7, 263, 265–67, 378
National Land Use Policy Act, 309
National parks, 8, 249, 263–65, 267–70, 271, 272, 308, 378

National Parks and Recreation Act, 368
National Priorities List, 353, 382
National Resources Inventory, 6, 10, 335, 340, 342
National Resources Planning Board, 275–78, 338
National seashores and lakeshores, 267–68, 269–70, 282, 308, 329, 378
National Wetlands Inventory, 9, 308, 328–29, 331
National Wildlife Refuge System, 263, 308, 329, 378
New Communities Program, 365
New Deal, 248, 274, 275–78, 300
New England towns, 44–45, 136–40, 259
New Jersey Pinelands Commission, 367–69, 373
New Lanark (Scotland), 101–3
New Orleans, 171, 215–16
New York
  parks, 92–96
  urban growth, 67–68, 70, 75, 173–74
  water supply, 88–91
New York City Zoning Ordinance, 165, 177
NIMBY, 204, 343, 346, 354, 382
Nonconforming use, 183, 191–93
Noncumulative zoning, 180–81
Nuisance, 53, 60, 117–19, 171–72, 196–98

Office of Ocean and Coastal Resource Management, 310–11
Office of Technology Assessment, 333, 353–54
Offshore oil and gas leasing, 272, 283, 295, 311–312
Olmsted, Frederick Law, Jr., 167, 169, 177
Olmsted, Frederick Law, Sr., 92–99, 106
Open space preservation, 189–92, 281–84, 303, 363–65, 378
Osborn, F. J., 106, 110
Outdoor Recreation Resources Review Commission, 281–83, 285
Owen, Robert, 72, 100, 101–3, 105, 106

Paris, 67–68, 78, 81–88, 248
Parks
  national, 8, 249, 263–65, 267–70, 271, 272, 308, 378
  state, 363–65
  urban, 76–88, 92–99, 236–37
Pesticides, 4, 6, 300, 349
Petaluma (CA), 230–33
Pinchot, Gifford, 7, 264
Pinelands (NJ), 367–69
Pittman-Robertson Act, 327
Planned unit developments, 188, 189–91
Planning
  city, 30–33, 59, 175–77
  regional, 251–53, 276–77, 288–93, 366–69
  state, 309–14, 365–73, 383
  and zoning, 201–4
Planning board, 179, 180, 181, 188, 201–2, 221, 223
Police power, 160, 171–73, 178, 195–99, 242, 244
Pope, Robert Anderson, 176
Potential Cropland Study, 339–40
Powell, John Wesley, 262–63, 264
Preemption Act of 1841, 259
Prescriptive easement, 125
Progressive Movement, 35, 166, 174–77
Property ownership, 17, 23–25, 51–53, 113–33

Property tax, 52, 119, 159, 250
Property values, 115, 125, 196, 249
Public, 23–25, 43–44, 49–50
Public Health Act of 1848, 65, 74, 166
Public housing programs, 285–88, 379
Public land, acquisition, 161, 242, 246–53, 327
Pullman, George M., 100, 103–6

Racial segregation, 204, 285–88, 377
Reagan, Ronald, 288, 289, 300, 309, 340, 353, 366, 378, 381
Real property, defined, 113–17
Reasonableness, 198, 199–201
Recreation land, 8–9, 266, 267–72, 281–84, 363–65
Redevelopment, urban, 65, 76–100, 166 (*See also* Urban renewal)
Redlining, 277–78
Regent's Park (London), 78–81, 248
Regional context, of zoning, 205–12, 220–21, 232–33
Regional Planning Association of America, 169
Regulatory power. *See* Police power
Reilly, William K., 365
Resettlement Administration, 277
Resource Conservation and Recovery Act, 350–51, 382
Resource management, 34, 264–66
Resource Recovery Act, 346
Restrictive covenant, 59–60, 122–23, 217, 223–24
Revenue, 159–60
Riparian rights, 36, 114, 115–16
River basin management, 276–77, 293
Roosevelt, Theodore, 7, 263, 264, 274, 275–78, 338

Sanitary landfill, 378
Sanitary Reform Movement, 35, 39, 61, 68–76, 85, 92
Savings and Loans crisis, 381–82, 384
Scale, geographic, 20–21
Scenic easement, 122, 249, 364
Sears Island (ME), 306
Section 404 Wetlands Program, 329–34, 372
Sewage disposal, 73–75, 85, 144–46, 347–48, 378
Sierra Club, 262, 264, 306–7
Single-family homes, 179, 213, 221, 278
Site plan review, 221
Skyscrapers, 173–74
Slum Clearance Program, 285
Soil and Water Resources Conservation Act, 340
Soil Conservation Service, 277, 292, 316, 334–35, 339–43
Soil erosion, 6–7, 275–77, 334, 341–43
Soils, 334–36
Solid waste management, 343–49
Sovereign immunity, 251
Special districts, 137, 145–46, 151–53, 157, 160, 362
Special permit, 179, 189
Spot zoning, 188, 235
Standard Planning Enabling Act, 178, 206, 221
Standard Zoning Enabling Act, 178, 192, 213
State Comprehensive Outdoor Recreation Plan, 283
State parks, 263, 363–65
Stratton Commission, 309

Street plans, 28–30, 82–86, 131–32, 222
Strip mining, 292, 344, 354–56
Subdivision plan, 132, 221–23
Subdivisions, 60, 188–92, 221–23, 277–78
Suburbs
  and federal policies, 277–78, 286, 377–78
  growth of, 47–48, 76, 144–45, 149–53, 227–29, 274, 377, 381
  zoning issues, 169–71, 207–13, 287–88
Superfund Act, 352–54, 382
Surface Mining Control and Reclamation Act, 292, 344, 355–56

Taking issue, 195–99, 214, 236–37, 242–43, 246–48
Task Force on Metropolitan and Urban Problems, 286
Tax rate, 159
Taylor Grazing Act, 8, 265
Tennessee Valley Authority, 9, 276–77, 292–93, 316, 341
Tidewater counties, 140–41
Timber industry, 7, 261, 265–67, 271
Timber rights, 114, 120, 121
Town and Country Planning Act (England), 109
Towns, 21, 136–40, 140–41, 259
Townships, 129–31, 142
Toxic Substances Control Act, 349–50
Toxic waste, 349–54
"Tragedy of the Commons," 40–42, 54–56
Transfer of development rights, 216, 234–37, 369, 370–71, 383
Trespass, 25, 53, 125

Udall, Stewart, 283
Urban areas. *See also* Central city
  expansion, 10–15, 65–68, 143–49, 338
  functions, 21
Urbanized Areas, 10, 12–13
Urban renewal, 214, 248, 278, 285–88, 379
U.S. Department of Agriculture, 305, 328–29, 334–43
U.S. Department of Commerce, 178, 206
U.S. Department of Housing and Urban Development, 10, 284, 305, 317, 383
U.S. Department of Transportation, 282, 303, 305
Use regulations, 179, 181, 183–85, 197
U.S. Forest Service, 7, 9, 264, 265–67, 271
U.S. Geological Survey, 262–63
Utopian settlements, 100–101

V zone, 323
Validity, presumption of, 173, 188, 200
Variances, 188–89, 192, 198, 199, 235
Veterans Administration, 278, 285

War on Poverty, 287–88, 290, 379, 383
Washington, D.C.
  historic preservation, 237
  L'Enfant plan, 86, 166–67
  urban renewal, 214, 248
Waste management, 290, 343–54, 378, 382

Water resource projects, federal, 9, 275–77, 293, 295, 302–3
Water Resources Council, 293, 325
Water Resources Planning Act, 276, 282, 293
Water rights, 36, 114, 115–16
Water supply, 55–56, 72–75, 77, 85, 88–92, 144–48, 157
Watt, James, 266, 272
Wenner-Gren Conference, 279–80
Wetlands, 6, 9, 181, 239–46, 308–9, 312, 327–34, 371–77, 378
White City, 86, 98–99

White, Gilbert F., 276, 317–18, 319, 325
Whyte, William H., 204, 281, 364, 381
Wilderness Act, 282, 378

Zero population growth, 13
Zoning
 court tests, 195–201, 206–217
 criticisms of, 219–20, 382–83
 Euclidean, 165–66, 177–93, 220, 228, 237–38
 and planning, 201–206
Zoning board of appeals, 161, 179, 188–89

LIST OF CASES CITED

A.M.G. Associates v. Springfield Township, 243
Appeal of Girsh, 207–9
Appeal of Kit-Mar Builders, 208
Associated Home Builders v. City of Livermore, 233
Associated Home Builders v. City of Walnut Creek, 225
Baker v. Carr, 274
Bartram v. Zoning Commission of Bridgeport, 188
Berman v. Parker, 214, 248, 286
Bersani v. EPA, 333
Boca Raton v. Boca Villas Corp., 234
Borough of Cresskill v. Borough of Dumont, 206–7
Citizens to Preserve Overton Park v. Volpe, 303
City of New Orleans v. Impastato, 215
City of New Orleans v. Pergament, 215–16
Construction Industry Assn. of Sonoma County v. City of Petaluma, 230–33
Courtesy Sandwich Shop Inc. v. Port of New York Authority, 248
Devereaux Foundation Inc. Zoning Case, 188
Duffcon Concrete Products, Inc. v. Borough of Cresskill, 206–7
Dupont Circle Citizens Association v. District of Columbia Zoning Commission, 236
First English Evangelical Lutheran Church v. County of Los Angeles, 199
Fred F. French Investing Co. v. City of New York, 236–37, 246–47
Gautreaux v. Chicago Housing Authority, 285, 288
Golden v. Township of Ramapo, 229–30
Hadacheck v. Sebastian, 171–72, 196–97
Hills Development Co. v. Somerset County, 211–12
J. W. Jones Companies v. City of San Diego, 225
John Donnelly and Sons, Inc. v. Outdoor Advertising Board, 213–14
Just v. Marinette County, 244, 246, 323
Keystone Bituminous Coal Assn. v. DeBenedictis, 198–99
Lawton v. Steele, 172

Lochner v. New York, 171
Maher v. City of New Orleans, 216
Metromedia Inc. v. City of San Diego, 215
Metropolitan Housing Development Co. v. Village of Arlington Heights, 287
Morris County Land Improvement Co. v. Parsippany-Troy Hills Twp., 242–44, 246, 322–23
Mugler v. Kansas, 171–72
Naegele Outdoor Advertising v. Village of Minnetonka, 213
National Land and Investment Co. v. Easttown Twp. Board of Adjustment, 208
Natural Resources Defense Council v. Callahan, 331
New York City Housing Authority v. Muller, 248
Oakwood at Madison v. Twp. of Madison, 210
Penn Central Transportation Co. v. City of New York, 216, 236
Pennsylvania Coal Co. v. Mahon, 196–99, 242
Powell v. Pennsylvania, 172
Reid v. Architectural Review Board of City of Cleveland Heights, 217
Robinson v. City of Boulder, 234
Secretary of the Interior v. California, 312
Sierra Club v. March, 306–7
Slaughterhouse cases, 171
South Burlington County NAACP v. Township of Mount Laurel, 209–12, 379, 384
Thomas Cusack Co. v. City of Chicago, 213
Tulk v. Moxhay, 59, 223
Turnpike Realty Co. & Town of Dedham, 242, 243–45, 323
Vernon Park Realty Inc. v. City of Mount Vernon, 200–201
Vickers v. Township Committee of Gloucester Township, 209
Village of Belle Terre v. Borass, 214
Village of Euclid v. Ambler Realty Co., 27, 165, 169, 179, 197–98, 201, 205, 213
Welch v. Swasey, 172–73, 196, 197
Zabel v. Tabb, 330